Springer
Tokyo
Berlin
Heidelberg
New York
Barcelona
Budapest
Hong Kong
London
Milan
Paris
Santa Clara
Singapore

K. Hayashi · A. Kamiya · K. Ono (Eds.)

Biomechanics

Functional Adaptation and Remodeling

With 180 Figures

Springer

Kozaburo Hayashi, Ph.D.
Department of Mechanical Engineering, Faculty of Engineering Science, Osaka University, 1-3 Machikaneyama-cho, Toyonaka, Osaka, 560 Japan

Akira Kamiya, M.D., Ph.D.
Institute of Medical Electronics, University of Tokyo School of Medicine, 7-3-1 Hongo, Bunkyo-ku, Tokyo, 113 Japan

Keiro Ono, M.D., Ph.D.
Osaka Koseinenkin Hospital, 4-2-78 Fukushima, Fukushima-ku, Osaka, 553 Japan

ISBN 4-431-70173-7 Springer-Verlag Tokyo Berlin Heidelberg New York

Printed on acid-free paper

Typesetting: Best-set Typesetter Ltd., Hong Kong
Printing & binding: Yokoyama, Japan

Preface

Biomechanics is a relatively new research area that seeks to understand the mechanics of living systems, and to develop approaches applicable to a wide variety of fields that include biomedical engineering, medical science, clinical medicine, applied mechanics, and engineering. For the past quarter-century, this area has grown rapidly and developed greatly, and is now recognized as one of the most important and interesting fields in basic science. Owing to such recent progress, the research project entitled Biomechanics of Structure and Function of Living Cells, Tissues, and Organs was selected as one of the most important national research projects in Japan. Between 1992 and 1995 this project was supported financially in part by a Grant-in-Aid for Scientific Research on Priority Areas (Biomechanics, Nos. 04237101-04237106) from the Ministry of Education, Science and Culture, Japan, with Kozaburo Hayashi as the principal investigator, and Hiroyuki Abe, Keiro Ono, Akira Kamiya, and Hiromasa Ishikawa as the principal co-investigators.

The project was composed of the following five programs: Mechanical Properties of Living Cells, Tissues, and Organs; Biomechanics of Orthopedic Systems and Motion Analysis; Biomechanics of the Circulatory System; Computational Biomechanics; and Functional Adaptation and Remodeling of Biological Tissues and Organs.

More than 70 biomechanics scientists who are also experts in the fields of engineering, basic science, medical science, clinical medicine, and dentistry participated in the project and obtained a great deal of invaluable new information.

The classic doctrine "function dictates structure" is reaffirmed in Wolff's law of the skeletal system. Whether this doctrine holds true for the cardiovascular system and connective tissues such as heart muscle, blood vessels, tendons, ligaments, and intervertebral discs is a central question of biomechanics. The multidisciplinary approach of our research project has thrown new light on the key issues of biomechanics. Biologists refer to remodeling as the continuous renewal of bone in the skeleton, while biomechanists consider remodeling to be the continuous adaptation of bone structure to its mechanical environment. Continuous renewal of bone is necessary for maintaining the shape and function of the skeleton during growth. The process takes place, for example, in the cranium and facial bone where accompanying tissue and organ development leads the skeletal anlage to modified growth (modeling). Here local influences such as the mechanical, chemical, and electrical environments alter the growth pattern and organization of the tissue and thereby produce macroarchitectural features.

Strain-adaptive bone remodeling is an example of continuous adaptation of bone structure and is often seen in the healing of fractures in juvenile patients, a representative case being the spontaneous correction of a broken tubular bone with angular deformity. The relevant sensor, signal, and activators in bone-mass regulation in this situation have been partially clarified. Recently, one of our project team discovered a similar response mechanism in vascular endothelial cells as a result of fluid shear stress. This opens a new frontier in biomechanics.

Another important question is whether remodeling processes underlie the regressive changes in mechanical properties often accompanying overuse or aging. As far as the skeletal system is concerned the answer appears to be yes; There must be an efficient remodeling process induced with advanced bone reduction in the vertebra, since the load-bearing vertical trabeculae in such vertebrae avoid resorption but the horizontal ones do not. Ligaments, tendons, and intervertebral discs lack a vigorous reparative or remodeling capacity after the growth period, hence, the cumulative effects of microdamage to the matrices of these structures changes from strengthening in the juvenile to weakening in the adult.

Functional adaptation and remodeling has proved not to be a universal phenomenon in living cells or tissues but to depend upon species, age, location, and loading conditions. The importance of nonlinear relationships should not be overlooked in mathematical models for studying the relationship between stimulus and remodeling.

This book was produced to document the unique and important results obtained from the above-mentioned research project, and to offer the most up-to-date information on the phenomena of functional adaptation and optimal remodeling observed in living organs and components for biomedical engineers, medical scientists, clinicians, and other biomechanics-related engineers and scientists.

This publication was supported financially in part by a Grant-in-Aid for Publication of Scientific Research Result (No. 77012) from the Ministry of Education, Science and Culture, Japan. Finally, we wish express our appreciation to the editorial and production staff of Springer-Verlag, Tokyo for their cooperation in producing this book.

Kozaburo Hayashi
Akira Kamiya
Keiro Ono

Contents

List of Contributors

Response of Endothelial Cells to Mechanical Stress

Response of Vascular Endothelial Cells to Flow Shear Stress: Phenomenological Aspect

MASAAKI SATO[1], NORIYUKI KATAOKA[1], and NORIO OHSHIMA[2]

Summary. Blood vessel walls are covered by an endothelial cell monolayer that is always exposed to blood flow. A fluid-imposed shear stress acting on endothelial cells is defined as a tangential force produced by blood viscosity and a velocity gradient. There are many reports that endothelial cells respond to shear stress and change their morphology and functions. Blood flow characteristics have received special attention because atherogenesis is found around arterial branches and curved regions. In this chapter the effects of fluid-imposed shear stress on endothelial cell morphology, which have been reported by ourselves and other researchers, are mainly summarized. We also focused on flow patterns such as reverse, secondary, pulsatile, and turbulent flows. This kind of complex flow would occur in the in vivo condition, especially around arterial bifurcations and curved sites, and would certainly affect cell morphology. From the importance of adhesive proteins in the role of connecting endothelial cells and subendothelial structures, we inferred the effects of extracellular matrices and cytoskeletal components on cell morphology from the changes in endothelial cell shape and actin filament structure. Because the cytoskeletal changes of endothelial cells after exposure to shear stress are closely correlated with mechanical properties, we applied a micropipette technique to examine stiffness and viscoelastic properties.

Key words: Atherosclerosis—Endothelial cell—Extracellular matrix—Shear stress—Mechanical property

[1] Biomechanics Laboratory, Graduate School of Mechanical Engineering, Tohoku University, Aoba, Aramaki, Aoba-ku, Sendai, 980-77 Japan
[2] Institute of Basic Medical Sciences, University of Tsukuba, Tsukuba, 305 Japan

1 Introduction

It is known that atherosclerotic lesions tend to localize around arterial branches and curved sites [1,2]. These regions would experience localized complex blood flows such as elevated or reduced wall shear stress, boundary layer separation, and secondary flows [3,4]. From these phenomenological results, hemodynamic forces are believed to be one of the important factors in the process of atherogenesis; however, the detailed mechanisms are still poorly understood and the precise role is uncertain. In addition, because the accumulation of lipid within arterial intima might be characteristic in the early stages of atherosclerosis, the process of transendothelial macromolecule transport and the influence of hemodynamic-related events, in particular shear stress, received early attention [5–7]. Schwartz et al. [2] have pointed out that lesion-prone or prelesion areas are delineated by the in vivo uptake of the protein-binding azo dye Evans blue, using experimental atherosclerosis animal models. In their experiments, the prelesion-area endothelial cells were relatively more polygonal or cobblestone shaped than nonlesion-prone areas, and the surface glycocalyx was some two- to fivefold thinner.

This background caused many researchers to be interested in the effects of blood flow states on endothelial cell morphology and functions. Hemodynamic forces have been found to affect the shape and orientation of endothelial cells studied both in vivo and in vitro [7–11]. Considering fluid-imposed hemodynamic forces, the normal component, i.e., arterial pressure, should be considered in addition to shear stress (Fig. 1). In living systems, arterial pressure is pulsatile and produces a cyclic stretch of the blood vessel wall. Although changes in the shape and functions of cultured endothelial cells can also be induced by a cyclic stretch of substrate [12], this subject is not referred to in this chapter.

A fluid-imposed wall shear stress, τ_w, is defined in the following equation:

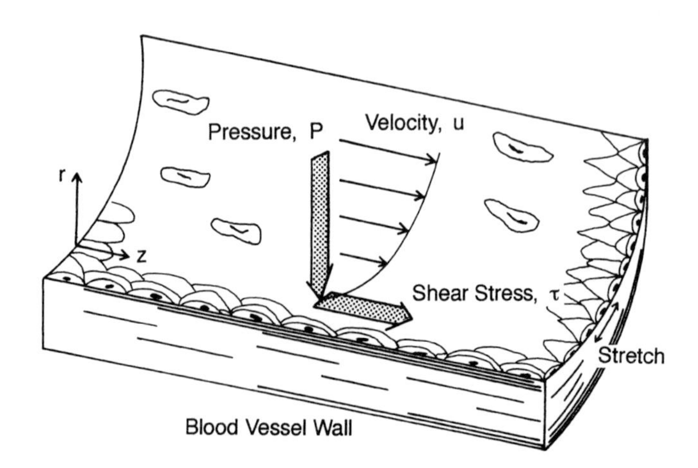

Fig. 1. Schematic diagram of hemodynamic forces acting on a blood vessel

$$\tau_w = \mu \cdot \left(du/dy\right)_w \tag{1}$$

where μ is the blood viscosity coefficient, u is the velocity component in the direction parallel to the wall, y is distance in the direction perpendicular to the wall, and $(du/dy)_w$ means the velocity gradient of blood flow or shear strain rate at the wall. Morphological changes of cultured endothelial cells caused by exposure to a shear stress are reported to depend on the magnitude and the time of exposure [13] and flow conditions such as turbulent [14] and pulsatile flow [15]. Furthermore, cell adhesion proteins and extracellular matrices affect endothelial cell morphology and functions. The areas where endothelial cells are polygonal or cobblestone shaped may correspond to regions of relatively low shear stress. Recent studies have shown that atherogenesis is frequently found in these regions. This low shear stress means that the time-averaged value is low, and complex blood flow conditions such as reverse, secondary, and turbulent flows will have occurred in these regions. Focusing on the effect of the local flow conditions on the endothelium in our study, we designed our experiments to investigate the effects of flow direction on the morphological responses of cultured bovine endothelial cells.

Flow experiments on endothelial cells are usually carried out under the confluent condition, where the configurational changes are analyzed. To understand a part of the mechanical stress mechanism, we have attempted flow experiments using subconfluent endothelial cells. Furthermore, the cytoskeleton (in particular the stress fibers), which is an important structure in membrane support and in the maintenance of cell shape, also appears to differ in regions of differing shear stress [16–18]. This could be reflected in the mechanical properties of the endothelial cells, and any change in these properties could be important to the deformation a cell undergoes as part of the adaptation process. For this reason, it was believed that a measurement of the mechanical properties of an endothelial cell would be instructive. Such properties would appear to be an important correlation between cell structure and function; however, to date no data are available on the mechanical properties of the endothelial cell.

2 Shape Changes of Endothelial Cells Exposed to Shear Stress

2.1 Endothelial Cell Shape in Arteries In Vivo

It is well known that endothelial cell shape is affected by in vivo blood flow condition. To examine this phenomenon, aortic tissue specimens were obtained from two normal adult male New Zealand white rabbits; their endothelial cells were observed by a scanning electron microscope [19]. Photomicrographs of the en face shape of endothelial cells around the bifurcation of the left subclavian artery are shown in Fig. 2. Our interests concentrated on the cell morphology near the flow divider, shown as a solid line in the figure. Because the flow

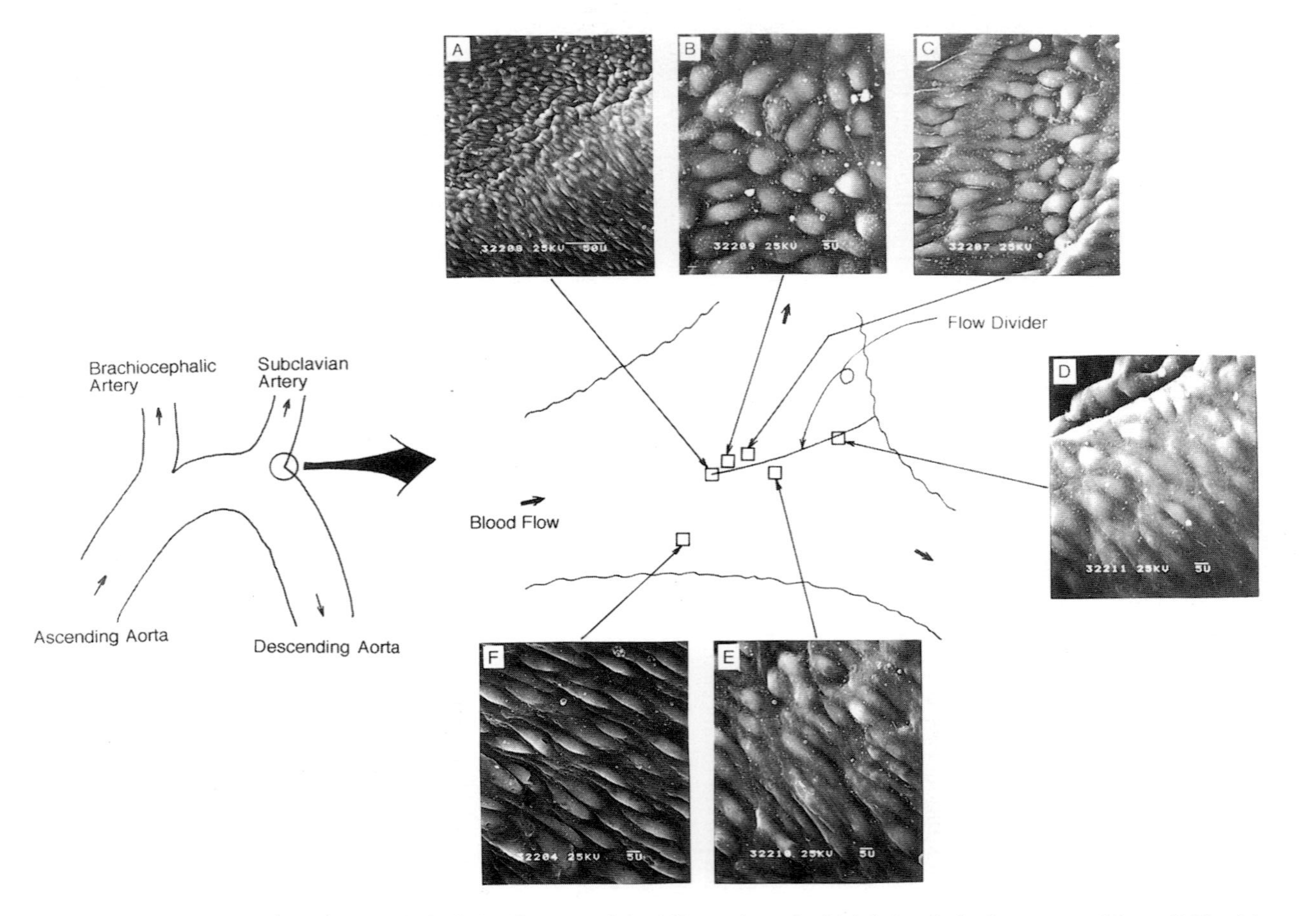

FIG. 2. En face shapes (*A–F*) of endothelial cells around the bifurcation of rabbit left subclavian artery. (From [19], with permission)

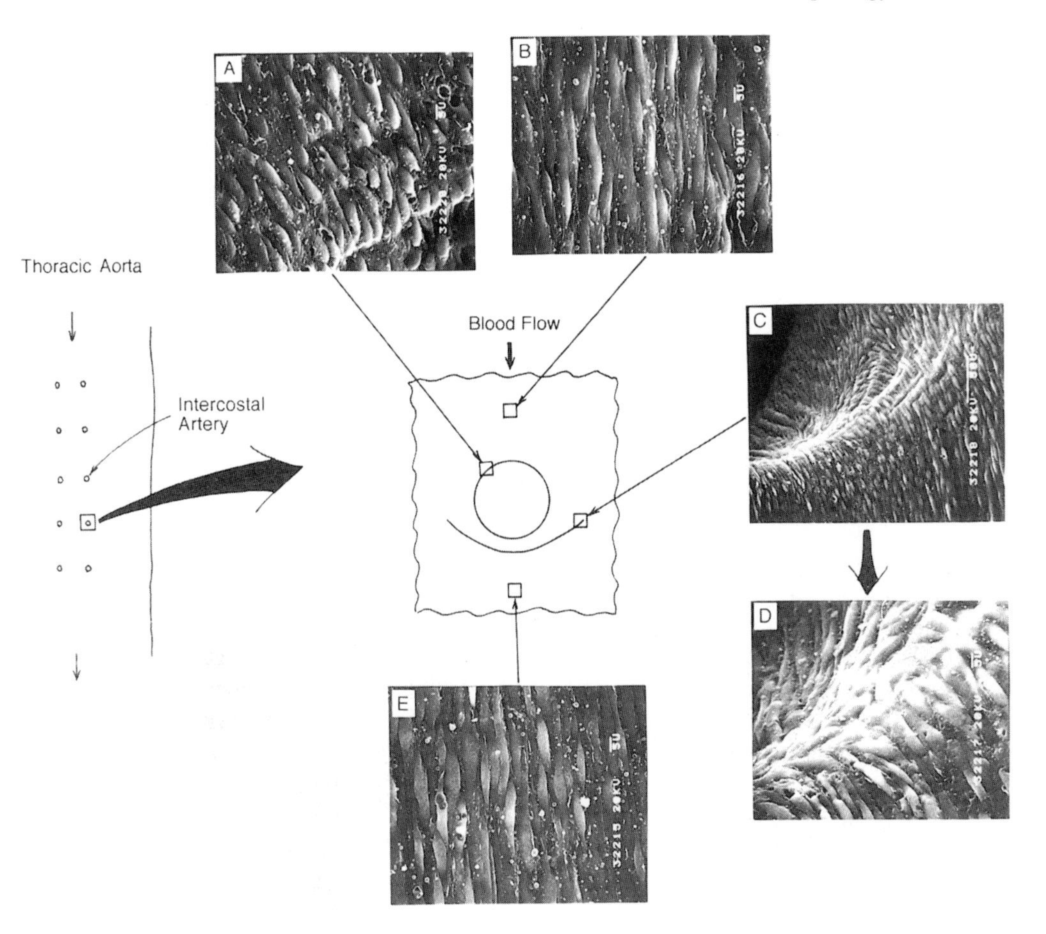

FIG. 3. En face shapes (*A–E*) of endothelial cells around a rabbit aortic intercostal ostium. (From [19], with permission)

condition around this flow divider was especially complex, the endothelial cells showed different shapes depending on location and flow conditions. Most cells appeared to align with flow direction and elongate by the fluid-imposed shear stress (panels D, E, and F in Fig. 2). Some cells (panels B and C) showed an almost polygonal shape and seemed to be exposed to relatively low shear stress or shear stress that fluctuated in direction.

En face cell shapes of the endothelial cells around the aortic intercostal ostium are shown in Fig. 3. Cell morphology in this location generally reflected well the blood flow condition; elongated cells aligned with the flow direction were seen. Some irregularly shaped cells were found on the flow dividers only (panel D, Fig. 3).

Flaherty et al. [8] measured the orientation, morphology, and population density of endothelial nuclei of the canine thoracic aorta to obtain evidence that blood flow patterns do indeed affect patterns of nuclear orientation. In a uniform vessel segment with stable blood flow, such as the middle and lower descending aorta, the nuclei were oriented parallel to the axis of the blood vessel. Flaherty et al. tried very interesting experiments in which a segment of the descending thoracic aorta was removed, opened longitudinally, and reclosed to form a tube with a new longitudinal axis 90° from the original axis. The results showed the nuclear pattern realigned in the direction of flow within 10 days after surgery, that is, nuclear orientation clearly depended on altered flow pattern.

Okano and Yoshida [20] measured endothelial cell shape at bifurcations of brachiocephalic and subclavian arteries. The stagnation point of flow and the leading edges of flow dividers were covered by round and long fusiform endothelial cells, respectively. The hips of flow dividers of both branches, where flow velocity was confirmed to be relatively slow, were covered by ellipsoidal cells. It would be suggested, by this result, that blood flow state is reflected in endothelial cell shape and orientation. These configurational changes of endothelial cells would appear in changes of the cytoskeletal structure.

Kim et al. [17] stained F-actin filaments in endothelial cells of rabbit aorta in vivo to confirm the flow effects on the microstructure. They clearly observed the actin microfilament bundles in endothelial cells along flow dividers at branch points. The prominent central microfilament bundles and the reduced peripheral microfilaments were seen at higher shear stress regions. This phenomenon was believed to be an adaptive response of endothelial cells to the mechanical environment.

2.2 Cultured Endothelial Cells

The effects of flow on the shape and cytoskeletal structure of cultured endothelial cells have been studied by many researchers, including the authors. We usually use porcine or bovine aortic endothelial cells harvested from thoracic and abdominal aortae and cultured in flasks. From the third to seventh subculture, endothelial cells were transferred onto a glass coverslip or a plain glass. A fully confluent endothelial cell monolayer was exposed to shear stress of 2Pa for 1–24h in a parallel-plate flow chamber (Fig. 4). This flow chamber has a flow section

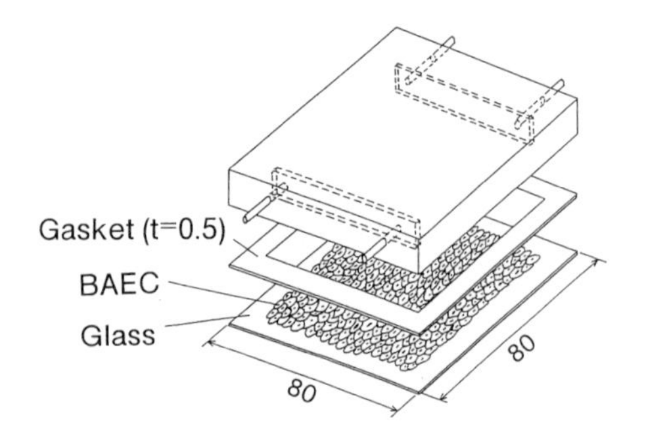

Fig. 4. Schematic diagram of flow chamber in which endothelial cells are cultured and exposed to fluid-imposed shear stress. *BAEC*, bovine aortic endothelial cells

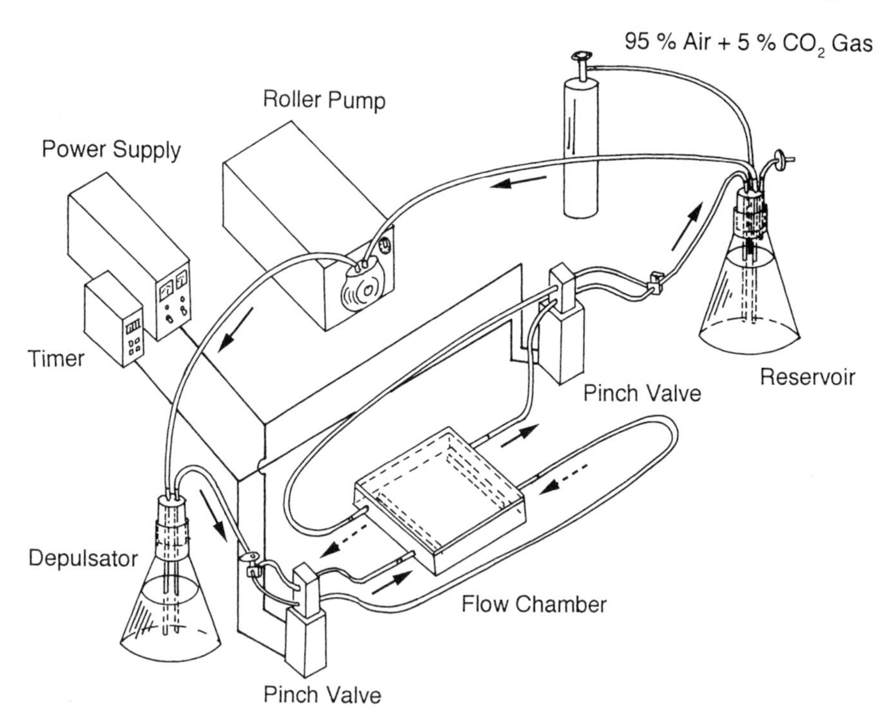

Fig. 5. Schematic diagram of flow circuit. In some experiments, pinch valves were used to control flow direction

that is 0.5 mm high, 50 mm wide, and 60 mm long. The flow circuit we have used (Fig. 5) is mainly composed of the chamber, two reservoirs (one for the depulsator), and a roller pump. When necessary, pinch valves were used to control flow direction in our experiments [21].

After exposure to flow, the endothelial cell monolayer was rinsed with phosphate-buffered saline (PBS) and fixed in 1% formaldehyde in PBS for 20 min. After fixation, cell samples were incubated in 0.1% Triton X-100 for 5 min. Then cells were rinsed in PBS and incubated with 150 n*M* rhodamine-phalloidin in PBS for 20 min to stain F-actin filaments. Endothelial cells were rinsed and mounted over glycerol/PBS (1:1). Their microstructure was observed and photographed with a photomicroscope equipped with epifluorescence optics (Olympus IMT-2, Tokyo, Japan), and the following geometrical parameters were measured: angle of cell orientation, defined as the deviation of the longest diagonal of the cell from the direction of the flow, and shape index, defined as $4\pi A/P^2$ (A, area; P, perimeter). The shape index is a nondimensional parameter defined such that a circular cell would have a shape index of 1.0, while a highly elongated cell would have a value near 0.0.

Endothelial cells responded to a fluid-imposed shear stress and changed their shape and structure. This depended on the level of applied shear stress and the exposure duration [13]. In general, the higher the level of shear stress and the

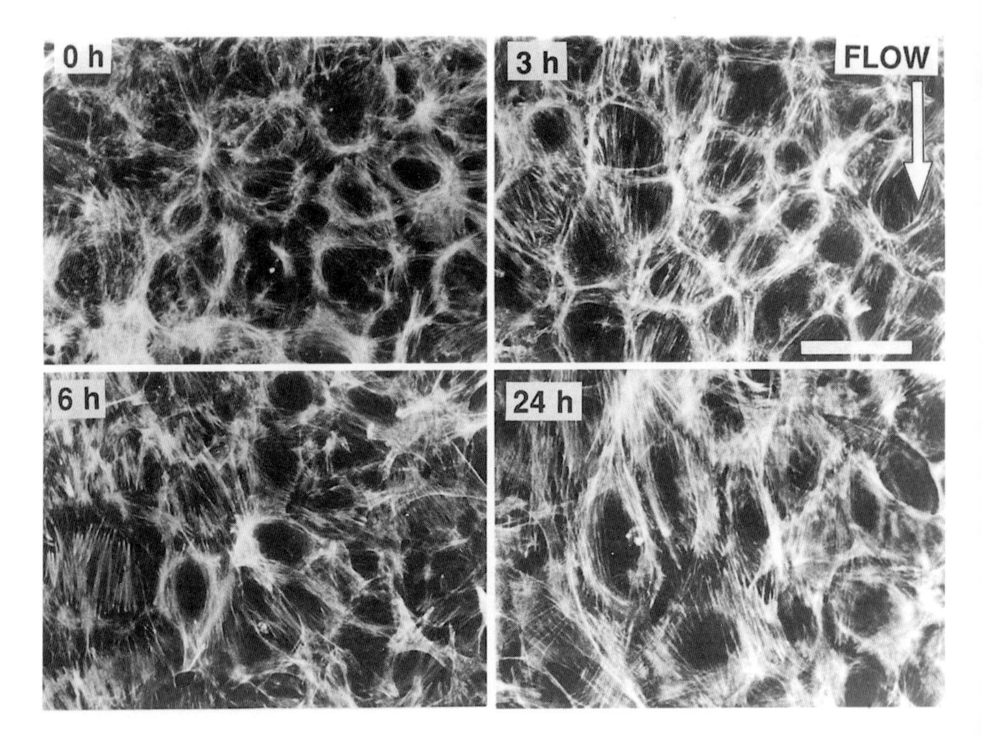

Fig. 6. A typical example of the time-course changes (0, 3, 6, and 24 h) in the cytoskeletal structure of F-actin filaments of porcine aortic endothelial cells stained with rhodamine phalloidin. Applied shear stress was 2 Pa. *White bar*, 50 μm; *arrow*, flow direction (3 h). (From [22], with permission)

longer the duration of exposure, the smaller the shape index, i.e., the more elongated is the endothelial cell. Because this cell shape change progresses with time, we have studied time-course changes in cell shape and in the patterns of microfilament distribution using cultured porcine endothelial cell monolayers after flow exposure [22]. Photomicrographs of cells in which F-actin filaments were stained are shown in Fig. 6. Under the no-flow condition, F-actin filaments were mainly distributed at random around the peripheral regions. After a 3-h exposure, F-actin bundles aligned clearly with flow direction were formed in the central part of some cells. However, cell shape itself did not significantly change at this time. After a 6-h exposure, we observed significant changes in cell shape and alignment of F-actin filament in most cells. After 24-h exposure, cell elongation and orientation were clearly observed with alignment of bright F-actin bundles, i.e., stress fiber, and with dense peripheral bands.

Time-course changes in the shape index values are shown in Fig. 7. At no-flow condition a value of shape index was 0.83 ± 0.01, indicating an almost polygonal or cobblestone shape. The value of shape indices after a 6-h and longer exposure became significantly smaller than the control value, which means that cultured

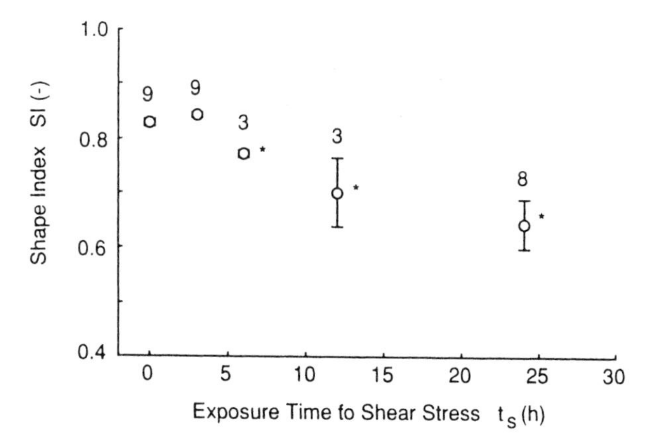

FIG. 7. Effect of the time of exposure to shear stress on the shape index of porcine aortic endothelial cells. Mean ± SE; n, number of measured coverslips; *, $P < 0.05$. (From [22], with permission)

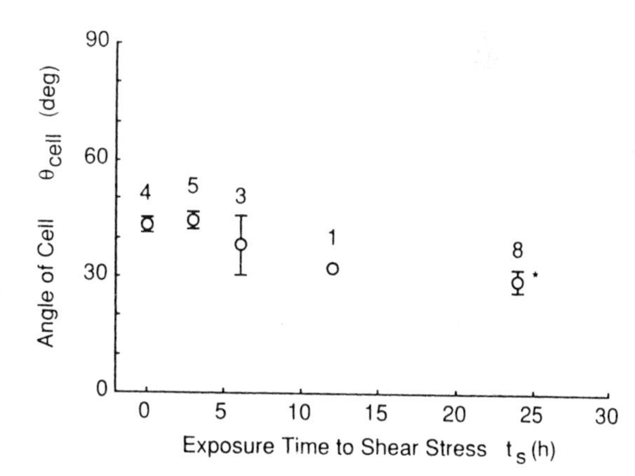

FIG. 8. Effect of time of exposure to shear stress on the angle of cell orientation of porcine aortic endothelial cells. Mean ± SE; n, number of measured coverslips; *, $P < 0.05$. (From [22], with permission)

porcine endothelial cells might start to change cell shape at about 6h after exposure to a 2-Pa shear stress. We espccially need to pay attention to phenomena at about a 3-h exposure, because some F-actin filaments are already aligned with flow direction without significant changes of cell shape. From this result, we can expect that the change of F-actin filament orientation would precede the change of cell elongation. In time-course changes in angle of cell orientation (Fig. 8), cells are distributed at random in the no-flow condition. Although some cells aligned with flow direction before 12h of exposure, the angle of most cells significantly changed only after 24-h exposure. This means that endothelial cells would start to change their cell shape first and then their orientation when they are exposed to fluid-imposed shear stress.

Endothelial cell shape also depends on the physicochemical properties of the substrate. For example, we cultured porcine aortic endothelial cells on extracel-

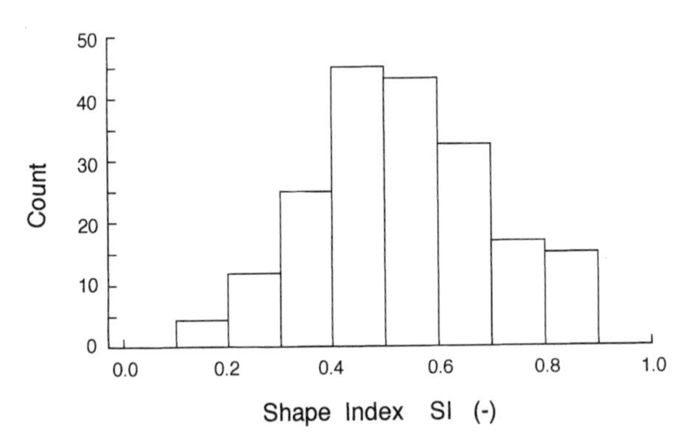

FIG. 9. Shape index (*SI*) values of endothelial cells cultured on mixed extracellular matrices under no-flow condition. (From [23], with permission)

lular matrices in which type IV collagen, heparan sulfate, chondroitin sulfate, and dermatan sulfate were mixed [23]. The cells cultured on the mixed extracellular matrices showed marked elongation and segmental orientation with randomly distributed cell axes even under a no-flow condition, and the F-actin filaments were mostly observed in parallel with the cell axis. Distribution of shape index in this culture condition (Fig. 9) indicates that some cells are polygonal and some other cells elongated.

Endothelial cells adhere to the substrate by means of a complicated mechanism in which transmembrane cell adhesion receptors interact with extracellular matrix proteins at focal contacts. One component of the focal contact complex linking F-actin filaments to the plasma membrane is vinculin. Girard et al. [10] examined the role of extracellular matrix proteins and cytoskeletal-associated focal contact proteins in the modulation of shear stress-related endothelial cell morphological changes. In a static culture condition, vinculin appeared to be located primarily at the periphery of each cell. Following exposure of cells to 5h of arterial levels of shear stress, vinculin staining with immunofluorescence appeared more streaked and seemed to undergo a reorganization concomitant with initial cell shape changes. After 24-h exposure, they observed a reorganization of vinculin with prominent localization at the upstream end of most cells. From this result, vinculin was suggested to play a role in directing flow-induced stress fiber assembly.

Fibronectin is one of the major components of extracellular matrices and plays an important role in adhering cells to substrate. When we culture endothelial cells on fibronectin-coated substrate, adhesion and proliferation of cells are enhanced and production of fibronectin from endothelial cells is controlled by flow. The pattern of localization of fibronectin associated with endothelial cells responding to shear stress was studied by Girard et al. [10] and by Gupte and Frangos [24]. When cells were exposed to relatively high levels of shear stress (3–5Pa) for 12h, the fibronectin appeared to be organized into a reticular pattern. At 24h prominence of the fibronectin reticular pattern was diminished, and the fibronectin clearly aligned with the direction of flow. However, after the onset of flow, a decrease in the amount of fibronectin was reported to occur initially [24].

This phenomenon, as Girard et al. pointed out, quite possibly is associated with the mobility required of endothelial cells if they are to change shape and align themselves with the direction of flow.

We have studied the effect of flow direction on bovine aortic endothelial cell morphology using the flow chamber and the flow circuit shown in Figs. 4 and 5 [21]. The flow direction to which endothelial cells were exposed was changed in addition to the usual one-way flow; i.e., reciprocating flow and alternately orthogonal flow were used with 30-min intervals. In the reciprocating and alternately orthogonal flow experiments, a medium flow was alternately changed in the opposite and orthogonal directions, respectively, by pinch valves controlled with a timer. Endothelial cells showed polygonal shape under the no-flow condition. F-actin filaments were mainly localized at the periphery of the cells, the same as the porcine endothelial cells shown in Fig. 6, and some filaments had starlike configurations around the central portion (Fig. 10).

After application of arterial level of shear stress (2 Pa) for 24 h, endothelial cells showed marked elongation and oriented with the flow direction under the one-way flow conditions. In the reciprocating flow, cells slightly elongated and aligned with the flow direction. Although the cells elongated under the alter-

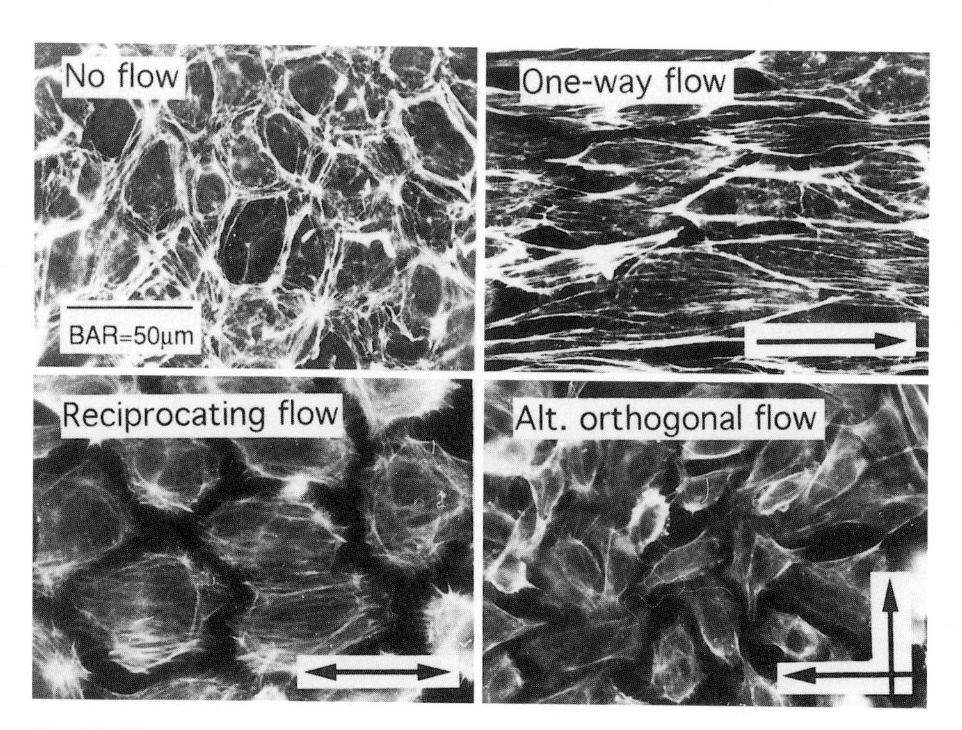

Fig. 10. Photomicrographs of rhodamine phalloidin-stained endothelial cells for four flow conditions: *no-flow*, *one-way flow* (2 Pa, 24 h), *reciprocating flow* (2 Pa, 30-min interval, 24 h), and *alternately* (*Alt.*) *orthogonal flow* (2 Pa, 30-min interval, 24 h). *Arrows* indicate flow direction; *bar*, 50 μm

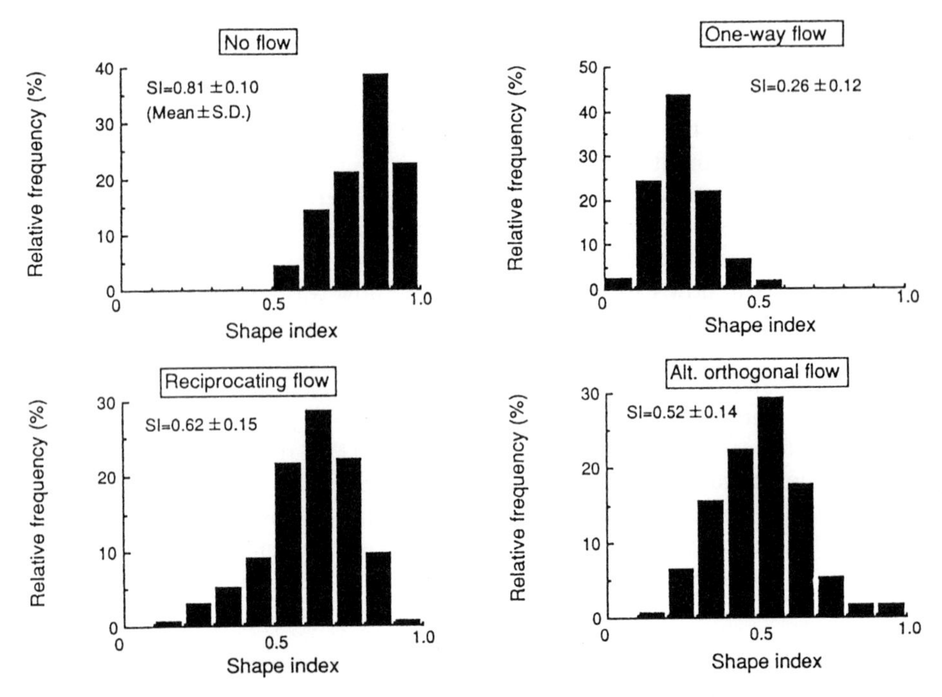

FIG. 11. Shape index values for four flow conditions: *no-flow, one-way flow* (2 Pa, 24 h), *reciprocating flow* (2 Pa, 30-min interval, 24 h), and *alternately (Alt.) orthogonal flow* (2 Pa, 30-min interval, 24 h)

nately orthogonal flow condition, the cells were oriented at random as in the case of the no-flow condition. Under one-way flow conditions, stress fibers were mainly formed in the central portion of cells and aligned with the flow direction. Also, the F-actin filaments around the upstream side of each cell were observed to be strongly stained. In contrast, the F-actin filaments in the cells exposed to the reciprocating flow were stained almost homogeneously.

Distribution of shape index for the four different experimental conditions is summarized in Fig. 11. The cell shape of the no-flow condition was mostly polygonal, having a shape index value of 0.81 ± 0.10. Most endothelial cells exposed to shear stress of the one-way flow were quite elongated; the mean value of the shape index was 0.26 ± 0.12, which was significantly lower than that of the no-flow condition. For reciprocating and alternately orthogonal flows, the values were 0.62 ± 0.15 and 0.52 ± 0.14, showing very similar values. Comparing these results with one-way flow, the shape index was significantly larger in the reciprocating and alternately orthogonal flows, even under the same shear stress and the same exposure duration. These shape index values were between those of the control and the one-way flow condition, indicating that the cells in the reciprocating and alternately orthogonal flow conditions were more polygonal in shape than for one-way flow even under flow conditions.

Distribution of cell orientation to the flow direction is summarized in the same manner in Fig. 12. At the no-flow condition, cell orientation was quite random and the SD value was 52.5°. Under the one-way and reciprocating flow conditions, most endothelial cells aligned with the flow direction. The SD value became smaller, being 8.0° for the one-way flow and 20.0° for the reciprocating flow. However, quite interestingly, in the case of the alternately orthogonal flow the SD value was 52.9° and the distribution was relatively random.

Davies et al. [14] also studied the effects of hemodynamic forces on vascular endothelial cell turnover by exposing contact-inhibited confluent cell monolayers to shear stresses of varying amplitude in either laminar or turbulent flow. In their results, turbulent shear stresses as low as 0.15 Pa for as short a period as 3 h stimulated substantial endothelial DNA synthesis in the absence of cell alignment. From the point of view of the in vivo condition in which endothelial cells are exposed to a pulsatile flow, Helmlinger et al. [15] investigated the influence of pulsatile flow on cell shape and orientation and on F-actin microfilament localization in confluent bovine aortic endothelial cell monolayers using a 1-Hz

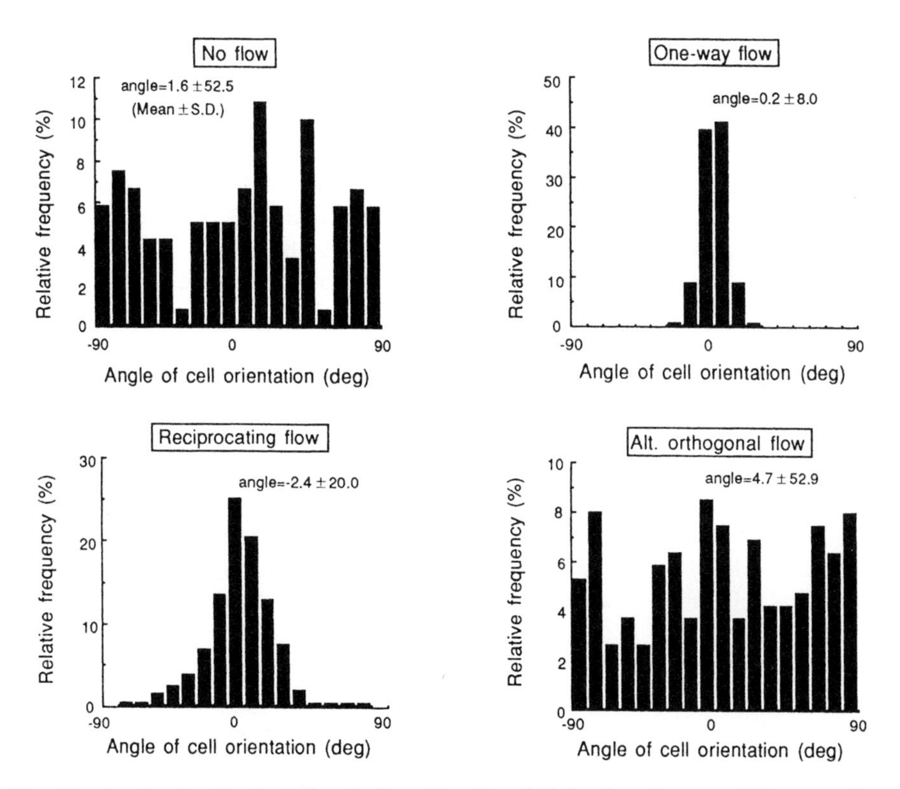

FIG. 12. Angle of cell orientation to flow direction (0°) for four flow conditions: *no-flow*, *one-way flow* (2 Pa, 24 h), *reciprocating flow* (2 Pa, 30-min interval, 24 h), and *alternately (Alt.) orthogonal flow* (2 Pa, 30-min interval, 24 h)

nonreversing sinusoidal shear stress, a 1-Hz reversing sinusoidal shear stress, and a 1-Hz oscillatory shear stress (mean flow rate = 0). Endothelial cells exposed to each different flow condition showed different response in cell shape, cell orientation, and F-actin filament structure depending on the flow state. Nonreversing sinusoidal flow resulted in less rapid cell shape change, but a more elongated cell shape. In the oscillatory flow, they did observe cells that remained polygonal in shape, as in the static culture, and no stress fibers developed in these cells.

Both in vivo and in vitro studies have revealed that vascular endothelial cell shape and cytoskeletal structure strongly reflect the environmental differences in blood flow. It has been demonstrated that cells in a high-shear environment have a more elongated shape and more prominent stress fiber formation, while those in the low-shear environment have a polygonal shape. On the other hand, the cell shape has been found to be polygonal in the regions of high permeability in in vivo studies [2,25]. Jo et al. [26] reported in their in vitro study that the permeability of cultured endothelial cells increased depending on the level of shear stress. In the atherosclerotic regions, such as arterial bends and branches, the blood flow is very complex with reverse and secondary flow. From these aspects, there is a possibility that the changes of the flow direction affect the morphology and function of the cells in the in vivo flow condition. In fact our experimental results showed that endothelial cells might recognize the flow direction and change their shape and cytoskeletal structure.

To examine the effects of the contact-inhibited confluent condition on cell morphology, we cultured bovine aortic endothelial cells, as described previously, and seeded a glass plate with a low cell concentration of 1×10^4 cells/cm^2. Flow experiments were started under the subconfluent conditions of endothelial cells using the same flow chamber. The area covered with endothelial cells or cell concentration in the subconfluent condition was almost one-tenth that in the confluent condition. Applied shear stress was controlled to 2 Pa for 1, 3, 6, 12, and 24 h. In some experiments, cell movement was recorded on videotape using a time-lapse video recorder and the cell motility in flow conditions was analyzed. After flow exposure, endothelial cells were fixed and their F-actin filaments were stained in the same way. Changes in the shape index of endothelial cells after exposure to the shear stress of 2 Pa are shown in Fig. 13. Even at the no-flow control conditions, the shape index was 0.47 in mean value, indicating relatively elongated cell shape. After exposure to shear stress for 24 h, the shape index was 0.32 (mean value). There was no significant difference between these two values: endothelial cells cultured in the sparse condition did not respond to a shear stress of 2 Pa.

This phenomenon was confirmed by observing the changes in cell angle after exposure to shear stress (Fig. 14). The angle of cell orientation at the no-flow condition was $-2° \pm 46°$ (mean ± SD), indicating that cells were almost randomly distributed. This randomness continued even after applying shear stress for 24 h: i.e., the angle of cell orientation was $-2° \pm 59°$. In the photomicrograph of F-actin filament microstructure after exposure to shear stress for 24 h (Fig. 15), the endothelial cells cultured in the sparse condition are observed to orient in ran-

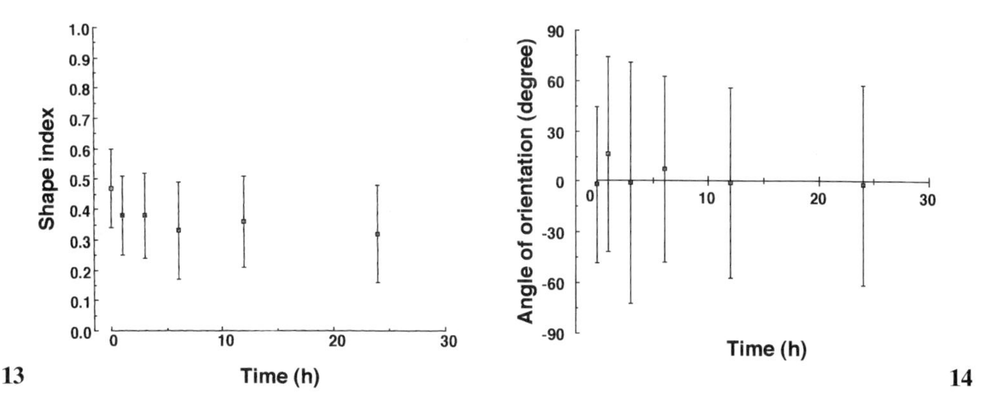

Fig. 13. Time-course changes of shape index values for endothelial cells cultured in sparse condition. Shear stress, 2 Pa

Fig. 14. Time-course changes of the angle of cell orientation for endothelial cells cultured in sparse condition. Shear stress, 2 Pa

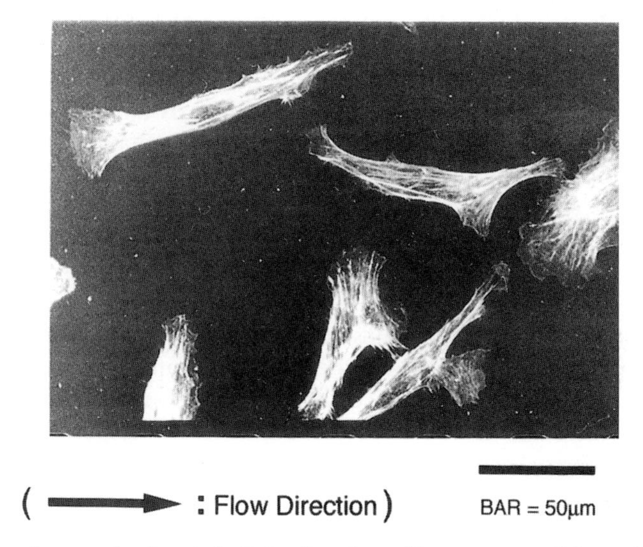

Fig. 15. Photomicrograph of endothelial cells cultured in sparse conditions after exposure to shear stress of 2 Pa for 24 h

dom directions. F-actin filaments were almost aligned with the long axis of the elongated cell. This was also observed for all the cells, independent of the duration of exposure to shear stress. As we have already seen, the endothelial cells cultured in the confluent condition responded to shear stress and changed their configuration. However, the cells cultured in the subconfluent condition did not show any response in configuration. What difference in response mechanisms to shear stress is considered to occur between these two cultured conditions? We have observed that intracellular calcium levels increased when applying fluid-imposed shear stress even for the cells cultured in the subconfluent conditions. Is it not true, then, that endothelial cells begin to respond to shear stress during the cell mitosis phase, do endothelial cells need to contact each other to respond to shear stress? We need to do further experiments to elucidate this phenomenon.

3 Mechanical Properties of Cultured Endothelial Cells

The viscoelastic deformation of endothelial cells cultured in the static condition and in shear stress-exposed conditions was measured using a micropipette technique [27,28]. Cultured endothelial cells were detached from the substrate using trypsin or a mechanical method in which a sterile wood stick was used to scratch the cells on the coverslip. The sheet of cell aggregates was detached further by aspiration and then suspended in the culture medium for observation through an inverted microscope. Each cell shape was spherical in appearance. A micropipette experiment was performed at room temperature (22°–24°C). The tip of the micropipette was caused to approach the surface of an endothelial cell by controlling the micropipette and the sample holder with two separate hydraulic micromanipulators. In this viscoelastic experiment, the desired negative pressure, produced by drawing water from the reservoir, was applied to the tip of the micropipette by opening the solenoid valve. The deformation process of the cell in the micropipette was observed through a TV camera, and the image and elapsed time were recorded on videotape.

3.1 Statically Cultured Endothelial Cells

Porcine aortic endothelial cells cultured in 12-well plastic plates were used for this experiment. In applying the micropipette technique to measurement of the elastic properties of cells, it was necessary to wait for several minutes before the aspirated length of the cell in the micropipette reached its asymptotic value. From this, it appeared that the endothelial cells exhibited a strong viscous behavior. Using cytoskeletal-disrupting agents, we investigated the influence of two different cytoskeletal elements, the F-actin filaments and the microtubules, on these viscoelastic properties.

The cells were divided into four groups: (1) cells detached by trypsin [control (T)]; (2) mechanically detached cells [control (M)]; (3) cells treated by $2\,\mu M$ cytochalasin B (CB); and (4) cells treated by $2\,\mu M$ colchicine. In these last two groups (3, 4), the cells were treated for about 14h before detachment, F-actin filament assembly of the cytoskeletal elements was disrupted by CB, and microtubule coils were depolymerized by colchicine. The cells treated by CB or colchicine were also detached from the substrate by a mechanical method and their viscoelastic properties measured using the micropipette technique.

A sequence of photographs of the progressive deformation of a trypsin-detached endothelial cell under a constant negative pressure, $\Delta p = 20\,\text{mmH}_2\text{O}$, is shown in Fig. 16. The diameter of the cells was approximately $16\,\mu$m and the internal diameter of the micropipette approximately $4\,\mu$m. The left photograph in the top row of Fig. 16 shows the cell made contact with the micropipette tip at time $t = 0$. The small arrowhead in each photograph indicates the aspirated portion of the cell, which is increasing with elapsed time. The magnitude of the deformation appeared to approach its maximum at about 4 min.

The time-course of the normalized aspirated length under constant aspiration pressure is shown in Fig. 17 for cells treated by the four different method. The cells already showed some deformation at time $t = 0^+$. After this initial deformation, which is considered to be an elastic response, the cell surface showed a creep displacement that was nonlinear with time. For example, in the case of the control (M) cell, the deformation reached an almost constant level after about 360 s. Almost the same results were obtained for the control (T) cell. In the case of the colchicine-treated cell, the time when the deformation became stable was about 240 s, and this time was slightly shorter than that of the control (M) cell.

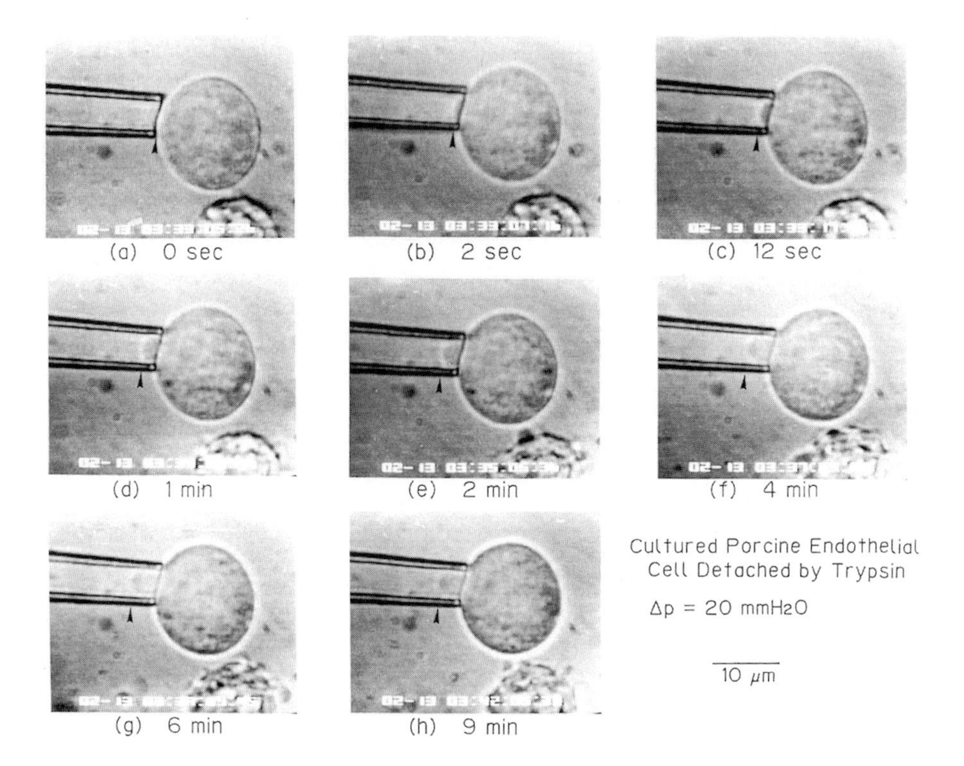

Fig. 16. Sequence of photomicrographs [(*a*)–(*h*)] of progressive deformation of a trypsin-detached endothelial cell into a micropipette tip. Aspiration pressure of $20\,\text{mmH}_2\text{O}$ was applied stepwise to the tip of the micropipette at time $t = 0$

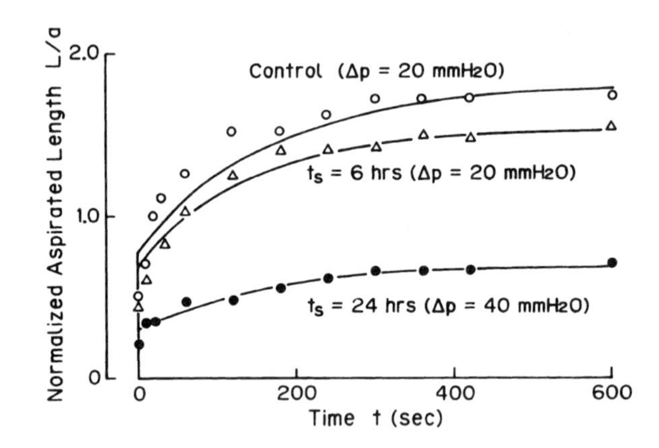

Fig. 17. Time-course changes (control, *open circles*; 6h, *triangles*; 24h, *solid circles*) in normalized aspirated length, *L/a*, for four differently treated cells cultured under static condition. *Solid lines* were obtained by fitting Eq. 2 to the data

Furthermore, because the deformation was larger even though the aspiration pressure was lower, $\Delta p = 10\,\mathrm{mmH_2O}$, the stiffness of this colchicine-treated cell would be smaller. The CB-treated cell showed a quite characteristic elastic deformation process; i.e., the deformation was rapid and large even for a relatively low aspiration pressure, $\Delta p = 5\,\mathrm{mmH_2O}$, when the time at which the deformation became stable was about 60s. From this result it appeared that the F-actin filaments were playing a very important role in the process of viscoelastic deformation.

To analyze the data, a standard linear viscoelastic half-space model of the endothelial cell [27,29] was used (Fig. 18). According to this model the aspirated length, L, is expressed as an exponential function of time, t:

$$L(t) = L_s\left[1 - k_2/(k_1 + k_2)\exp(-t/\tau)\right]h(t) \tag{2}$$

$$L_s = 2a\Delta P/\pi k_1 \tag{3}$$

where k_1 and k_2 are elastic constants, τ is time constant, a is the radius of the micropipette, and $h(t)$ denotes the unit step function; k_1 is determined from Eq. 3 by measuring the asymptotic aspirated length L_s. The coefficient of viscosity, μ, is expressed as follows:

$$\mu = \tau k_1 k_2/(k_1 + k_2) \tag{4}$$

Fitting Eq. 2 to the data of each group by a least-squares method, we then obtained the values of k_2, τ, and μ. Calculated curves are shown in Fig. 17 as solid lines.

Elastic constants k_1 and k_2, the coefficient of viscosity μ, and the time constant τ of endothelial cells treated with the four different methods were obtained by

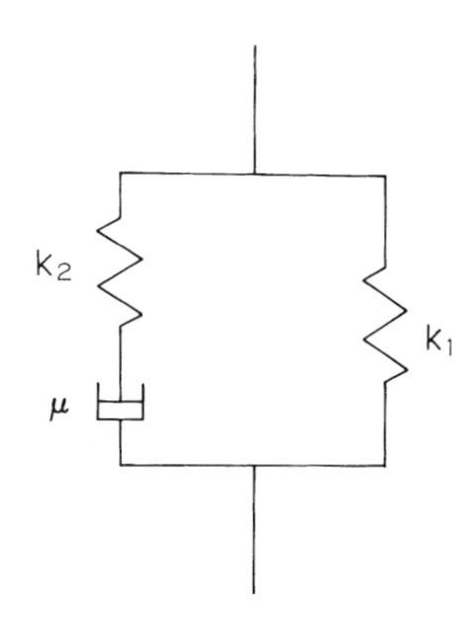

FIG. 18. Standard linear model for a viscoelastic solid. k_1, k_2, elastic constants; μ, coefficient of viscosity. (From [27], with permission)

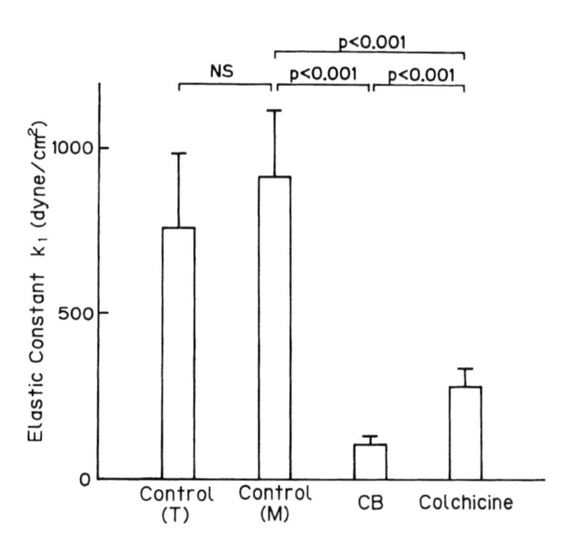

FIG. 19. Elastic constant k_1 obtained for four groups of statically cultured endothelial cells detached by trypin, *Control (T)*; cells detached mechanically, *Control (M)*; cells treated by 2 μM cytochalean *CB*; cells treated by 2 μM colchicine, *Colchicine*. Mean ± SD; 10 dyne/cm^2 = 1 Pa

applying these equations to experimental data. As an example of the obtained data, values of k_1 and k_2 are shown in Figs. 19 and 20. The k_1 value in the standard linear viscoelastic model for control (T) cells (76 ± 23 Pa, mean ± SD) was slightly lower than that for control (M) cell (92 ± 20 Pa). The values for CB-treated (11 ± 3 Pa) and colchicine-treated (28 ± 5 Pa) cells were significantly lower, approximately one-tenth and one-third of the control (M) value. Although we could not find statistically significant differences among the values of the other elastic constant, k_2, for the four different groups of cells, the mean k_2 values of CB-treated (44 ± 40 Pa) and colchicine-treated (65 ± 44 Pa) cells were considerably lower than those of the control cells (210 ± 140 Pa in control (T) and 190 ± 150 Pa in control (M)).

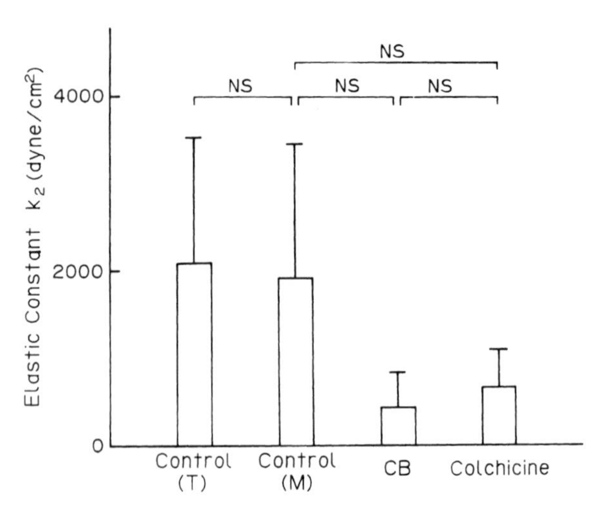

FIG. 20. Elastic constant k_2 obtained for four groups of statically cultured endothelial cells. Mean ± SD; 10 dyne/cm^2 = 1 Pa; *NS* not significant

The values of the coefficient of viscosity μ also showed results similar to those for the k_1 value. The coefficient of viscosity of control (T) cell ($8.3 \pm 4.0 \times 10^3$ Pa·s) had no significant difference with control (M) data ($7.2 \pm 2.7 \times 10^3$ Pa·s). Quite significant differences were found for the cells treated by CB ($1.3 \pm 1.1 \times 10^2$ Pa·s) and colchicine ($2.3 \pm 0.6 \times 10^3$ Pa·s). No significant difference in time constant was found comparing control [160 ± 50 s in control (T) and 130 ± 10 s in control (M)] and colchicine-treated cells (130 ± 30 s), and only the CB-treated cells exhibited a significantly different time constant (18 ± 15 s). This might come from the fact that the three viscoelastic parameters, k_1, k_2, and μ, affect the time constant in a rather comple manner.

In this study, the micropipette technique was applied to measure the viscoelastic behavior of cultured endothelial cells. The data were analyzed using the standard linear viscoelastic model and a homogenous, half-space geometric model. It should be stressed that the cultured endothelial cells exhibited a quite long time constant for deformation under a constant applied stress. This characteristic viscoelastic time constant appeared to mainly depend on the F-actin filaments, one of the components of a cell's cytoskeletal structure. Another cytoskeletal component, the microtubules, also exhibited a viscous behavior, with μ and the elastic constants k_1 and k_2 all being altered by the use of the agent colchicine. However, the time constant, being a ratio of viscosity to elastic modulus, was not significantly influenced by the use of colchicine. This was probably fortuitous.

3.2 Endothelial Cells Exposed to Shear Stress

Endothelial cells are oriented to the direction of flow and are more elongated in regions of higher shear stress. The cytoskeletal structure, especially the microfilaments, also becomes aligned with flow direction and develops in the central region of each cell, as previously noted. Our previous study [30] indicated that these cells become more resistant to micropipette aspiration, indicating a change in cell mechanical properties. Furthermore, from the viscoelastic properties of endothelial cells cultured in no-flow condition, the F-actin filaments are expected to play some important roles for such viscoelastic behavior. Viewing these points, we investigated the viscoelastic property of cultured endothelial cells exposed to fluid-imposed shear stress [28].

Porcine aortic endothelial cells cultured on glass coverslips were used in the fourth and fifth passage. These cells were divided into two groups, one being the cells cultured under no-flow, control conditions and the other cells cultured under the shear flow conditions of 2 Pa for 6 or 24 h. Fluid-imposed shear stress was applied using a parallel-plate flow chamber. After exposure to shear stress, the cells were detached from the substrate by scratching with a wood stick and then the micropipette technique was used to measure the viscoelasticity. Applying the half-space model mentioned previously to the experimental data, we obtained viscoelastic parameters such as elastic constants, viscosity, and time constant in the same way.

Cells exposed to shear stress for 6 h became spherical after detachment from the substrate. In the case of a 24-h exposure, about the half the detached cells

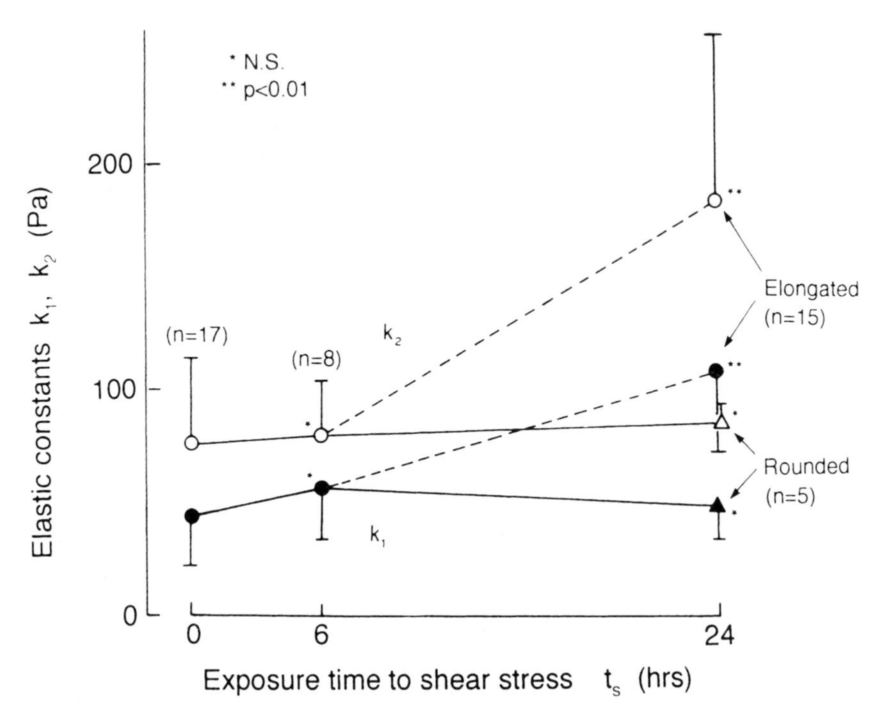

FIG. 21. Time-course changes in elastic constants, k_1 and k_2, after exposure to shear stress of 2 Pa. Mean ± SD; *, N.S.; **, $P < 0.01$. (From [28], with permission)

retained an elongated shape on detachment, with the others taking on a spherical shape. The k_1 and k_2 values as a function of exposure time to shear stress are shown in Fig. 21. Both values of 6-h exposure cells were not significantly different from the control cells, but 24-h-exposure cells with elongated shapes had significant, approximately twofold, higher values of elastic parameters. Cells exposed for 24h to shear stress, but which became spherical on detachment, did not show any significant change in either k_1 or k_2. The values of the coefficient of viscosity μ also became larger after exposure to shear stress (data not shown). The μ value of 6-h-exposure cells had no significant difference with control data; however, a significant difference was found for the cells exposed for 24h. No significant difference was found for the time constant τ in comparing control and shear-exposed cells.

The cultured porcine endothelial cells exposed to shear stress for 24h had high viscosity and stiff elastic properties in comparison to those cultured in no-flow conditions. Observation of the cytoskeletal structure suggests that this change in viscoelastic property is closely relate to the development and alignment of F-actin filaments. Because we do not have any information about the role of microtubules, we need to investigate this other important component of the cytoskeleton.

4 Events Found in Endothelial Cells After Exposure to Flow

After exposure to a fluid-imposed shear stress, cultured endothelial cells will show many different reactions, as was summarized by Girard et al. [10] (Table 1). We would recognize that some kinds of chain reaction might occur in endothelial cells at the onset of flow. At this relatively early stage, second messengers such as Ca^{2+} and inositol-1,4,5-trisphosphate (IP_3) are first activated [31,32]. Although it is phenomenologically known that a fluid flow certainly stimulates endothelial cells and induces the second-messenger activation, the recognition events and the signal transduction pathway in the cells are still not known. We do not even know what physical factors in flow, shear stress, shear strain, the secondary effects, or other factors are important and are dominant in the responses of cells [33,34].

Possible mechanisms of stress transmission and signal transduction were reviewed by Davies and Tripathi [35]. F-actin filaments are anchored in the plasma membrane at several sites and have been shown to play roles in stress transmission. These sites are (1) focal adhesions on the ventral surface, (2) intercellular adhesion proteins at the cell periphery, (3) integral membrane proteins at the apical cell surface, and (4) the nuclear membrane. These locations are considered as the likely sites of signal transduction [35]. After these events, disappearance of dense peripheral F-actin bands begins at about 30min. Morphological changes are then found in rearrangement of the F-actin filaments in cells, followed by cell elongation and orientation. During these phases, production of extracellular

TABLE 1. Time-course of events following onset of flow for bovine aortic endothelial cells cultured on a Thermanox or mylar surface exposed to moderate or high shear stress. (From [10], with permission).

Time after onset of flow (h)	Biological event
0.1	Second-messenger activation: Ca^{2+}, IP_3
0.2–0.5	Translocation of protein kinase C
0.5	Disappearance of dense peripheral F-actin bands is initiated; after peaking, IP_3 level by now has decreased to below initial, control value
2	Cell shape changes first observed
4	Cell orientation initiated
6	Amount of cell-associated fibronectic decreased
12	Extensive F-actin stress fibers aligned with the flow direction already are formed
24	Dense peripheral F-actin bands reappear; amount of cell-associated fibronectin increasing; intracellular IP_3 level now returning to initial, control value
36–48	Process of cell elongation-orientation complete; extensive F-actin stress fibers as well as dense peripheral bands; amount of cell-associated fibronectin comparable to, perhaps even greater than, initial control values

matrices from endothelial cells might be controlled to make cell shape changes easier. After an approximately 12-h exposure to flow, F-actin stress fibers are extensively formed in both the central and peripheral portions and aligned with the flow direction.

Production of fibronectin is reported to increase after the process of cell elongation and orientation has been completed [10]. These phenomena would seem to indicate that the endothelial cells had finished their morphological changes and would firmly adhere to the substrate, trying to adapt to the mechanical environment. Recent studies have clarified the phenomenological process of cell morphological changes and the events associated with the response of endothelial cells to the hemodynamic environment. However, the mechanisms of stress transmission and signal transduction in the cell are still unclear, and the process of F-actin filament rearrangement should be elucidated in understanding the mechanisms of cell shape change. This research, using cultured endothelial cells, has usually been approached by cellular biology and cellular engineering, but not physiology. In the near future we need to investigate the response mechanisms of endothelial cells from more physiological points of view to clearly show the relationship between atherogenesis and blood flow.

Acknowledgment. This work was supported in part by research grants from the Ministry of Education, Science and Culture, Japan (Grants-in-Aid for Scientific Research on Priority Areas, No. 04237104 and Grants-in-Aid for Scientific Research (C), No. 04670513). The authors thank S. Ujita and K. Kimura for their efforts in performing flow experiments and data analysis at Biomechanics Laboratories in Tohoku University.

References

1. Nerem RM, Cornhill JF (1980) The role of fluid mechanics in atherogenesis. J Biomech Eng 102:181–189
2. Schwartz CJ, Valente AJ, Sprague EA, Kelley JL, Nerem RM (1991) The pathogenesis of atherosclerosis: an overview. Clin Cardiol 14:I-1–I-16
3. Batten JR, Nerem RM (1982) Model study of flow in curved and planar arterial bifurcations. Cardiovasc Res 16:178–186
4. Yamamoto T, Tanaka H, Jones CJH, Lever MJ, Parker KH, Kimura A, Hiramatsu O, Ogasawara Y, Tsujioka K, Caro CG, Kajiya F (1992) Blood flow velocity profiles in the origin of the canine renal artery and their relevance in localizaton and development of atherosclerosis. Arterioscler Thromb 12:626–632
5. Fry DL (1968) Acute vascular endothelial changes associated with increased blood velocity. Circ Res 22:165–197
6. Caro CG, Firz-Gerald JM, Schroter RC (1971) Atheroma and arterial wall shear—observation, correlation and proposal of a shear-dependent mass-transfer mechanism for atherogenesis. Proc R Soc Lond (Biol) 177:109–159
7. Cornhill JF, Levesque MJ, Herderick EE, Nerem RM, Kilman JW, Vasco JS (1980) Quantitative study of the rabbit aortic endothelium using vascular casts. Atherosclerosis 35:321–337
8. Flaherty JT, Pierce JE, Ferrans VJ, Patel DJ, Tucker WK, Fry DL (1972) Endothelial nuclear patterns in the canine arterial tree wtih particular reference to hemodynamic events. Circ Res 30:23–33
9. Dewey CF Jr (1984) Effects of fluid flow on living vascular cells. J Biomech Eng 106:31–35
10. Girard PR, Helmlinger G, Nerem RM (1993) Shear stress effects on the morphology and cytomatrix of cultured vascular endothelial cells. Academic Press, San Diego, pp 193–222
11. Silkworth JB, Stehbens W (1975) The shape of endothelial cells in en face preparations of rabbit blood vessels. Angiology 26:474–487
12. Shirinsky V, Antonov A, Birukov K, Sobolevsky A, Romanov Y, Kabaeva N, Antonova G, Smirnov V (1989) Mechanochemical control of human endothelium orientation and size. J Cell Biol 109:331–339
13. Levesque MJ, Nerem RM (1989) The study of rheological effects on vascular endothelial cells in culture. Biorheology 26:345–357
14. Davies PF, Remuzzi A, Gordon EJ, Dewey CF Jr, Gimbrone MA Jr (1986) Turbulent fluid shear stress induced vascular endothelial cell turnover in vitro. Proc Natl Acad Sci USA 83:2114–2117
15. Helmlinger G, Geiger RV, Schreck S, Nerem RM (1991) Effects of pulsatile flow on cultured vascular endothelial cell morphology. J Biomech Eng 113:123–131
16. Rogers KA, Kalnins V (1983) Comparison of the cytoskeleton in aortic endothelial cells in situ and in vitro. Lab Invest 49:650–654
17. Kim DW, Langille BL, Wong MKK, Gotlieb AI (1989) Patterns of endothelial microfilament distribution in the rabbit aorta in situ. Circ Res 64:21–31
18. Wong AJ, Pollard TD, Herman IM (1983) Actin filament stress fibers in vascular endothelial cells in vivo. Science 219:867–869
19. Sato M, Ohshima N (1994) Flow-induced changes in shape and cytoskeletal structure of vascular endothelial cells. Biorheology 31:143–153

20. Okano M, Yoshida Y (1992) Endothelial cell morphometry of atherosclerotic lesions and flow profiles at aortic bifurcations in cholesterol-fed rabbits. J Biomech Eng 114:301–308
21. Kataoka N, Ujita S, Ogawa Y, Sato M (1994) Flow direction affects shape and cytoskeletal structure of cultured bovine aotric endothelial cells. In: Biofluid mechanics, proceedings of the 3rd international symposium, vol 17, VDI, Düsseldorf, pp 551–556
22. Ookawa K, Sato M, Ohshima N (1992) Changes in the microstructure of cultured porcine aortic endothelial cells in the early stage after applying a fluid-imposed shear stress. J Biomech 25:1321–1328
23. Ookawa K, Sato M, Ohshima N (1993) Morphological changes of endothelial cells after exposure to fluid-imposed shear stress: differential responses induced by extracellular matrices. Biorheology 30:131–140
24. Gupte A, Frangos JA (1990) Effects of flow on the synthesis and release of fibronectin by endothelial cells. In Vitro Cell Dev Biol 26:57–60
25. Gerrity RG, Richardson M, Somer JB, Bell FP, Schwartz CJ (1977) Endothelial cell morphology in area of in vivo Evans blue uptake in the aorta of young pigs: II. Ultrastructure of the intima in area of differing permeability to proteins. Am J Pathol 98:313–334
26. Jo N, Dull RO, Hollis TM, Tarbell JM (1991) Endothelial albumin permeability is shear dependent, time dependent, and reversible. Am J Physiol 26:H1992–H1996
27. Sato M, Theret DP, Wheeler LT, Ohshima N, Nerem RM (1990) Application of the micropipette technique to the measurement of cultured porcine aortic endothelial cell viscoelastic properties. J Biomech Eng 112:263–268
28. Sato M, Ohshima N (1995) Viscoelastic properties of cultured endothelial cells. J Biomech (in press)
29. Theret DP, Levesque MJ, Sato M, Nerem RM, Wheeler LT (1988) The application of a homogeneous half-space model in the analysis of endothelial cell micropipette measurements. J Biomech Eng 110:190–199
30. Sato M, Levesque MJ, Nerem RM (1987) Micropipette aspiration of cultured bovine aortic endothelial cells exposed to shear stress. Arteriosclerosis 7:276–286
31. Ando J, Komatsuda T, Kamiya A (1988) Cytoplasmic calcium responses to fluid shear stress in cultured vascular endothelial cells. In Vitro Cell Dev Biol 24:871–877
32. Nollert MU, Eskin SG, McIntire LV (1990) Shear stress increases inositol trisphosphate levels in human endothelial cells. Biochem Biophys Res Commun 170:281–287
33. Ando J, Kamiya A (1993) Wall shear stress rather than shear rate regulates cytoplasmic Ca^{++} responses to flow in vascular endothelial cells. Biochem Biophys Res Commun 190:716–723
34. Ando J, Ohtsuka A, Katayama Y, Korenaga R, Ishikawa G, Kamiya A (1994) Intracellular calcium response to directly applied mechanical shearing force in cultured vascular endothelial cells. Biorheology 31:57–68
35. Davies PF, Tripathi SC (1993) Mechanical stress mechanisms and the cell—an endothelial paradigm. Circ Res 72:239–245

Responses of Vascular Endothelial Cells to Fluid Shear Stress: Mechanism

Akira Kamiya[1] and Joji Ando[2]

Summary. The endothelial cells (ECs) that cover the inner surface of every vessel wall as a monolayer have a variety of cell activities and play important roles in maintaining the vascular wall morphology and function, showing both self-adaptive actions and local interactive responses with other cells such as vascular smooth muscle cells and blood corpscules. Recent evidence suggests that these functions are almost all affected by fluid shear stress on the endothelial wall induced by blood flow. In vivo studies have shown that wall shear stress regulates adaptive vessel growth and angiogenesis and is essentially involved in the pathogenesis of atherosclerosis. In vitro evidence indicates that shear stress modulates the production of a variety of such vasoactive substances as histamine, nitric oxide, prostacyclin, and endothelin and controls permeability and endocytosis of macromolecules. More recent studies have revealed that this stress regulates endothelial mRNA levels of proteins including tissue plasminogen activator, platelet-derived growth factor, and adhesion molecules, through a core element at the promoter of genes. These facts suggest that EC has a mechano-sensor to detect changes in shear stress and to transmit the signal to intracellular organelles. It is established that the cytoplasmic Ca^{2+} response is involved in this mechanism as the second messenger of the internal signaling system. Theoretical analyses also suggest that these shear-dependent EC activities functionally optimize the entire vascular system for oxygen transport.

Key words: Vascular endothelial cells—Fluid shear stress—Cytoplasmic calcium ion—Gene expression—Optimum models

1 Introduction

In the past two decades, much attention has been paid to the function and biomechanics of the vascular endothelial cells (ECs). The reason is that many studies have successively offered evidence indicating various important roles of

[1] Institute of Medical Electronics and [2] Department of Cardiovascular Biomechanics, Faculty of Medicine, University of Tokyo, 7-3-1 Hongo, Bunkyo-ku, Tokyo, 113 Japan

ECs in maintaining the physiological functions of the cardiovascular system. It is now well recognized that the EC monolayer is not merely an anticoagulant or antipermeable passive barrier between flowing blood and vascular tissue, but an active layer controlling vascular growth and remodeling and regulating vascular tonus via secretion of both relaxant and contractile factors. ECs also interact with other cells by means of various cell growth factors, adhesion proteins, and other chemical mediators. In addition, many reports have revealed that ECs respond not only to chemical stimuli such as hormones, autacoids, and neurotransmitters, but also to mechanical stimuli such as shear stress elicited by blood flow or tensile stress generated by blood pressure.

Investigation of the influence of mechanical stresses on EC functions, i.e., EC biomechanics, is expected to shed light on the physiological and molecular biological basis of vascular functions and also the pathogenesis of various vascular diseases including atherosclerosis. This category of studies, interdisciplinarily extending over biology, medicine, and engineering, is now forming a new frontier in biomechanics.

We outline here recent findings concerning the effects of fluid shear on EC functions, mainly focusing on our own studies. More detailed reviews are found elsewhere [1,2].

2 In Vivo Adaptive Responses of Vascular Wall to Fluid Shear

Statistical analyses of vascular branching systems [3–5] suggest that when blood flow (f) through a vessel is related to its internal radius (r) as

$$f/r^n = \text{constant} \tag{1}$$

the best-fitted value of n is nearly equal to 3. The mechanism inducing such a flow–radius (f–r) relationship has long been suspected to be the adaptation of vessel wall to flow rate. About 100 years ago, Thoma [6] documented that vascular branching (angiogenesis) and growth in the chick embryo were observed to be rapid in vessels with high flow. Later, Rodbard [7] put forward the hypothesis that the viscous drag generated by blood flow, which is now commonly called the wall shear stress, may regulate the vascular caliber. His hypothesis was based on many clinical findings, such as vessel dilatation at arteriovenous fistulas or dilatation and atrophy of the maternal uterine artery during pregnancy and after delivery, and so on.

Kamiya and Togawa [8] tried to verify this mechanism experimentally, because they noticed that this is a persuasive mechanism which can adaptively autoregulate the vascular system to hold the statistical f–r relationship, because wall shear stress (τ) for Poiseuille flow is written as

$$\tau = \mu\dot{\gamma} = 4\mu \mathrm{f}/\left(\pi \mathrm{r}^3\right) \tag{2}$$

where μ and $\dot{\gamma}$ are blood viscosity and wall shear rate (the perpendicular gradient of flow velocity profile at the wall surface), respectively. Consequently, when flow is increased or decreased and if the concomitant changes in shear stress induces enlargement or reduction of radius, the stress τ is eventually kept constant, because the third power of radius (r^3) may exert a potent negative feedback effect on τ. Constant τ in Eq. 2 is consistent with the f-r relationship in Eq. 1 for $n = 3$.

An in vivo experimental study was performed by constructing an arteriovenous shunt between the canine carotid artery and the external jugular vein (Fig. 1a), and by measuring changes in arterial blood flow rate and internal diameter at various times after the shunt operation. Fig. 1b,c summarizes the results, showing gradual alterations in the shunted/control ratio of the arterial radius and changes in wall shear stress plotted against the flow ratio. From these results it is evident that the vascular wall responded to sustained changes in blood flow as expected and regulated the radii to maintain the stress constant.

This finding was later confirmed by several research groups [9–12]. They examined the adaptive vessel diameter changes to the shear stress loads at several arteries in various animals, by increasing blood flow with a similar arteriovenous shunt or by decreasing flow with a flow-restricting clip. Langille and O'Donnell [10] found that the reduction of arterial diameter to diminished blood flow was abolished when the endothelium was removed from the vessel, indicating that the vascular wall response to fluid shear is endothelium dependent. More recently, Masuda's group [12] reported that, at the EC-desquamated portion of the shunted artery, the arterial dilatation to the flow load did not occur, and showed evidence that vascular response to the increased flow is also EC dependent.

It is also well documented that blood flow or shear stress is one of the regulatory factors of capillary angiogenesis [13,14]. According to Hudlicka et al. [14], sustained increase in peripheral blood flow in the skeletal and cardiac muscles induced by repeated electrical nerve stimulation or by chronic vasodilator administration stimulated capillary angiogenesis to augment capillary density; Wang and Prewitt [15] observed that the size and number of arterioles was attenuated by reducing peripheral blood flow.

These in vivo studies clarified the important roles of EC responses to fluid shear in the adaptive remodeling of large vessels and capillary angiogenesis. In vivo studies were followed by in vitro studies for examining the influences of fluid shearing force on EC activities in more detail, by virtue of developed cell culture techniques and molecular biological approaches to cell functions. The findings from such in vitro experiments using cultured ECs and flow-loading devices are described in the following sections together with related in vivo data.

3 Effects of Fluid Shear on EC Proliferation and Migration

One of the most striking characteristics of EC function is that ECs always construct a confluent monolayer in the culture dish, as they do in vivo. If a portion of such a monolayer is denuded, surrounding ECs migrate and proliferate to cover

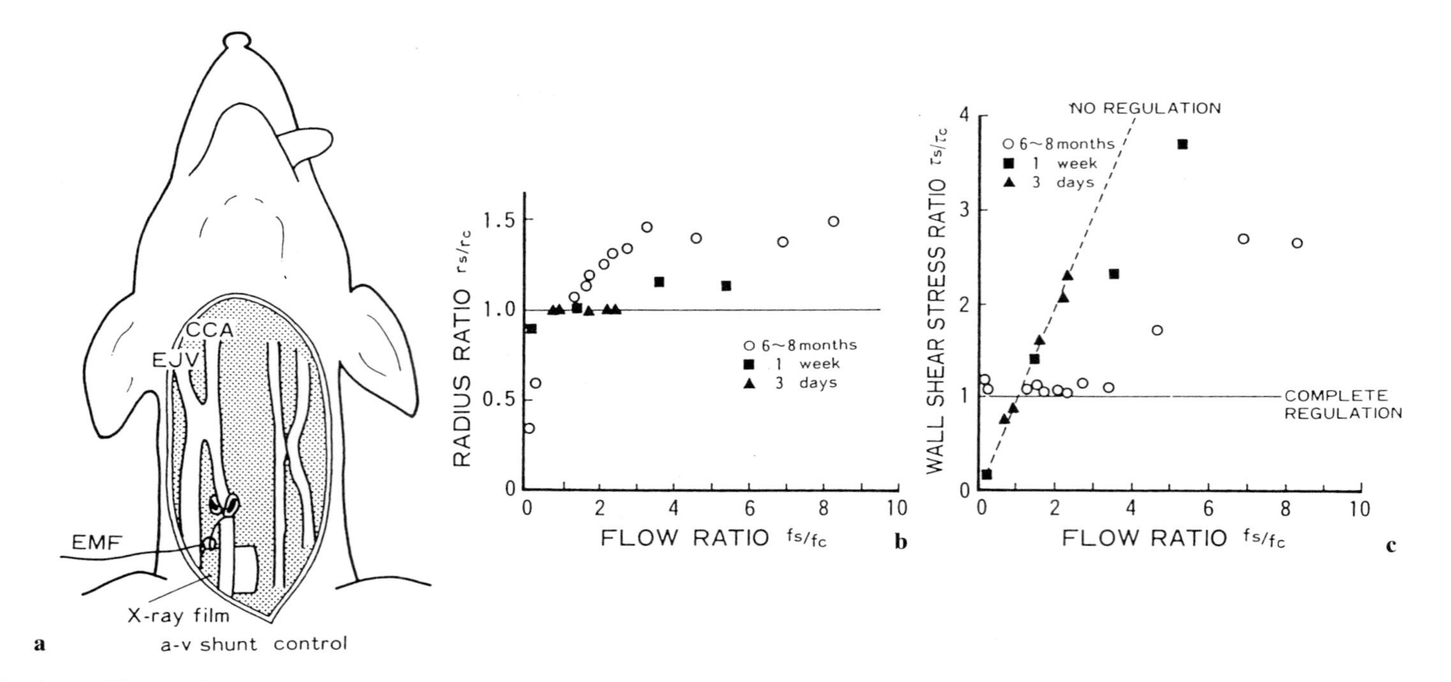

Fig. 1a–c. The arteriovenous (*a-v*) shunt experiment. The results show the adaptive responses of vascular internal radius to flow regulated by wall shear stress. **a** The a-v shunt was constructed between the canine common carotid artery (*CCA*) and the external jugular vein (*EJV*). After a flow-loading period (3 days, 1 week, or 6–8 months), the blood flow rate (f_s) and the internal radius (r_s) of the shunted artery were measured using electromagnetic flow metering (EMF) and angiography (X-ray film) to compare them with those (f_c, r_c) of the ipsilateral control artery. **b** The obtained radius ratio (r_s/r_c) plotted against flow ratio (f_s/f_c). **c** The relationships of shear stress ratio (τ_s/τ_c), calculated from radius ratio (r_s/r_c) by Eq. 2 versus flow ratio (f_s/f_c). Note that for the chronic data (6–8 months) with flow ratio less than 4, the values of shear stress ratio had shifted close to the line of unity, indicating complete regulation of radius by wall shear stress. (From [8], with permission)

the denuded area. Ando et al. [16,17] examined the effect of fluid shear application on EC regeneration in this repairing process. Following mechanical denudation of a part of a cultured confluent monolayer [16], they exposed ECs to shear stress (0.4–1.4 dyne/cm^2) for 24 h in a flow-loading device (a rotating-disk type) and counted the number of cells that migrated to and proliferated in the denuded area. The number of cells in the recovered area was much greater under fluid shear than under the static condition, suggesting that shear application stimulates EC regeneration in the repairing process. In addition, the area covered by ECs in the downstream site of the flow was larger than that in the upstream site, indicating a vector effect of fluid shear on EC. In the long-term observation of single EC motility using a video recording, Masuda and Fujiwara [18] found a similar vector effect that the individual cell moves at the same speed in all directions immediately after flow application but thereafter tends to go downstream.

Figure 2 shows the shear effects on EC proliferation during the repair of mechanical denudation observed in the DNA histogram [17]. It shows that the cells subjected to wall shear stress (1.3–2.1 dyne/cm^2, 24 h) contain a smaller relative population of cells in the resting state (left peak; G_1/G_0 phases) than the static control and a higher population in the mitotic state (right peak; S_1, G_2, and M phases of the cell cycle). It is obvious that fluid shear enhances DNA synthesis and stimulates ECs to divide.

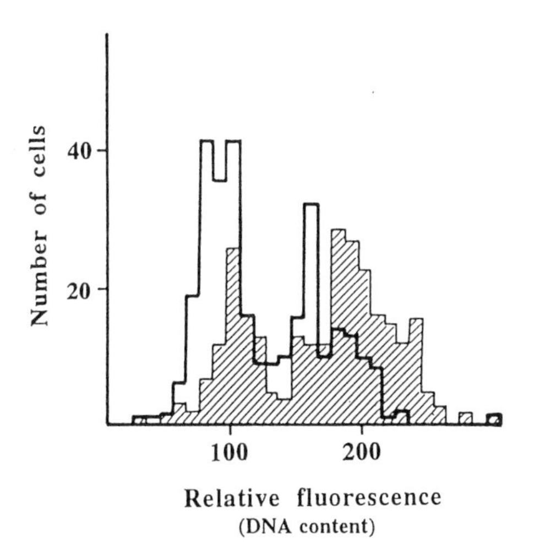

Fig. 2. The shear effects on endothelial cell (EC) DNA synthesis during repair of mechanical denudation. Cell nuclei were stained with propidium iodide; their DNA histogram was obtained by photometric fluorescence microscopy. In the shear-loaded cultures, the number of cells decreased in the G_0 and G_1 phases of the cell cycle (*left peak*), while it increased in the G_2 and M phases, which suggests a stimulatory effect of shear stress on DNA synthesis in ECs. *Open bars*, static control; *shaded bars*, shear stress (1.3–2.1 dynes/cm^2, 24 h). (From [17], with permission)

In the confluent monolayer, however, the effects of shear application on cell proliferation are still controversial. Eskin et al. [19] applied a shear stress of 3.3 dyne/cm^2 to bovine cultured EC for 1–2 weeks and observed a 1.5- to 5-fold increase in cell density compared with static controls; Dewey [20] however reported neither a change in cell density after a 1-week application of shear stress of 5 dyne/cm^2 nor an increase in [^{3}H]thymidine uptake under the shear stress application of 1–8 dyne/cm^2 for 24 h. On the other hand, Davies et al. [21] showed that turbulent shear stress as low as 1.5 dyne/cm^2 stimulated substantial DNA synthesis of ECs, although laminar shear stress (ranging from 8 to 15 dyne/cm^2) revealed no effect. They suggested that unstable characteristics of shear stress, rather than its magnitude, are the major determinant of the effect of shear in EC growth. In the in vivo study concerning the shear effects on EC growth, Masuda et al. [22] showed that EC density increased twofold in a month when the wall shear stress had been increased up to fivefold by making an arteriovenous shunt in a canine carotid artery.

In addition to shear stress, ECs in vivo are subjected to stretch tension generated by blood pressure. This mechanical stress seems to stimulate EC growth as well. The uptake of [^{3}H]thymidine by ECs cultured on an elastic membrane was reported to increase markedly following cyclic stretching [23].

4 Effects of Shearing Force on EC Morphology

The shape and orientation of EC are known to be influenced by blood flow in vivo; in regions where blood flows unidirectionally at high speed, ECs are ellipsoidal in shape, with their long axis aligned in the direction of flow, while in regions where blood flows at low speed or stagnates they are rounded and show no definite orientation. Flaherty et al. [24] surgically removed a segment of canine thoracic aorta, where ECs were aligning along the flow stream, and reimplanted it in a vertical direction from its original orientation. They observed that the EC nuclear pattern was realigned along the flow direction within 10 days after the surgery. In the in vitro experiments, Dewey et al. [25] applied 8-dyne/cm^2 shear stress to cultured calf ECs in a flow-loading device of the cone-plate type. No significant cell reorientation was observed until 24 h, but by 48 h, the cells had uniformly reassumed an ellipsoidal configuration with their major axis aligned in the flow direction. It has now been confirmed that the onset of cell shape change and orientation is sensitive to both the duration and magnitude of the shear stress applied and that these shear-induced changes in morphology are reversible.

From the sensitive effect of flow on the EC shape and orientation, it is anticipated that the cytoskeleton, forming cell shape and structure, can be affected by flow. Actually, actin filaments in ECs exposed to a high flow rate are aligned parallel to the flow direction and form striking rich stress fibers [26,27]. When shear stress was experimentally increased in a canine carotid artery by an arteriovenous shunt, the number of bundles of 6- to 7-nm microfilaments in the ECs were increased [28]. Kim et al. [29,30] observed in experimentally coarcted aortas that the stress fibers in a high-shear region upstream of the coarctation became markedly thicker and longer than in control aortas. Similar findings were obtained in vitro.

Franke et al. [31] applied shear stress to cultured human umbilical vein ECs (HUVEC) in a cone-plate apparatus and observed that stress fibers, which were not seen in static controls, were formed in the shear-loaded ECs. Wechezak et al. [32,33] also reported that under the influence of shear stress, F-actin stress fibers in cultured bovine ECs were aligned with flow direction; that the actin reorganization was accompanied by redistribution of the underlying fibronectin matrix; and that following the stress application, focal contacts that coincided with the termini of stress fibers at the basal plasma membrane were concentrated in the downstream regions of the cell.

5 Fluid Shear Effects on Protein and Nitric Oxide Synthesis

Shear stress modulates the synthesis and secretion of various substances by ECs. Concerning histamine, which increases endothelial permeability, it was reported that when cultured bovine ECs were subjected to shear stress of 2.8–6.2 dyne/cm^2 for 1.5 h in a flow chamber, the activity of histidine decarboxylase in ECs increased 2.8- to 3.7 fold more than the static control [34]. By applying mean shear stress ranging from 22 to 109 dyne/cm^2 with pulsatile perfusion of rabbit aortas with platelet-free blood, DeForest and Hollis [35] found a high correlation between the shear level and the EC histamine-forming capacity.

The fluid shear effect on production of prostacyclin, a potent vasodilator and endogenous inhibitor of platelet aggregation, was first reported by Reeves et al. [36]. When the perfusion flow to an isolated rat lung was enhanced, the minute production of prostacyclin measured from its stable metabolite, 6-keto-prostaglandin $F_1\alpha$, increased linearly with the flow rate. Frangos et al. [37] applied a shear stress of 0.016–10 dyne/cm^2 to HUVEC and observed that the onset of shear stress led to a sudden burst of prostacyclin production that decreased to a steady state within several minutes. The rate of steady-state production by cells subjected to pulsatile shear stress (range, 1.2–8 dyne/cm^2; frequency, 1 Hz) was more than twice that of cells exposed to steady shear stress (10 dyne/cm^2) and 16 times greater than that of cells in stationary culture. Grabowski et al. [38] reported, concerning the prostacyclin production burst induced by fluid shear in cultured bovine ECs, that only the peak rate of the burst increased in a flow-rate-dependent manner, and that once stressed, cells did not produce additional prostacyclin in response to repeated stimulation.

The production and release of endothelium-derived relaxing factor (EDRF), a potent vasodilator and inhibitor of platelet aggregation like prostacyclin, are also found to be modulated by fluid shear. In blood-perfused coronary artery, Holtz et al. [39] found EDRF, not prostacyclin, to be the sustained mediator of the flow-dependent, endothelium-mediated dilation of the artery. This finding was confirmed by Rubanyi et al. [40] and by Miller and Vanhoutte [41]. In an in vitro study using cultured ECs, Cooke et al. [42] showed that when bovine aortic ECs cultured on microcarrier beads were subjected to vortical flow in a bath by magnetic stirring, relaxation of vascular rings in the same bath occurred in a flow-rate-dependent manner. The authors also indicated that the mediator of this EDRF response is nitric oxide (NO) or NO-related substances, because the flow-rate-dependent relaxations were inhibited by methylene blue and hemoglobin, which interfere with biosynthesis of NO. More recently, Korenaga et al. [43] clearly demonstrated shear-dependent NO production in ECs by measuring concomitant responses in cGMP, instead of directly detecting the NO, which is quite unstable. The working hypothesis underlying this study is that the cGMP level of ECs is regulated by NO in a concentration-dependent manner through cyclic guanylate activation. If cGMP levels in ECs depend on flow rate, and if these cGMP responses are completely abolished by a selective inhibitor of NO [e.g., N^{ω}-monomethyl-L-arginine (L-NMMA)], then cGMP levels reflect NO production rates in ECs. As shown in Fig. 3, the results of the in vitro experiments of Korenaga et al. verified the validity of their hypothesis. The authors also demonstrated that NO response depends on both applied shear level and extracellular ATP concentration in its magnitude and is reversible.

The effects of shear on endothelin (ET) release by ECs remain poorly defined. Yoshizumi et al. [44] measured ET released from cultured porcine ECs by using an enzyme-linked immunosorbent assay (ELISA) following the applica-

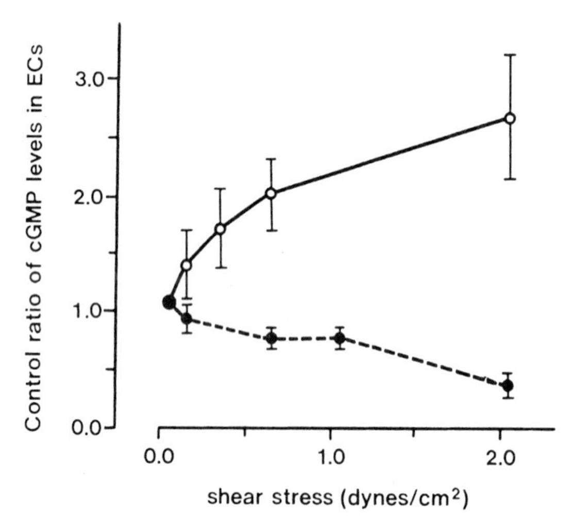

Fig. 3. The effect of fluid shear on nitric oxide (NO) production in ECs. NO production rate was monitored from cyclic guanosine monophosphate (cGMP) levels, induced through guanilate cyclase activation by NO and completely abolished by a specific NO inhibitor, L-NMMA (N^{ω}-monomethyl-L-arginine). Shear stress dependency of NO production in ECs was confirmed. *Open circles*, control; *solid circles*, L-NMA (ATP, 1 μM). (From [43], with permission)

tion of a shear stress of 5 dyne/cm^2 for 24 or 48 h in a cone-plate apparatus. The amount of ET increased from 0 for the control to 3.5 ng/ml after 24 h to 5 ng/ml after 48 h, and the level of ET mRNA showed a peak 2–4 h after the initiation of shear application. The shear-induced increase in the production of ET was confirmed by Milner et al. [45] in rabbit ECs. However, contradictory results were reported by Sharefkin et al. [46]. They tested the effect of a shear stress of 25 dyne/cm^2 on ET-1 precursor mRNA and ET peptide release in HUVEC in parallel-plate flow chambers and found that application of shear stress for 24 h reduced the levels of this mRNA and also decreased ET-1 peptide release. The reason for the discrepancy is not clear, but it may be the difference in the intensity of shear stress applied or the phase-dependent expression of ET to the load.

Shear application also modulates the production of tissue plasminogen activator (tPA) in ECs. Diamond et al. [47] observed that low (venous) shear stress (4 dyne/cm^2) had no effect on tPA secretion but that at high (arterial) shear stresses of 15 and 25 dyne/cm^2, the tPA secretion rate was 2.1- and 3.0 fold greater than that of the basal tPA secretion, respectively. The increase in tPA production was not observed in 4–6 h but did occur 10–30 h after shear application, indicating that the tPA response to shear is very slow. tPA mRNA levels were also found to increase in ECs exposed to fluid shear [48].

The production of platelet-derived growth factor (PDGF), which stimulates smooth muscle cell (SMC) growth and constricts vessels, is affected by fluid shear as well. The mRNA levels of both PDGF-A and PDGF-B in HUVEC increased shear dependently [49]. Ono et al. [50] observed that SMC migration factors were released from ECs by fluid shear loading and surmised that PDGF might be related to the phenomenon. They also showed the enhanced synthesis of collagen and an extracellular matrix produced by ECs because of flow load.

In some cases, however, flow application to ECs downregulates their synthetic activities. For instance, EC production of fibronectin, one of the typical adhesive proteins, is decreased by fluid shear [51]. The production of proteoheparan sulfate, a glycoprotein, has also revealed a downregulatory response to applied shear [52]. By utilizing immunostain with monoclonal antibodies and flow cytometry, Ohtsuka et al. [53] recently found that flow application to mouse lymph node ECs markedly reduced the expression of vascular cell adhesion molecule-1 (VCAM-1) but induced no change in another adhesion molecule, CD44. This VCAM-1 response to flow was both fluid shear- and loading time dependent and was also reversible. The detailed nature of this downregulatory response is described later.

6 Other Shear-Dependent EC Functions

It has been known since the work of Fry [54] that shear stress exerts an influence on endothelial permeability. Jo et al. [55] have reported data showing that EC permeability is sensitive to shear stress. Bovine aortic ECs cultured on a filter membrane were subjected to flow in an in vitro flow system, and simultaneous changes in permeability for albumin were measured. The results showed that permeability increased 4.3 and 10.3 fold in an hour at 1 and 10 dyne/cm^2 shear stress, respectively. After removal of the shear stress, permeability returned to preshear levels within 2h. By using chromium ethylenediaminetetraacetate (Cr-EDTA) as a test tracer having a molecular weight comparable to that of sugar, Shibata and Kamiya [56] observed about a twofold increase in transcapillary permeability in the rabbit tenuissimus muscle as capillary flow increased from 500 to 2500 μm/s, and suggested from these in vivo data that increased capillary wall shear induced this permeability enhancement.

Pinocytosis, as measured by uptake of horseradish peroxidase by cultured bovine ECs, was found to increase in response to fluid shear, time and amplitude dependently [57]. Shear stress also enhances receptor-mediated low-density lipoprotein internalization and degradation by ECs [58].

Adhesive interaction between blood cells and ECs is modulated by shear stress as well. Shear stress values less than 2 dyne/cm^2 decrease the rate of adherence of neutrophils to ECs [59], suggesting that the shearing force might be an important determinant of the number of neutrophils which adhere to and migrate under the endothelium in postcapillary venules or inflammatory sites. Further studies are required to elucidate whether shear stress acts as a physical force to detach neutrophils or modulates EC adhesiveness by changing the expression of adhesive molecules on their surface, as suggested by VCAM-1 downregulation as described previously [53].

Changes in mechanical properties of ECs from shear stress have been reported by Sato et al. [60]. After cultured bovine aortic ECs were exposed to shear stress, the authors measured cell stiffness by aspirating cells with a micropipette. The results showed that the mechanical stiffness of shear-loaded cells was significantly greater than that of control cells.

7 Transduction of Flow Signal

ECs have the ability to change their morphology and functions in response to fluid shear as described. This strongly suggests the presence of a mechanism by which ECs recognize the shear level and mediate the signal to intracellular organelles. Recent findings concerning this signal-transduction process are summarized in this section.

It has been generally accepted that a wide variety of cell types have an internal signaling system. When external stimuli arrive at the cell membrane, they are recognized by specific receptors on the membrane and are transferred into changes in second messengers such as G protein, inositol triphosphate, diacylglycerol, cyclic adenosine monophosphate (cAMP), cGMP, and Ca^{2+}. These second messengers are known to regulate cell functions via many biochemical processes. Ca^{2+} response in ECs to shear stress was first reported by Ando et al. [61]. Nollert et al. [62] also found flow-induced inositol triphosphate transients in ECs. In addition, it is reported that shear stress increases the frequency of opening of the K^+ channel stress dependently [63], leading to hyperpolarization of the cell membrane [64]. This response of the K^+ channel might be associated with the cytoplasmic Ca^{2+} increase because ECs are known to have Ca^{2+}-dependent K^+ channels.

The mechanism of intracellular Ca^{2+} response was further analyzed in our studies by using cultured monolayers of bovine aortic ECs loaded with the fluorescent Ca^{2+} probe Fura-2 [65]. The cell layer was exposed to a laminar flow in a flow chamber, and change in intracellular Ca^{2+} concentration, $[Ca^{2+}]_i$ was simultaneously monitored by a photometric fluorescence microscope. As is shown in Fig. 4, the onset of flow induced an immediate increase in $[Ca^{2+}]_i$, followed by a gradual decline to attain a plateau level, and when the flow was stopped the concentration returned to the control level. When extracellular Ca^{2+} concentration was reduced to zero by adding 2 m*M* ethyleneglycoltetraacetic acid (EGTA) to the medium, only the initial peak response remained and the later slow component disappeared, indicating that the initial response is from intracel-

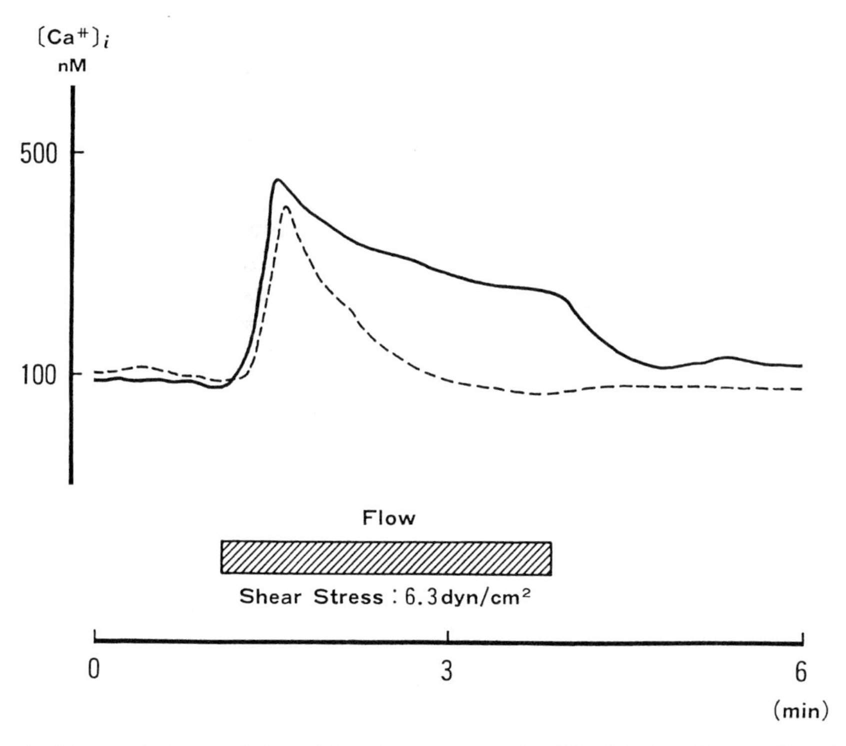

FIG. 4. Changes in intracellular calcium ion concentration $[Ca^{2+}]_i$ induced by medium flow. When Ca^{2+} in the external medium was depleted by 2 m*M* ethyleneglycotetraacetic acid (EGTA), only the initial peak response remained, indicating that the early components were from the intracellular calcium stores and the rest was the result of calcium influx from the external medium. *Solid line*, control; *dashed line*, 2 m*M* EGTA. (From [65], with permission)

lular Ca^{2+} stores such as mitochondria or endoplasmic reticulum and the later component was caused by Ca^{2+} influx from the external medium.

The magnitude of such a flow-induced $[Ca^{2+}]_i$ response was found to be strongly modified by the extracellular ATP level [65]. When ATP concentration in the external medium was less than 100 n*M*, no noticeable responses to flow change were observed. In the presence of 500 n*M* ATP, a stepwise elevation of flow rate elicited corresponding increases in $[Ca^{2+}]_i$; for an ATP level greater than 1 m*M*, however, every step response was nearly saturated and unstable, often associated with a large initial peak.

Such high sensitivity of flow-induced Ca^{2+} responses to ATP, which has been observed by several groups [65–68], caused a debate about the EC flow sensing mechanism. The context of the arguments and the experimental evidence that identified the mechanism are outlined next.

8 Crucial Verification of Shear Stress-Dependency of $[Ca^{2+}]_i$ and Other Responses

To explain the flow-induced $[Ca^{2+}]_i$ response, two possible mechanisms were proposed. One mechanism is based on the conventional concept that EC has some mechanoreceptors, which can directly detect mechanical cell deformation elicited by wall shear stress (the shear stress hypothesis). The sensitivity of the receptors may be modulated by chemical agonists such as ATP. Another model, introduced later, postulates that diffusional transport of ATP to the cell surface is the main mechanism of the $[Ca^{2+}]_i$ response; the basis is that ATP itself induces Ca^{2+} transients in ECs concentration dependently [69] and its transport from the flowing medium to the cell surface is known to be hindered by the diffusion boundary layer, which is built up near the surface and the thickness of which is modulated by wall shear rate (the shear rate hypothesis). According to Caro and Nerem [70], the thickness of the diffusion boundary layer (δ) and the accumulation rate at the surface ($\dot{m}$) are related to the shear rate ($\dot{\gamma}$) as

$$\delta = \alpha\left(D/\dot{\gamma}\right)^{1/3}$$
$$\dot{m} = \left(D/\delta\right)\Delta C/0.0595 \quad (3)$$

where D and ΔC indicate ATP diffusivity in the medium and its concentration difference beyond the diffusion boundary layer (α, a constant).

As a matter of practice, it is not easy to prove which hypothesis is true by discriminating the effect of shear stress (τ) and that of shear rate ($\dot{\gamma}$), because they are closely related each other ($\tau = \mu\dot{\gamma}$) (see Eq. 2). One possible way is to change viscosity (μ) and to see (1) whether the same τ at different μ and γ induces the same $[Ca^{2+}]_i$ response (to confirm the shear stress hypothesis), or (2) whether Ca^{2+} response diminishes when μ is increased at the same $\dot{\gamma}$ (to verify the shear rate hypothesis), because the accumulation rate ($\dot{m}$) of ATP is, from Eq. 3 and the Stokes–Einstein equation, expressed as

$$\dot{m} = \kappa \dot{\gamma}^{1/3} \mu^{-2/3} \quad (4)$$

where κ is a constant that does not depend on $\dot{\gamma}$ or μ. Figure 5a shows a typical example of $[Ca^{2+}]_i$ response in our in vitro experiments recently performed by altering medium viscosity from 0.765 cP of Hanks balanced salt solution (HBSS) to 3.372 cP by adding 5% dextran [71]. The obtained results (Fig. 4b) deny the shear rate hypothesis because the $[Ca^{2+}]_i$ response to the higher viscosity was always significantly larger than that to the lower viscosity at any shear rate level. In contrast, clear one-to-one correspondence between Ca^{2+} response and shear stress (Fig. 4c), regardless of differences in viscosity and shear rate, verifies the validity of the shear stress hypothesis. It has been confirmed that flow-induced $[Ca^{2+}]_i$ responses in ECs are elicited by mechanical deformation of the cell by wall shear stress, rather than by ATP accumulation modulated by the shear rate-dependent diffusion boundary layer. These findings are supported by the fact that when ECs are rubbed lightly with a latex balloon and exposed to frictional force, a transient Ca^{2+} increase occurs in an ATP-free medium and even in air [72], and the fact that shear stress exerts vector effects on EC morphology [24,25] and proliferative activity [16]. ATP (or other chemical ligands) may be regarded as the sensitivity modulator of EC responses to fluid shear stress.

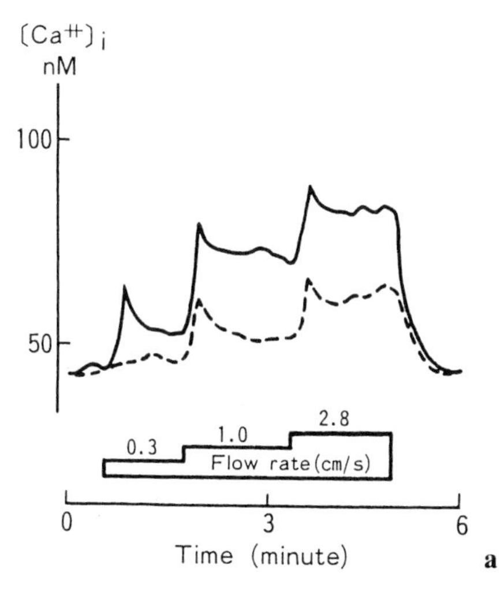

FIG. 5a–c. Flow-induced Ca^{2+} responses obtained by perfusing a low-viscosity medium (Hank's solution, HBSS) and a high-viscosity medium (HBSS with 5% dextran) (**a**) (*solid line*, high viscosity; *dashed line*, low viscosity). The magnitude of the responses are plotted against wall shear rate (**b**) and wall shear stress (**c**). Note, in **b** (*open circles*, HBSS; *solid circles*, HBSS + 5% Dextran), that the levels of Ca^{2+} response to the high-viscosity medium are always greater than those to the low-viscosity medium at any shear rate level; in **c**, both high-viscosity and low-viscosity data are clustered around a single curve against the shear stress. These results demonstrated the validity of the shear stress hypothesis rather than the shear rate hypothesis. (From [71], with permission)

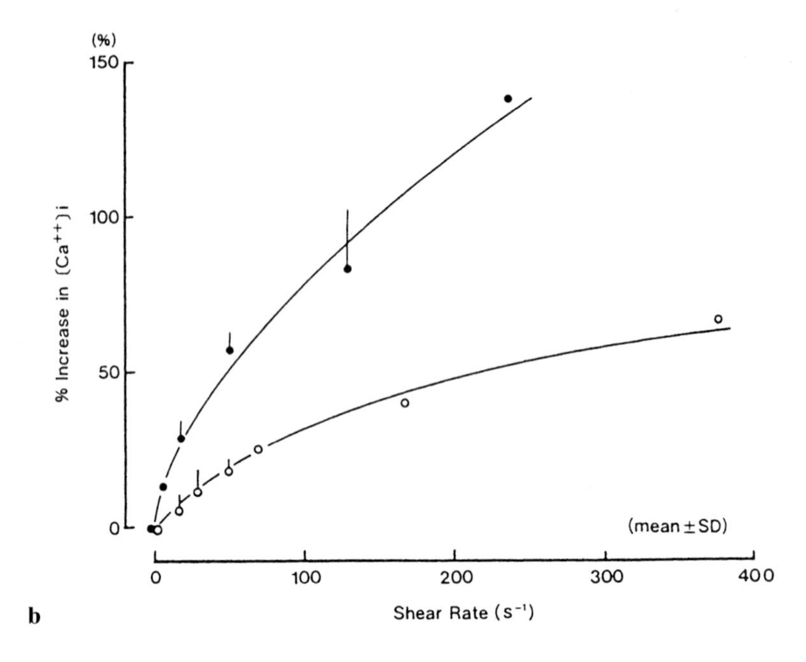

b

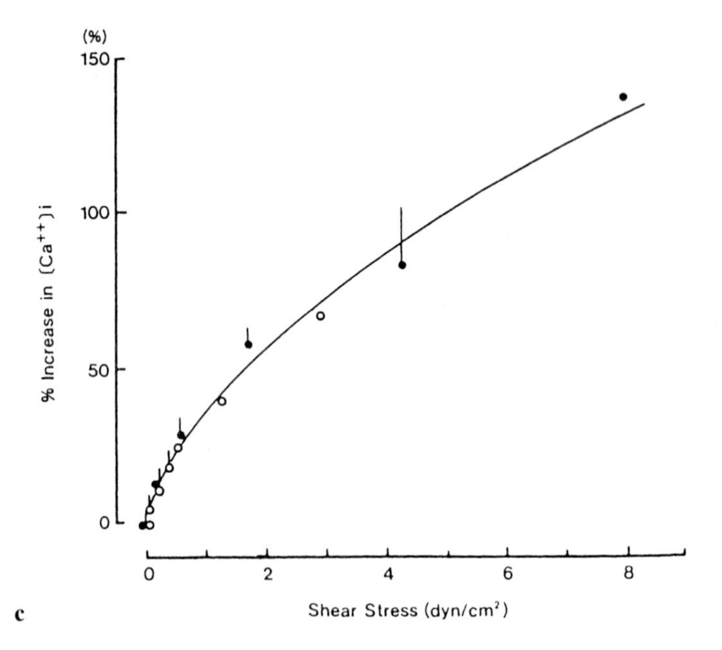

c

FIG. 5. *Continued*

Concerning protein production of ECs, the same examination of shear stress dependency as $[Ca^{2+}]_i$ response was performed, using media of high or low viscosity for VCAM-1 by Ando et al. [73] and for ICAM-1 (an intracellular adhesion molecule) by Tsuboi et al. [74]. Figure 6 depicts flow-induced changes in VCAM-1 mRNA level plotted against shear rate (Fig. 6a) and those against shear stress (Fig. 6b). It is apparent that down regulation of VCAM-1 expression is not shear rate dependent but strictly shear stress dependent. Similar results were observed for ICAM-1 mRNA, although the gene expression of this molecule is upregulated by the stress. These results have demontrated that the flow-induced EC responses are essentially stimulated by mechanical shear stress to the cell.

9 Frontiers of EC Shear-Related Studies

The intra- or extracellular mechanism of EC for sensing its own shear deformation has not been well identified, even where the sensor exists. Furthermore, in addition to shear stress, pulsatile stretching of ECs cultured on a flexible membrane also induces an immediate increase in intracellular Ca^{2+} concentration, probably through stretch-activated Ca^{2+} channels, as found by Letsou et al. [75], and stimulates adenylate cyclase activity [76]. Because physiological vascular responses to shearing stress and to tangential stress are apparently different, it is of interest to see how ECs discriminate these stresses. Ingber [77] has suggested that integrins, transmembrane molecules mediating cell attachment by physically interlinking the extracellular matrix with the intracellular cytoskeleton, may act as mechanochemical transducers. Including such a possibility, further analyses are necessary to visualize the sensing mechanisms in EC to physical stresses.

With respect to EC protein production, Resnik et al. [78] found that the EC genes upregulated by shear stress, such as those for the PDGF-B chain, tPA, ICAM-1, etc., conserve a common core sequence (GAGACC) in the promoter, which is termed the shear stress-responsive element (SSRE). It has been confirmed, however, that this element is not present in the promoter of downregulated genes like those of VCAM-1 and ET. The common element in the promoter of these genes should be identified. It is also suspected that there must exist a number of unknown proteins, gene expressions of which are up- and downregulated in the vascular cells according to the shear stress signal, that participate in all adaptive remodeling of the vessel wall to the stress. The cellular and molecular mechanisms by which shear stress regulates the vascular morphology remain unclear in many aspects.

Functional characteristics of ECs are substantially different in the arterial and venous side of microvasculature. The physiological level of shear stress at arterioles (15–25 dyne/cm^2) is greater by nearly fivefold than that of venules [79]. This suggests that arterial ECs have a much higher setpoint for shear-induced adaptive remodeling than venous ECs. Capillary angiogenesis induced by enhanced peripheral blood flow in the skeletal and cardiac muscles is also known to start by offshooting only from the venular end [13], implying that the angiogenetic

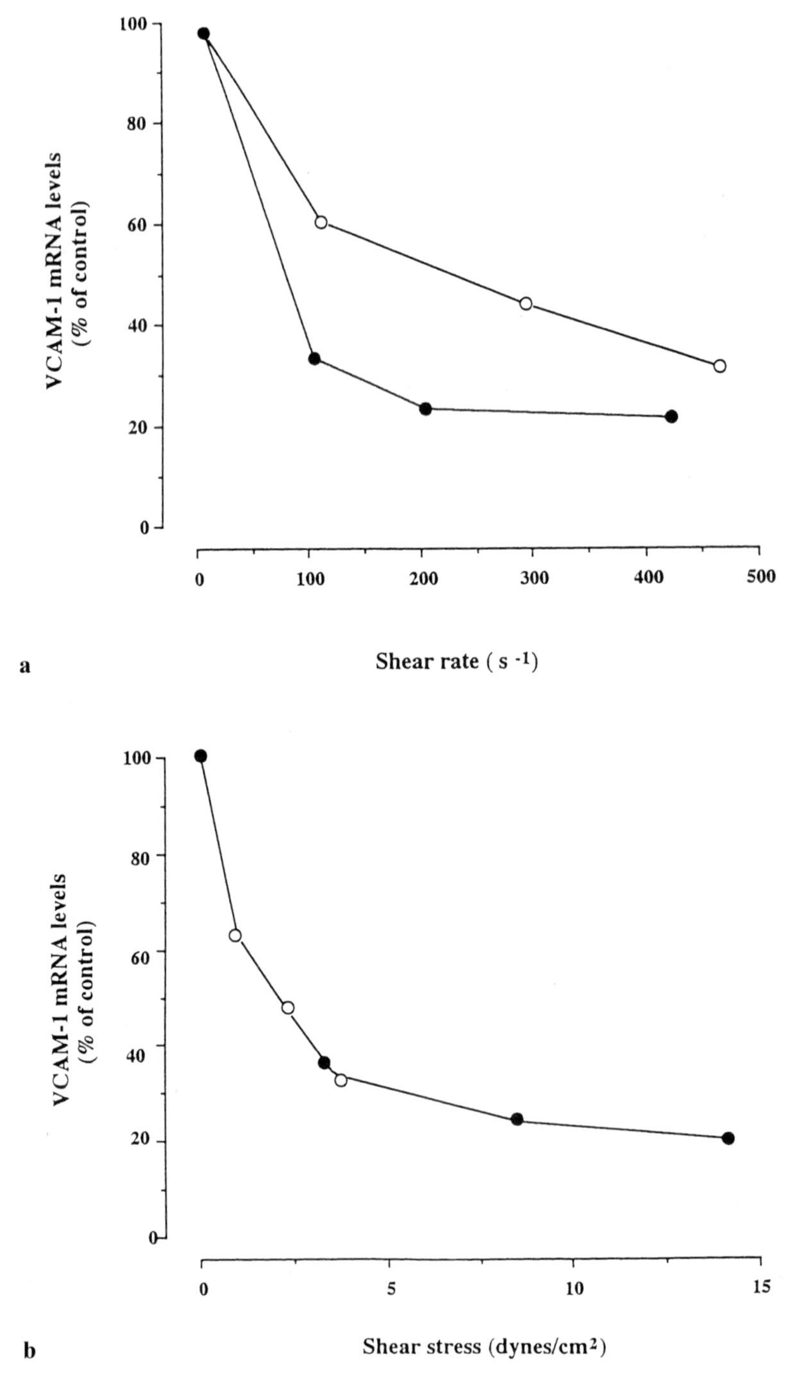

FIG. 6a,b. Relative changes in VCAM-1 mRNA level induced by perfusion of high- and low-viscosity medium. When the changes are plotted against shear rate, the high- and low-viscosity curves are separated (**a**); when plotted against shear stress, however, they form a single curve (**b**). *Open circles*, low viscosity; *solid circles*, high viscosity

activity of capillary ECs is more potent at the venular end. Furthermore, transcapillary permeability of plasma proteins is nearly tenfold greater at the venular end than at the arteriolar end in the skeletal muscle (unpublished observation). Because these discrepancies are observed at ECs very closely located, within 1 mm of one another, or 20–30 cell lengths, it is difficult to believe that they are essentially different in their inherent nature. Instead, it is more likely that some environmental factor such as oxygen tension in blood or in adjacent tissue, which always differs between the arteriolar and venular end of a capillary channel, adaptively modulates such EC activities as the shear stress sensitivity. Well-established morphological findings that the capillary density and structure of intercellular junctions are organ or tissue specific [80] also suggest significant inductory influences of surrounding cells on EC functions. It is of interest to clarify these synergistic effects of environmental factors on the shear-sensitive features of ECs.

It is now generally accepted that one of the earliest detectable events in human and experimental atherosclerosis is adherence of circulating monocytes and lymphocytes to the arterial endothelial lining [81]. It is followed by their transendothelial migration and accumulation in the intima, transformation of monocytes into foam cells, and secretion of cytokines and growth factors, eventually to form atherosclerotic plaques. Recent evidence indicates that the expression of the adhesion molecule VCAM-1 is selectively observed in a localized fashion in the aortic ECs that overlie early foam cell lesions [82]. Because EC expression of VCAM-1 is downregulated by shear stress [73] and the lesions of atherosclerosis seem to be localized at low-shear portions of the aorta and its branches [83], the shear-dependent EC–leukocyte interaction must be a key factor in understanding atherogenesis. It is of interest to know the mechanism by which VCAM-1, which ordinarily plays a physiological role in blood–lymph leukocyte circulation at the capillaries, is involved in the pathogenesis of this disease, particularly at large arteries only.

10 Optimum Regulation of Circulatory Systems by EC Functions

As described hitherto, ECs have various cell functions, almost all of which are shear stress dependent. The physiological significance of such activities is discussed here with the aid of system physiological model analyses of the circulatory systems for oxygen (O_2) transport to tissue.

Model analysis concerning the efficiency of blood as an O_2 carrier was performed by Chien [84]. As depicted in Fig. 7, the viscosity of blood (μ) non-linearly rises as the relative content of red cells in it (hematocrit, Ht in %) is increased. This implies that the amount of O_2 carried by a unit flow increases in proportion to Ht, although the elevated μ augments the mechanical work to drive a unit flow. Thus, the ratio (Ht/μ) can represent the efficiency of blood as an O_2 carrier, provided that the vessel structure remains unchanged. When this ratio is calcu-

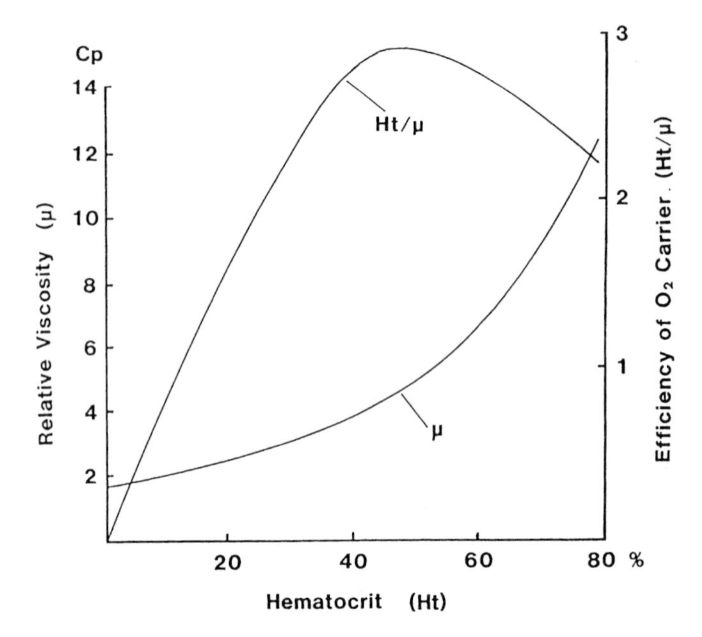

Fig. 7. Changes in viscosity (μ) against hematocrit (*Ht*) in human blood and the calculated efficiency of blood as O_2 carrier (Ht/μ). Note that the Ht/μ curve has a peak at Ht 40%–45%

lated from a Ht–μ relation in normal human blood (Fig. 7), the curve has a peak at a Ht level about 40%–45%, which is consistent with the physiological Ht value in humans. It is evident from this analysis that normal blood is maintained in the optimum state as the O_2 carrier.

The level of Ht in blood is mainly regulated by a glycoprotein hormone called erythropoietin (EPO), which is selectively synthesized in peritubular capillary ECs in the kidney and stimulates erythrocyte production in the bone marrow when released into blood circulation [85]. These EPO-producing ECs are reported to have a sensor for hypoxia composed of heme protein [86] and are sensitively activated under anemic and hypoxic conditions. However, it is also likely that shear-dependent EC activity participates in the regulation of EPO production, because renal blood flow is known to be controlled at a constant level through the autoregulatory mechanism regardless of hemodynamic situations, and the peripheral vascular bed is able to detect changes in blood viscosity (μ) from wall shear stress alterations. As suggested in the preceding section to explain the functional differences between arterial and venous ECs, it is suspected that some synergistic influence of oxygen tension and shear stress is also exerted on EPO regulation. Further studies are necessary to elucidate this interactive mechanism.

The most direct coupling between the optimal model and the shear-dependent EC character is seen in the branching structure of the vascular system. The minimum work model for the system by Murray [87] explains that if the cost function (CF) of a vessel branch with a certain flow rate (f) is given as a sum of mechanical energy loss caused by pressure drop (ΔP) and chemical energy cost in proportion to blood volume (V)

$$\mathrm{CF} = \Delta P f + \lambda V \tag{5}$$

where λ is a constant representing the energy expenditure for maintenance of blood per unit volume per unit time, the condition of the minimum CF is expressed by the flow–radius relation

$$f/r^3 = (\pi/4)(\lambda/\mu)^{1/2} \tag{6}$$

by assuming Poiseuille flow. Thus, wall shear stress (τ) is written from Eq. 2 as

$$\tau = (\lambda\mu)^{1/2} \tag{7}$$

This indicates that shear stress should be constant for any vascular branch to satisfy the optimum condition of this model. The in vivo and in vitro findings, showing that ECs respond to shear stress so as to regulate the stress constant by adaptive responses [8–17], just fit the prediction in Eq. 7. The stress level calculated from the estimated values of μ (0.03 dyne s cm^{-2}) and λ (7100 to 12000 dyne cm^{-2} s^{-1} by Zamir [88]) is approximately 14–19 dyne cm^2, which is comparable with the physiological level of the artery [79].

It is known that the capillary–tissue arrangement in the skeletal muscle is nearly uniform for mammals, irrespective of large body weight differences ranging from a few grams to several tons [89]. With respect to the architecture of the capillary network, our recent model simulation [90,91] suggests that it is optimally constructed for O_2 transport to tissue during exercise and that the regulatory mechanism is the shear-dependent EC activity for capillary angio-genesis. The simulation is based on the allometric relationships of skeletal muscle blood flow (F_m) and O_2 consumption rate (qO_{2m}) to body weight (W_b) of mammals (Fig. 8). When these scale-dependent metabolic parameters are determined against W_b, it is possible to estimate, for any animal, the maximum tissue mass to which a network with capillary number (n) can furnish sufficient O_2 by employing the Krogh cylinder model. The maximum mass $M_m(n)$ is enhanced hyperbolically as n increases.

It is also possible to assess the cost function of the supplying vascular bed by using the minimum volume model by Kamiya and Togawa [92], an optimum vascular model derived for a multiterminal system. The cost function CF(n) is given as a parabolic function of n. The efficiency (η) of the entire system is calculated from the ratio [$\eta(n) = qO_{2m}M_m(n)/\mathrm{CF}(n)$], of which examples for three animals (rat, dog, and horse) are shown in Fig. 9 for the resting and exercising states. The optimal capillary number (n_O) is determined from the peak point of the individual efficiency curve. The results of this simulation (Fig. 10) indicate a good agreement of the actual morphological data with those of the optimum model during exercise in terms of muscle mass and tissue radius. They suggest close resemblance between the two situations.

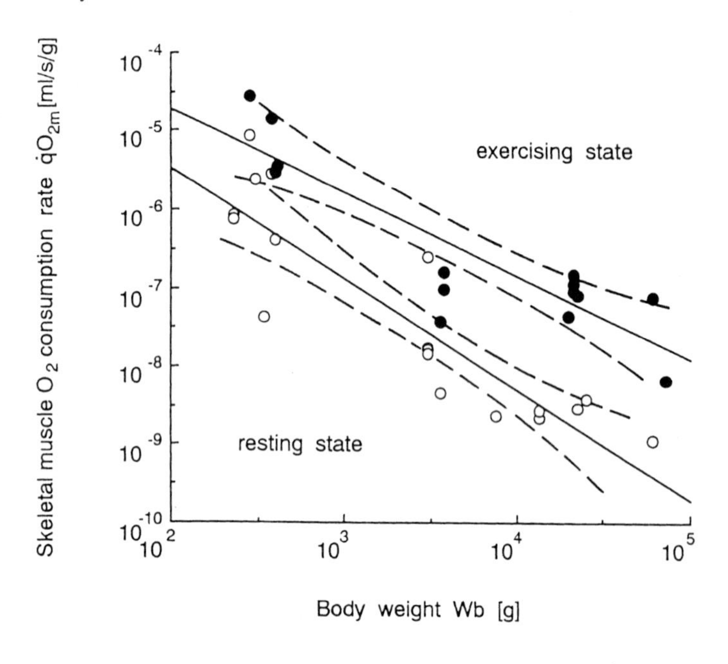

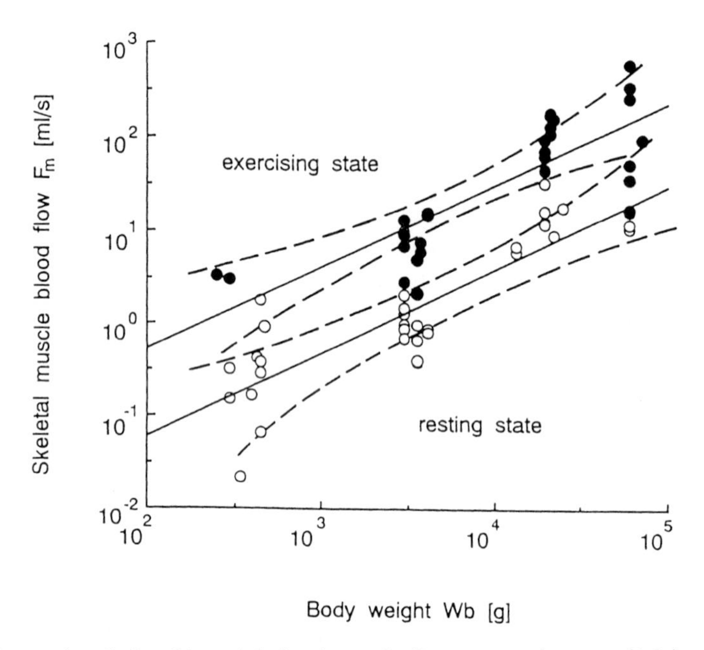

FIG. 8. Allometric relationships of skeletal muscle O_2 consumption rate ($\dot{q}O_2$) and blood flow (F_m) to body weight (W_b) in mammals. *Solid circles*, exercising state; *open circles*, resting state

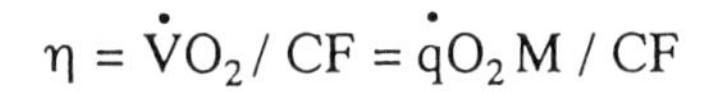

$$\eta = \dot{V}O_2 / CF = \dot{q}O_2 M / CF$$

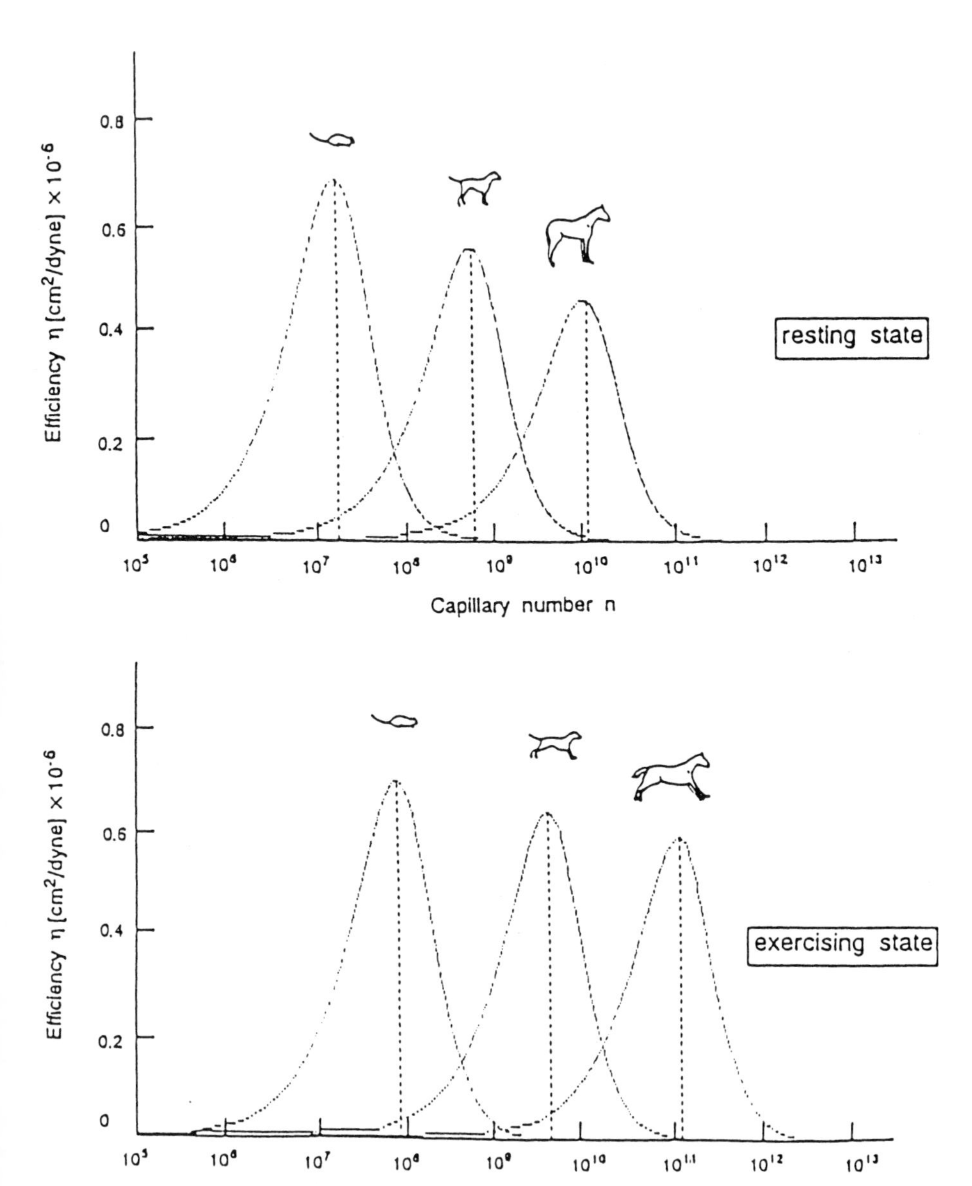

FIG. 9. Efficiency curves simulated for three mammals with different body weights (rat, 300 g; dog, 30 kg; horse, 700 kg) in the resting (*top*) and exercising (*bottom*) states, respectively

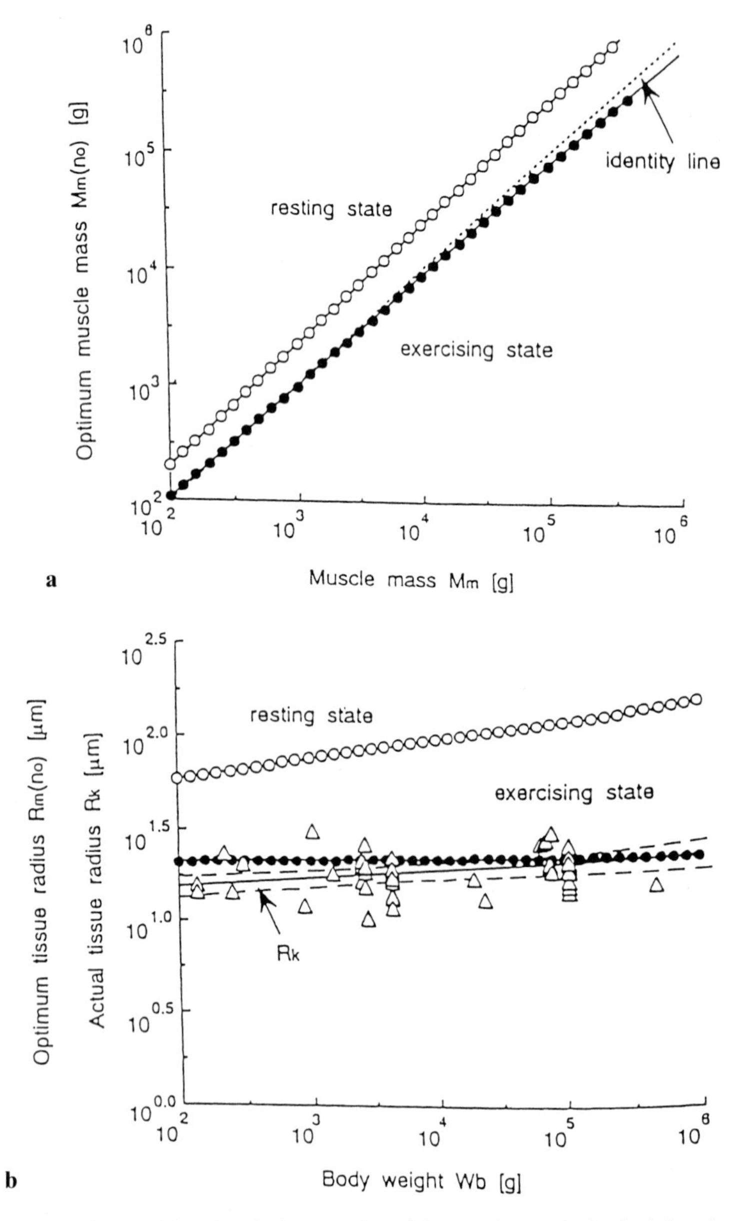

FIG. 10. Comparison of the simulation results with actual morphological data in terms of muscle mass (**a**) and the tissue radius surrounding a single capillary (**b**). Note that the optimum mass ($M_m(n_O)$) and tissue radius ($R_m(n_O)$) both in the exercising state (*solid circles*) are very close to the actual mass (M_m) and radius (R_k). *Open circles*, resting state; *solid circles*, exercising state; *open triangles*, reported values of R_k

If it is assumed that the capillary network in skeletal muscle is regulated to optimize O_2 transport to tissue during exercise, the responsible mechanism must be the shear-induced angiogenetic responses of venular ECs, as described previously. This is because this adaptive response acts to keep capillary flow constant despite peripheral blood flow change, and the simulated optimum state is achieved by regulating capillary flow at a certain level. The simulation study indicates that the optimum capillary flow rates for the mammals ranging from 100 g to 1000 kg in W_b vary only from 160 to 120 μl/s, which yield a capillary red cell velocity of 750–600 μm/sec. Meanwhile, actual red cell velocity observed during muscle exercise ranges from 500 to 1000 μm/s. Accordingly, capillary angiogenesis or atrophy corresponding to increased or decreased regional blood flow to regulate capillary flow (or the wall shear stress) at an appropriate level can be the optimizing mechanism of the network. Although the stress level at which venular ECs initiate the angiogenetic responses has not been confirmed, the consistent capillary flow velocity between the predicted and the observed suggests that the setpoint of the stress seems to be properly selected by the venular ECs for this purpose.

A common feature of these three model analyses is that EC responses to O_2 tension or to shear stress form a negative feedback loop to the stimulus. In addition, the setpoint of each response is appropriately selected, so that the controlled systems are all optimized for the function of O_2 transport. This kind of autoregulatory response may help to achieve "homeostasis" in biological systems. The current results suggest that homeostasis, which usually means simply to keep a certain equilibrated state, actually implies maintaining a strictly selected "optimum" state by finely organized mechanisms, at least concerning functions essential for life.

References

1. Ando J, Kamiya A (1993) Blood flow and vascular endothelial cell function. Front Med Biol Eng 5:245–264
2. Davies PF, Tripatchi S (1993) Mechanical stress mechanism and the cell—an endothelial paradigm. Circ Res 72:239–245
3. Suwa N, Takahashi T (1993) Morphological and morphometrical analysis of circulation in hypertension and ischemic kidney. Urban and Schwarzenberg, Munchen
4. Hutchins GM, Milner MM, Boitnott JK (1976) Vessel caliber and branch angle of human coronary arterial branch-points. Circ Res 38:572–576
5. Seki J (1994) Flow pulsation and network structure in mesenteric microvasculature of rats. Am J Physiol 266:H811–H821
6. Thoma R (1893) Histogenese und Histomechanik des Gefassystems. Enke, Stuttgart
7. Rodbard S (1975) Vascular caliber. Cardiology 60:4–49
8. Kamiya A, Togawa T (1980) Adaptive regulation of wall shear stress to flow change in the canine carotid artery. Am J Physiol 239:H14–H21
9. Guyton JR, Hartley CJ (1985) Flow restriction of one carotid artery in juvenile rats inhibits growth of arterial diameter. Am J Physiol 248:H540–H546
10. Langille BL, O'Donnell F (1986) Reductions in arterial diameter produced by chronic decreases in blood flow are endothelium-dependent. Science 231:405–407

11. Zarins CK, Zatina MA, Giddens DP, Ku DN, Glagov S (1987) Shear regulation of arterial lumen diameter in experimental atherogenesis. J Vasc Surg 5:413–420
12. Tohda K, Masuda H, Kawamura K, Shozawa T (1992) Difference in dilatation between endothelium-preserved and -desquamated segments in the flow loaded rat common carotid artery. Arterioscler Thromb 12:519–528
13. Hudlicka O (1984) Developement of microcirculation: capillary growth and adaptation. In: Renkin EM, Mitchel CC (eds) Handbook of physiology, vol 4. American Physiological Society, Bethesda, pp 165–216
14. Hudlicka O, Brown M, Egginton S (1992) Angiogenesis in skeletal and cardiac muscle. Physiol Rev 72:369–417
15. Wang DW, Prewitt RL (1991) Microvascular developement during normal growth and reduced blood flow: introduction of a new model. Am J Physiol 260:H1966–H1972
16. Ando JH, Nomura H, Kamiya A (1987) The effect of fluid shear stress on the migration and proliferation of cultured endothelial cells. Microvasc Res 33:62–70
17. Ando J, Komatsuda T, Ishokawa C, Kamiya A (1990) Fluid shear stress enhanced DNA synthesis in cultured endothelial cells during repair of mechanical denudation. Biorheology 27:675–684
18. Masuda M, Fujiwara K (1991) Morphological responses of single endothelial cells exposed to physiological levels of fluid shear stress. Med & Biol Eng & Comput 29(suppl):268
19. Eskin SG, Sybers HD, O'Bannon O, Navarro LT (1982) Performance of tissue cultured endothelial cells in a mock circulatory loop. Artery 10:159–171
20. Dewey CF Jr (1984) Effects of fluid flow on living vascular cells. J Biomech Eng 106:31–35
21. Davies PF, Remuzzi A, Gordon EJ, Dewey CF Jr, Gimbrone MA (1986) Turbulent fluid shear stress induces vascular endothelial cell turnover in vitro. Proc Natl Acad Sci USA 83:2114–2117
22. Masuda H, Kawamura K, Tohda K, Shozawa T, Sageshima M, Kamiya A (1989) Increase in endothelial cell density before artery enlargement in flow-loaded canine carotid artery. Arteriosclerosis 9:812–823
23. Sumpio BE, Banes AJ, Levin LG, Johnson G Jr (1987) Mechanical stress stimulates aortic endothelial cells to proliferate. J Vasc Surg 6:252–256
24. Flaherty LT, Pierce JE, Ferrans VJ, Patel DJ, Tucker WK, Fry DL (1972) Endothelial nuclear patterns in the canine arterial tree with particular reference to hemodynamic events. Circ Res 30:23–33
25. Dewey CF Jr, Bussolari SR, Gimbrone MA Jr, Davies PF (1981) The dynamic response of vascular endothelial cells to fluid shear stress. J Biomech Eng 103:177–184
26. Wong AJ, Pollard TD, Herman IM (1983) Actin filament stress fibers in vascular endothelial cells in vivo. Science 219:867–869
27. White GE, Fujiwara K (1986) Expressin and intracellular distribution of stress fibers in aortic endothelium. J Cell Biol 103:63–70
28. Masuda H, Shozawa T, Hosoda S, Kanda M, Kamiya A (1985) Cytoplasmic microfilaments in endothelial cells of flow loaded canine carotid arteries. Heart Vessels 1:65–69
29. Kim DW, Langille BL, Wong MKK, Gotlieb AI (1989) Patterns of endothelial microfilament distribution in the rabbit aorta in situ. Circ Res 64:21–31
30. Kim DW, Gotlieb AI, Langille BL (1989) In vivo modulation of endothelial F-actin microfilaments by experimental alterations in shear stress. Arteriosclerosis 9:439–445

31. Franke RP, Grafe M, Schnittler H, Seiffge D, Mittermayer C, Drenckhahn D (1984) Induction of human vascular endothelial stress fibres by fluid shear stress. Nature 307:648–649
32. Wechezak AR, Viggers RF, Sauvage LR (1985) Fibronectin and F-actin redistribution in cultured endothelial cells exposed to shear stress. Lab Invest 53:639–647
33. Wechezak AR, Wight TN, Viggers RF, Sauvage LR (1989) Endothelial adherence under shear stress is dependent upon microfilament reorganization. J Cell Physiol 139:136–146
34. Rosen LA, Hollis TM, Sharma MG (1974) Alterations in bovine endothelial histidine decarboxylase activity following exposure to shear stresses. Exp Mol Pathol 20:329–343
35. DeForrest LM, Hollis TM (1978) Shear stress and aortic histamine synthesis. Am J Physiol 234:H701–H705
36. Reeves JT, Grondelle AV, Voelkel NF, Walker B, Lindenfeld J, Worthen S, Mathias M (1983) Prostacyclin production and lung endothelial cell shear stress. In: Hypoxia, exercise, and altitude: proceedings of the third Banff international hypoxia symposium. Liss, New York, pp 125–131
37. Frangos JA, Eskin SG, McIntire LV, Ives CL (1985) Flow effects on prostacyclin production by cultured human endothelial cells. Science 227:1477–1479
38. Grabowski EF, Jaffe EA, Weksler BB (1985) Prostacyclin production by cultured endothelial cell monolayers exposed to step increases in shear stress. J Lab Clin Med 105:36–43
39. Holtz J, Forstermann U, Pohl U, Giesler M, Bassenge E (1984) Flow-dependent, endothelium-mediated dilation of epicardial coronary arteries in conscious dogs: effects of cyclooxygenase inhibition. J Cardiovasc Pharmacol 6:1161–1169
40. Rubanyi GM, Romero JC, Vanhoutte PM (1986) Flow-induced release of endothelium-derived relaxing factor. Am J Physiol 250:H1145–H1149
41. Miller VM, Vanhoutte PM (1988) Enhanced release of endothelium-derived factor(s) by chronic increases in blood flow. Am J Physiol 255:H446–H451
42. Cooke JP, Stamler J, Andon N, Davies PF, McKinley G, Loscalzo J (1990) Flow stimulates endothelial cells to release a nitrovasodilator that is potentiated by reduced thiol. Am J Physiol 259:H804–H812
43. Korenaga K, Ando J, Tsuboi H, Yang W, Toyo-oka I, Kamiya A (1994) Laminar flow stimulates ATP- and shear stress-dependent nitric oxide production in cultured bovine endothelial cells. Biochem Biophys Res Commun 198:213–219
44. Yoshizumi M, Kurihara H, Sugiyama T, Takaku F, Yanagisawa M, Masaki T, Yazaki Y (1989) Hemodynamic shear stress stimulates endothelin production by cultured endothelial cells. Biochem Biophys Res Commun 161:859–864
45. Milner P, Bodin P, Loesch A, Burnstock G (1990) Rapid release of endothelin and ATP from isolated aortic endothelial cells exposed to increased flow. Biochem Biophys Res Commun 170:649–656
46. Sharefkin JB, Diamond SL, Eskin SG, McIntire LV, Dieffenbach CW (1991) Fluid flow decreases preproendothelin mRNA levels and suppresses endothelin-1 peptide release in cultured human endothelial cells. J Vasc Surg 14:1–9
47. Diamond SL, Eskin SG, McIntire LV (1989) Fluid flow stimulates tissue plasminogen activator secretion by cultured human endothelial cells. Science 243:1483–1485
48. Diamond SL, Sharefkin JB, Dieffenbach C, Frasier-Scott K, McIntire LV, Eskin SG (1990) Tissue plasminogen activator messenger RNA levels increase in cultured human endothelial cells exposed to laminar shear stress. J Cell Physiol 143:364–371

49. Hsieh H-J, Li N-Q, Frangos JA (1991) Shear stress increases endothelial platelet-derived growth factor mRNA levels. Am J Physiol 260:H642–H646
50. Ono O, Ando J, Kamiya A, Kuboki Y, Yasuda H (1991) Flow effects on cultured vascular endothelial and smooth muscle cell functions. Cell Struct Funct 16:365–374
51. Gupte A, Frangos JA (1990) Effects of flow on the synthesis and release of fibronectin by endothelial cells. In Vitro Cell Dev Biol 26:57–60
52. Grimm J, Keller R, Groot PG (1988) Laminar flow induced cell polarity and leads to rearrangement of proteoglycan metabolism in endothelial cells. Thromb Haemostasis 60:437–441
53. Ohtsuka A, Ando J, Korenaga R, Kamiya A, Toyama-Sorimachi N, Miyasaka M (1993) The effect of flow on the expression of vascular adhesion molecule-1 by cultured mouse endothelial cells. Biophys Biochem Res Commun 193:303–310
54. Fry DL (1968) Acute vascular endothelial change associated with increased blood velocity gradients. Circ Res 22:165–169
55. Jo H, Dull RO, Hollis TM, Tarbell JM (1991) Endothelial albumin permeability is shear dependent, time dependent, and reversible. Am J Physiol 260:H1992–H1996
56. Shibata M, Kamiya A (1992) Blood flow dependence of local capillary permeability of Cr-EDTA in the rabbit skeletal muscle. Jpn J Physiol 42:631–639
57. Davies PF, Dewey CF Jr, Bussolari SR, Gordon EJ, Gimbrone MA Jr (1984) Influence of hemodynamic forces on vascular endothelial function: in vitro studies of shear stress and pinocytosis in bovine aortic cells. J Clin Invest 73:1121–1129
58. Sprague EA, Steinbach BL, Nerem RM, Schwartz CJ (1987) Influence of a laminar steady-state fluid-imposed wall shear stress on the binding, internalization, and degradation of low-density lipoproteins by cultured arterial endothelium. Circulation 76:648–656
59. Worthen GS, Smedly LA, Tonnesen MG, Ellis D, Voelkel NF, Reeves JT, Henson PM (1987) Effects of shear stress on adhesive interaction between neutrophils and cultured endothelial cells. J Appl Physiol 63:2031–2041
60. Sato M, Levesque MJ, Nerem RM (1987) Micropipette aspiration of cultured bovine aortic endothelial cells exposed to shear stress. Arteriosclerosis 7:276–286
61. Ando J, Komatsuda T, Kamiya A (1988) Cytoplasmic calcium response to fluid shear stress in cultured vascular endothelial cells. In Vitro Cell Dev Biol 24:821–877
62. Nollert MU, Eskin SG, McIntire LV (1990) Shear stress increases inositol trisphosphate levels in human endothelial cells. Biochem Biophys Res Commun 170:281–287
63. Olesen S-P, Clapham DE, Davies PF (1988) Haemodynamic shear stress activates a K^+ current in vascular endothelial cells. Nature 331:168–331
64. Nakache M, Gaub HE (1988) Hydrodynamic hyperpolarization of endothelial cells. Proc Natl Acad Sci USA 85:1841–1843
65. Ando J, Ohtsuka A, Korenaga R, Kamiya A (1991) Effect of extracellular ATP level on flow-induced Ca^{++} response in cultured vascular endothelial cells. Biochem Biophys Res Commun 179:1192–1199
66. Nollert MU, Diamond SL, McIntire LV (1991) Hydrodynamic shear stress and mass transport modulation of endothelial cell metabolism. Biotechnol Bioeng 38:588–602
67. Mo M, Eskin SG, Schilling WP (1991) Flow-induced changes in Ca^{2+} signalling of vascular endothelial cells: effect of shear stress and ATP. Am J Physiol 260:H1698–H1707
68. Dull RO, Davies PF (1991) Flow modulation of agonist (ATP)-response (Ca^{2+}) coupling in vascular endothelial cells. Am J Physiol 261:H149–H154

69. Luckhoff A, Busse R (1986) Increased free calcium in endothelial cells under stimulation with ademine nucleotides. J Cell Physiol 126:414–420
70. Caro CC, Nerem RM (1973) Transport of ^{14}C-4-cholesterol between serum and wall in the perfused dog common carotid artery. Circ Res 32:187–205
71. Ando J, Ohtsuka A, Korenaga R, Kawamura T, Kamiya A (1993) Wall shear stress rather than shear rate regulates cytoplasmic Ca^{++} responses to flow in vascular endothelial cells. Biochem Biophys Res Commun 190:716–723
72. Ando J, Ohtsuka A, Korenaga R, Kamiya A (1992) Intracellular calcium response to mechanical shearing force in cultured vascular endothelial cells. In Vitro Cell Dev Biol 28:152
73. Ando J, Tsuboi H, Korenaga R, Toyama-Sorimachi N, Miyasaka M, Kamiya A (1994) Shear stress inhibits adhesion of cultured mouse endothelial cells to lymphocytes by downregulating VCAM-1 expression. Am J Physiol 267:C679–C687
74. Tsuboi H, Ando J, Korenaga R, Takada Y, Kamiya A (1995) Flow stimulates ICAM-1 expression time and shear stress dependently in cultured human endothelial cells. Biochem Biophys Res Commun 206:988–996
75. Letsou GV, Rosales O, Maitz S, Vogt A, Sumpio BE (1990) Stimulation of adenylate cyclase activity in cultured endothelial cells subjects to cyclic stretch. J Cardiovasc Surg 31:634–639
76. Lansman JB, Hallam TJ, Rink TJ (1987) Single stretch-activated ion channels in vascular endothelial cells as mechanotransducers? Nature 325:811–813
77. Ingber D (1991) Integrins as mechanochemical transducers. Curr Opin Cell Biol 3:841–848
78. Resnik N, Collins T, Atkinson W, Banthron DT, Dewey CF Jr (1993) Platelet-derived growth factor B chain promoter contains a *cis*-acting fluid shear stress-responsive element. Proc Natl Acad Sci USA 90:4591–4595
79. Kamiya A, Bukhari R, Togawa T (1984) Adaptive regulation of wall shear stress optimizing vascular tree function. Bull Math Biol 46:127–137
80. Simionescu M, Simionescu N (1984) Ultrastructure of the microvascular wall: functional correlations. In: Renkin EM, Michel CC (eds) Handbook of physiology, vol 4. American Physiology Society, Bethesda, pp 41–102
81. Ross R (1986) The pathogenesis of atherosclerosis—an update. N Engl J Med 314:488–500
82. Cybulsky MI, Gimbrone MA Jr (1991) Endothelial expression of a mononuclear leukocyte adhesion molecule during atherogenesis. Science 251:788–791
83. Yoshida Y, Okano M, Wang S, Kobayashi M, Kawasumi M, Hagiwara H, Mitsumata M (1995) Hemodynamic force-induced difference of interendothelial junctional complex. In: Numano F, Wissler RW (eds) Recent advances in atherosclerosis researches (Annals of New York Academy of Sciences, vol 748). New York Academy of Sciences, New York, pp 104–121
84. Chien S (1971) Present state of blood rheology. In: Messmer K, Schmidt-Schonbein H (eds) Hemodilution. Theoretical basis and clinical application. Karger, Basel, pp 1–45
85. Koury ST, Koury MJ, Bondurant MC, Caro J, Graber SE (1989) Quantitation of erythropoietin-producing cells in the kidneys of mice by in-situ hybridization: correlation with hematcrit, renal erythropoietin mRNA and serum erythropoietin concentration. Blood 74:645–651
86. Gordberg MA, Dunning SP, Bunn HF (1988) Regulation of erythropoietin gene: evidence that the oxygen sensor is a heme protein. Science 242:1412–1415

87. Murray CD (1926) The physiological principle of minimum work. I Proc Natl Acad Sci USA 12:204–214
88. Zamir M (1977) Shear forces and blood vessel radii in the cardiovascular system. J Gen Physiol 69:449–461
89. Schmidt-Nielsen K (1984) Scaling: why is animal size so important? Cambridge University Press, Cambridge
90. Kamiya A, Ando J, Shibata M, Wakayama H (1990) The efficiency of the vascular-tissue system for oxygen transport in the skeletal muscles. Microvasc Res 39:163–179
91. Baba K, Kawamura T, Schibata M, Sohirad M, Kamiya A (1995) Capillary-tissue arrangement in the skeletal muscle optimized for oxygen transport in all mammals. Microvasc Res 49:163–179
92. Kamiya A, Togawa T (1972) Optimal branching structure of the vascular tree. Bull Math Biophys 34:431–438

Functional Adaptation and Optimal Control of the Heart and Blood Vessels

Responses of the Heart to Mechanical Stress

Hitonobu Tomoike

Summary. Adaptation of the heart to mechanical load is seen during normal development. A harmonized coordination between structure and function is observed on the integration of right and left ventricles as a low-pressure and a high-pressure pump, respectively. Both ventricles eject the same amount of blood to their respective peripheral circulation, but differ in intracavitary pressure, wall thickness, and shape. When coronary perfusion pressure is decreased below the level (critical pressure) of coronary flow autoregulation, regional wall motion deteriorates. The level of critical pressure is determined by the level of afterload, and is lower in the right than in the left coronary artery. The spatial distribution of the infarcted area following coronary branch ligation was quite different between the right and left ventricles, which is largely explained by the different level of inherent mechanical load between the respective ventricles. Abnormalities of mechanical load in mature heart also alter the structure and function not only of the vascular system but also of the four chambers of the heart. Although hypertrophic changes of the heart as a consequence of adaptation to mechanical overload are beneficial at the early phase for mitigating hemodynamic burdens noted in the diseased heart, chronic overload results in maladaptation or remodeling. In inevitable sequelae, changes in structure do not provide amelioration of cardiac pump performance, ultimately resulting in death. Recent clinical trials have provided clues to long-term structural improvement in the heart as a bonus of the reduced ventricular preload and afterload. These beneficial effects improve prognosis in heart failure. Altered gene expression and phenotypic alteration have drawn attention as being causative of hypertrophy and remodeling.

Key words: Mechanical overload—Adaptation—Remodeling—Hypertrophy—Heart failure

---

The First Department of Internal Medicine, Yamagata University School of Medicine, 2-2-2 Iidanishi, Yamagata, 990-23 Japan

1 Introduction

The way in which the heart and circulatory system respond to mechanical overload depends partly on the functional state of cardiac performance as well as on the history and character of the increase in workload [1]. Functional and structural alterations of the heart as a result of sudden hemodynamic overload have been rigorously studied in patients with heart disease as well as experimental animal models based on organ physiology [2,3]. However, at least two lines of study are exploring a new horizon on the mechanisms involved in adaptation and failure of the heart resulting from an increase in workload. One aspect is the application of recombinant DNA technologies and sophisticated biological methods to cells of the cardiovascular system; the other is long-term clinical trials concerning the prognosis as well as symptoms of patients with heart failure [4]. This chapter therefore focuses on the immediate and chronic effects of mechanical overload on the structure and function of the heart.

2 Adaptive Mechanisms of the Heart to Inherent Mechanical Load

Differences in geometry and pump performance between the right and left ventricles accord well with the given hemodynamic requirements. Normal development of the heart is a programmed process and is different from phenotypic changes in the mature heart resulting from stress and disease. However, to elucidate the physiological characteristics involved in normal growth and its function, we may learn part of the given blueprints for maturation or adaptation of the cardiovascular system to an increase in workload. The right and left ventricles are the two major pumps of the heart. The right ventricle has been stated to operate primarily as a volume pump, in that ejection pressure is low, whereas the left ventricle is more of a pressure pump. The architecture of the two ventricles differs in shape and wall thickness in accord with their functional specializations; however, they have the same cardiac output, operating on the Frank–Starling law.

3 Immediate Responses of the Heart to Stress

To cope with a sudden increase in the impedance of ejection (pressure overload), an acute valvular regurgitation (volume overload), or a partial destruction of the wall, many adaptive mechanisms are available for the heart to ensure a flow that is compatible with its survival. In the intact organism, the heart and the peripheral circulation interact in a dynamic manner. Experimental models have been proposed, using small and large mammals to which mechanical stresses mimicking human diseases have been applied.

3.1 Effects of Afterload on Pressure–Function–Flow Relations

We examined the effects of graded reductions in regional coronary perfusion pressure on regional segment shortening in the right-ventricular free wall [5]. Because the level of afterload alters the extent of systolic wall motion as well as that of myocardial oxygen consumption, segment shortening was measured with and without an increased level of right-ventricular pressure. The level of afterload to the right ventricle can easily be elevated, by constricting the pulmonary artery, without altering right coronary perfusion pressure. Thus, the right ventricle behaves like a naturally designed model for elucidating the relationship among perfusion pressure, coronary flow, pump performance, and afterload.

Thirteen mongrel dogs (19–23 kg) were sedated with intramuscular morphine sulfate (10 mg) and then anesthetized with intravenous α-chloralose (45 mg/kg) and urethane (450 mg/kg). After endotracheal intubation, a left thoracotomy at the third intercostal space was performed under positive pressure respiration and the pericardium was opened. A hydraulic vascular occluder (ϕ = 16 mm) was placed around the main pulmonary artery. The pericardium was loosely closed and the chest was closed. A right thoracotomy then was performed through the fourth intercostal space, and the heart was supported in a pericardial cradle. The autoperfusion system from the right carotid artery to the proximal portion of the right coronary artery was established using a large-bore cannula. Right coronary blood flow and pressure were monitored by a cannulating-type electromagnetic flow probe (ϕ = 3 mm, SP7517, Statham, Cleveland, OH, USA) and Statham P23Db transducer, respectively. Systemic arterial pressure was monitored continuously at the ascending aorta with a P23Db pressure transducer through a 8-Fr catheter, and both left- and right-ventricular pressures were measured by catheter-tip pressure transducers (PC350, Millar, Houston, TX, USA). For pressure measurements, zero level was taken at the middle of chest.

For assessment of regional wall motion in the right ventricle, two pairs of miniature (1.5–1.7 mm in diameter) piezoelectric crystals (5 MHz, Murata, Kyoto, Japan) were positioned approximately midway between endocardium and epicardium in the right-ventricular free wall. One pair was implanted 0.8–1.2 cm apart, parallel to the atrioventricular groove, and the other pair was oriented perpendicular to the first pair. The correct location of these crystals relative to the vascular bed of the right coronary artery was verified by postmortem x-ray examination.

End-systolic segment length was measured within 20 ms before the peak negative first derivative of left-ventricular pressure (LV dP/dt). Regional shortening was calculated by the formula [(end-diastolic length–end-systolic length)/end-diastolic length] × 100, and was termed "percent shortening." We recorded the aortic pressure, right-ventricular pressure, coronary flow, and regional segmental lengths along with left-ventricular pressure and its first derivative, dP/dt, on a direct-writing pen recorder (eight-channel Rectigraph, Sanei Sokuki, Tokyo, Japan).

Figure 1 shows representative tracings of regional wall motion at various levels of right coronary perfusion pressure before (upper panel) and after (lower panel) pulmonary artery (PA) banding. Before PA banding, regional wall motion remained almost constant at the perfusion pressure of 48 mmHg and then decreased progressively depending on the level of perfusion pressure. When right-ventricular systolic pressure was increased by 25 mmHg without affecting

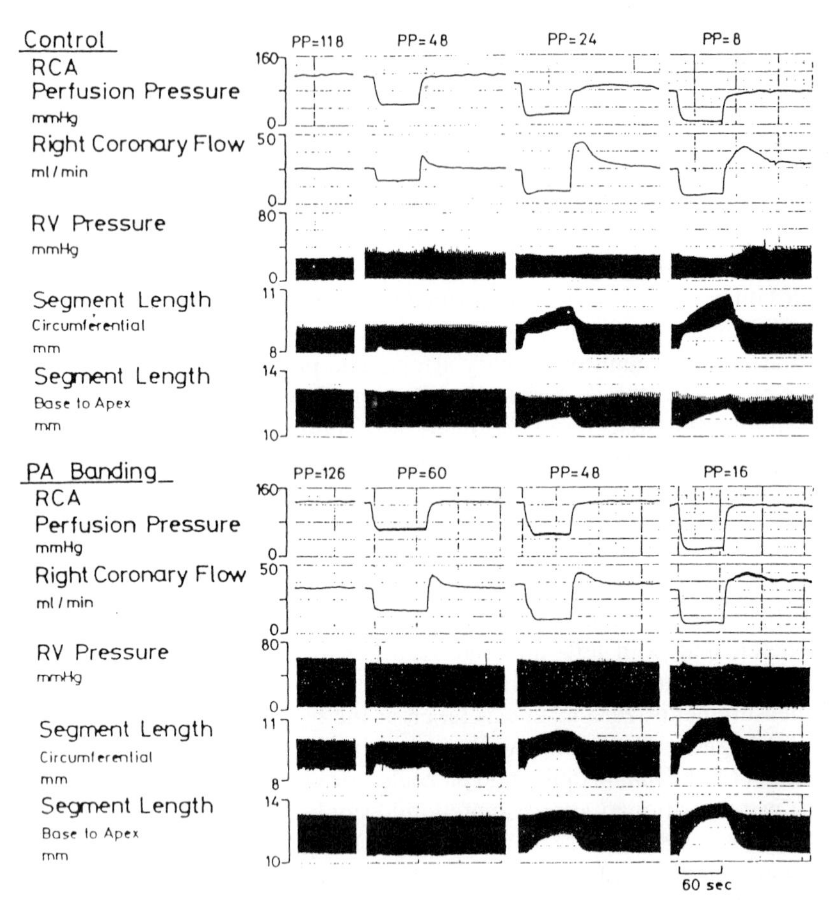

FIG. 1. Effects of various levels of perfusion pressure on right-ventricular pressure and segment length from a representative experiment. *Upper panel*, tracings taken at the normal right-ventricular pressure as a control; *lower panel*, tracings taken during pulmonary artery (*PA*) banding. After abrupt changes in coronary perfusion pressure, coronary blood flow changed depending on the pressure level and was always maintained constant during the procedure; thus, this phenomenon indicates the absence of autoregulation in the right coronary vasculature. Right-ventricular pressure was increased by constricting the PA with a cuff occluder. A reactive hyperemia was always noted, and the degree of the reactive hyperemia depended on the reduced level of coronary perfusion pressure. *RCA*, right coronary artery; *PP*, perfusion pressure (in mmHg); *RV*, right ventricle. (From [5] with permission)

mean arterial and left-ventricular peak systolic pressures, regional hypokinesis was first noted at a perfusion pressure of 60 mmHg. This means that the perfusion pressure critical for maintaining regional wall motion depends on the level of afterload.

In Fig. 2, all data points for each dog are replotted to demonstrate the pressure-shortening relation (Fig. 2a) and flow-shortening relation (Fig. 2b) in the right ventricle. The inflection points, below which the regional shortening was dependent on perfusion pressure, were 30 ± 9 and 45 ± 16 mmHg before and after PA banding, respectively ($P < 0.01$), in the base-to-apex segment and 32 ± 6 and 42 ±

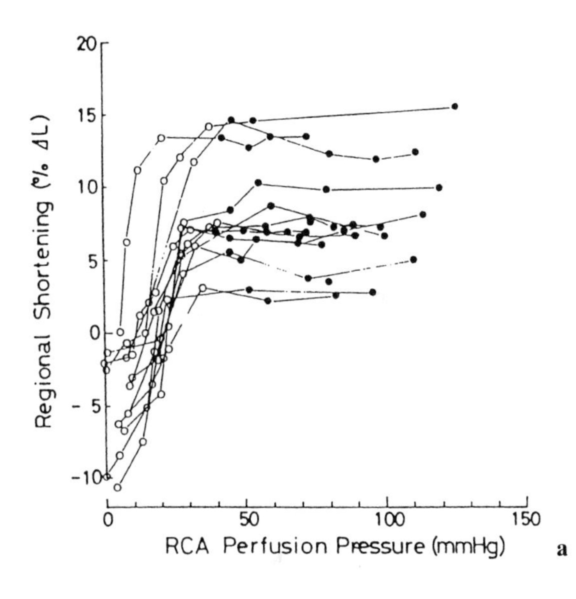

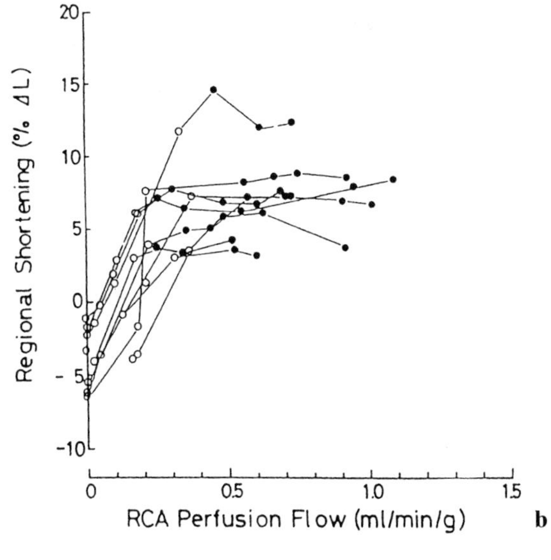

Fig. 2a,b. Relation between right-coronary perfusion pressure (**a**) or perfusion flow (**b**) and regional shortening in circumferential direction. *Open* and *closed circles* represent the relation below and above the critical perfusion pressure in 13 dogs in **a** or the critical perfusion flow in 9 dogs in **b**, respectively. Similar relations also were obtained in the base-to-apex direction, as well as in both directions during PA banding, but are not shown in this figure. (From [5] with permission)

11 mmHg before and after PA banding, respectively ($P < 0.01$), in the circumferential segment. Accordingly, (1) the critical perfusion pressure was lower in the right than in the left ventricle, and (2) an increase in right ventricle (RV) afterload elevated the critical perfusion pressure below which regional function deteriorated.

The difference in the critical perfusion pressure between the right and left ventricles may be explained by: (1) a marked difference in afterload for performing systolic wall motion, (2) lower oxygen consumption of the right than the left ventricle because of the lower level of right-ventricular pressure and work [6], (3) the lesser wall thickness, and (4) the lower extravascular compressive force of the right than the left ventricle.

3.2 Load-Dependent Shift in Critical Level of Perfusion Pressure

Independent control of ventricular cavity pressure and perfusion pressure is feasible in the right ventricle in situ. The higher level of critical perfusion pressure during PA banding may be attributable to the elevation of right-ventricular pressure. An increase of afterload inevitably increases myocardial oxygen consumption. The wall tension calculated from a pressure and an area at end-diastole by Laplace's law increased significantly, suggesting an augmentation of preload during PA banding. A monoexponential curvilinear relation between perfusion pressure or flow rate and percent shortening was noted before and after PA banding. It is noteworthy that the relationships shifted rightward after PA banding. Increases in RV systolic pressure during PA banding were directly related to the level of critical perfusion pressure. This presumably reflects increases in wall stress during PA banding. As Ellis and Klocke [7] described a rightward shift of the pressure–flow relation of the left coronary artery during diastole when the preload was increased, the rightward shift of the pressure–wall motion relation during PA banding may be explained in part by the increases in preload and the augmentation of oxygen consumption.

3.3 Difference in Autoregulation Between Right and Left Coronary Arteries

Coronary blood flow is controlled by many different mechanisms such as metabolic control, myogenic control, and neurohumoral control. The metabolic hypothesis proposes that coronary flow remains constant at a given level of metabolic demand because microvascular tone is governed by tissue concentrations of a metabolic substrate or product. Coronary blood flow is indeed closely matched to myocardial metabolism so that there is little change in coronary venous oxygen tension.

With regard to myocardial perfusion, coronary flow remains relatively constant over a wide range of perfusion pressure. When mean coronary driving pressure decreases to less than 50–70 mmHg, left coronary flow begins to fall sharply. The level of perfusion pressure for maintaining normal regional wall motion in the left ventricle is greater than 40–60 mmHg [8–10]. Thus, autoregulation adjusts coronary vascular resistance to maintain relatively constant myocardial perfusion as well as myocardial performance. The inflection point of the curvilinear relation between perfusion pressure and contractile force was considered as the limit of autoregulatory reserve [11]. Although the inflection point of the curvilinear relation between right coronary perfusion pressure and regional wall motion was easy to detect, it was difficult to determine the inflection between perfusion pressure and coronary flow, especially during right-ventricular hypertension. Thus, autoregulation (1) is a different category in flow and wall motion with regard to perfusion pressure and (2) is not so evident in the right coronary artery as in the repeatedly documented left coronary artery. Mechanisms of reduced level of autoregulation in the right coronary artery remain unclear.

3.4 Spatial Distribution of Myocardial Necrosis in Right and Left Ventricle

In experimental studies, the development of myocardial infarction after ligating a branch of coronary artery is determined by (1) time after flow cessation, (2) oxygen consumption of the heart, (3) the level of collateral blood flow, and (4) the size of the area at risk. A consistent linear relation between infarct area and the area at risk, or perfusion size of the ligated coronary artery, has been repeatedly noted in the left ventricle [12]. Using this normalization of infarct size by the area at risk, we attempted to define whether there is any topological difference in the distribution of myocardial necrosis through different transmural layers between the right (RV) and left ventricle (LV) [13]. Such an analysis may provide a quantitative basis for evaluating the role of hemodynamics and geometry of the ventricles as a determinant of infarct size and its distribution.

Fifty-one adult mongrel dogs of either sex (weight, 16–29 kg) were anesthetized with sodium pentobarbital (25 mg/kg iv) and ventilated by room air and supplemental oxygen. The chest was opened in the left fourth intercostal space. The largest obtuse marginal branch of the left circumflex coronary artery and the main trunk of the right coronary artery were ligated to produce myocardial infarction.

Five minutes after coronary occlusion, regional myocardial blood flow to the purely necrotic area decreased from 0.67 ± 0.03 to 0.06 ± 0.03 ml/g per minute in the RV and from 1.47 ± 0.09 to 0.19 ± 0.03 ml/g per minute in the LV. The blood flow to the total risk or central risk areas also decreased but was greater ($P < 0.05$) than that at the purely necrotic site in either the RV or LV. Between 5 min and 24 h after coronary occlusion, transmural myocardial blood flow toward the total or central risk area did not change significantly in either the RV or the LV, but that to the purely necrotic area of the LV decreased significantly from 0.19 ± 0.03 to 0.01 ± 0.01 ml/g per minute ($P < 0.01$). This phenomenon is explained by (1) disruption of the microvasculature in the region of severe ischemia [14], (2) a redistribution of collateral blood flow from the necrotic to the salvaged myocardium [15], and (3) the ischemia-induced vasoconstriction along the border of the ischemic zone [16].

Transmural distribution of myocardial blood flow inside the risk area differed between the RV and LV, both at 5 min and at 24 h after coronary occlusion. The subendocardial to subepicardial flow ratio in the central risk area of the RV was reduced slightly, from 1.19 ± 0.06 before coronary occlusion to 0.85 ± 0.05 at 5 min and, significantly, to 0.69 ± 0.06 at 24 h ($P < 0.01$). This ratio in the LV decreased markedly from 1.29 ± 0.08 before occlusion to 0.51 ± 0.09 at 5 min ($P < 0.01$ compared to before occlusion) and 0.09 ± 0.02 at 24 h of coronary occlusion ($P < 0.01$ compared to 5 min).

The mean infarct size of the risk area was similar between the right coronary artery and the obtuse marginal branch; 11.3 ± 0.5 g and 10.3 ± 0.9 g, respectively. The size of the risk area was more variable in the case of occlusion of the obtuse marginal branch than in that of the right coronary artery occlusion ($P < 0.05$). The mean infarct size and the ratio of the infarct size to the risk area (IS:RA) were not significantly different between occlusions of these two vessels. There was a linear relation between infarct size and risk area for both vessels. The infarct area was always located inside the risk area in both the RV and the LV.

The relation between the necrotic area (percent of the risk size, y) and transmural collateral flow (x) 24 h after coronary occlusion was inversely related in the LV and RV. The slope of the relationship was steeper in the RV than in the LV, and the x-axis intercept was lower in the RV than in the LV. The correlation coefficients between necrotic area and collateral flow remained unchanged between 5 min and 24 h of occlusion in the RV (−0.87 vs. −0.75), but increased from −0.47 to −0.78 during 24 h in the LV.

In the LV, the subepicardial layer tended to be salvaged in most dogs. On the other hand, myocardial necrosis in the RV was distributed rather homogeneously from the subendocardium to the subepicardium. In the RV, data from the subepicardial layers and subendocardial layers are distributed between the x-axis and a line of identity ($y = x$) of an infarct–risk relation, and there is no significant difference in variance between these layers; in the LV, however, data from the subendocardial layers are grouped along the line of identity and data from the subepicardial layers are scattered randomly. The ratios of subepicardial IS:RA to subendocardial IS:RA in the RV and LV were 0.76 ± 0.06 and 0.28 ± 0.05, respectively ($n = 31$, $P < 0.01$ between ventricles). Such a marked difference in the regional distribution of myocardial necrosis between the RV and LV suggests that the risk area is not the sole determinant of infarct size. Differences in hemodynamic burden and wall thickness between the two ventricles may affect the distribution of coronary flow and then determine the extent of necrosis.

A transmural difference in flow level in the normal LV is explained by a difference in myocardial oxygen consumption between layers. Weiss et al. [17]

demonstrated that regional myocardial oxygen consumption was 20% higher in the subendocardial layer than in the subepicardial layer. The tissue pressure gradient is also a dominant determinant of flow distribution when the peripheral coronary pressure of the collateral channels is decreased by coronary ligation [18]. However, the regional difference in susceptibility of myocytes to ischemia remained unanswered.

In general, the subendocardial region of the RV is also prone to be underperfused as in the LV. However, a transmural difference in flow or necrotic size was not so prominent in the RV as in the LV. Accordingly, transmural flow distribution after abrupt coronary ligation may explain the variations in infarct size. The relation between collateral flow and infarct size was similar in the subendocardial and subepicardial halves of the RV and the subendocardial layer in the LV. The salvage of the subepicardial layer in the LV suggest a tissue-sparing effect in the case of left coronary artery occlusion; namely, sacrificing the injured myocardium may be appropriate for maintaining a viable epicardium at the better condition.

In summary, the regional distribution of myocardial necrosis was determined by the level of regional myocardial blood flow inside the risk area. The slope of a necrosis–area at risk relation was similar between the subendocardium and subepicardium in the RV but in the LV it was larger in the subendocardium than in the subepicardium. Thus, in the dog, the inherent characteristics in the regional distribution of coronary collateral blood flow, the thickness of wall, intracavitary pressure, and the level of oxygen consumption are important modifiers in the development of myocardial infarction.

4 Responses of the Heart to Chronic Mechanical Overload

One of the pathophysiological features of the heart, common in chronic mechanical overload, is an increase in chamber volume. It has been believed that an increase in chamber volume reflects an increase in sarcomere length and that this effect enhances cardiac pump performance, on the basis of the Frank–Starling mechanism. Sonnenblick and Strobeck wrote in 1977 that "hypertrophy may help to compensate for excessive afterload by increasing the myocardial mass and in turn reducing the stress per unit area of muscle" [19]. However, the effect of chronic volume overload on the heart is noted mainly at the structural level, such as increases in LV chamber volume and muscle mass, proliferation of interstitial fibrosis, vascular proliferation, and myocyte hypertrophy [20]. An important consequence of these alterations are LV dysfunction. The distinction between compensatory or adaptive, and decompensatory or maladaptive, processes remains somewhat elusive. The term remodeling is now used to address pathological alterations of chamber volume and shape not related to a preload-mediated increase in sarcomere length [21]. Thus, the following review focuses on the ventricular adaptation and remodeling that occur in the presence of chronic mechanical overload.

4.1 Animal Model of Hypertrophy and Remodeling

A critical component of the goal targeted at elucidating complex disorders related to adaptation and remodeling of the heart is a reliable animal model that is the partial or complete reproduction of changes found in human diseases. The main difficulties are related to the differences between animal species and humans, including their dimensions, weight, longevity, metabolism, and signal transduction system. The animals most often used previously are the dog and pig; more recently, the mouse, rat, rabbit, and hamster have been used. A mechanical overload is created genetically or surgically, as has been described in a detailed review [22]. Three types of mechanical stresses are known as a cause of cardiac hypertrophy: (i) volume or diastolic overloads such as in aortic insufficiency, mitral insufficiency, or aortocaval fistula; (ii) pressure overloads caused by an increase in impedance of ejection caused by aortic constriction, systemic arterial hypertension, or pulmonary arterial hypertension; and (iii) functional or anatomical loss of contractile units as the result of myocardial lesions caused by infarction, myocardiopathy, or chronic tachypacing.

When an increased diastolic volume is imposed chronically on the RV or LV, a progressive augmentation of the end-diastolic volume (eccentric hypertrophy) is documented. In a model of chronic diastolic overload, created in the dog by means of an arteriovenous fistula for several months, LV dilation occurs, accompanied by an increase in LV end-diastolic pressure; the diastolic pressure–volume curve is shifted to the right, i.e., the volume rises at any level of pressure, and the LV becomes more globular in shape [3].

In contrast, with sustained systolic overload of the ventricle in patients with conditions such as aortic stenosis, the ventricular wall becomes hypertrophied but there may be relatively little change in the size and shape of the LV chamber, or so-called concentric hypertrophy [23]. Similar concentric hypertrophy has been observed in animals with systolic pressure overload [24]. In the presence of concentric hypertrophy, the diastolic pressure tends to be higher at any given intraventricular volume [2]. Sasayama et al. [25] clearly showed that the increase in wall thickness relative to chamber radius "compensates" for the pressure overload by normalizing systolic wall stress and "preserves" systolic shortening and ejection fraction. Izumo et al. [26] showed that work overload-induced hypertrophy produces a myosin heavy chain (MHC) transition from the normal adult (α-MHC) to the fetal (β-MHC) isoform and that this transition correlates with changes in contractility [26]. Phenotypic conversions characterized by protein isoform switches and the modulation of individual gene expression have been intensively studied at the myocyte level [27]. Observations that long-standing cardiac hypertrophy represents a progressive and lethal growth abnormality have accumulated [28].

Rapid ventricular pacing for 4–7 weeks results in progressive decline of myocardial contractility and chamber dilatation along with overt signs of congestive heart failure, such as ascites, dyspnea, and edema [29]. This model is characterized by reduced responsiveness to β-adrenergic receptor stimulation, as is observed in chronic human heart failure. The pathogenesis of the contractile abnormalities in this diseased state depends on changes in excitation–contraction coupling and potentially on the sarcoplasmic reticulum calcium release channel [30].

4.2 Time-Dependent Processes of Restructuring

A sudden hemodynamic overload such as that caused by aortic valve rupture results in cardiac hypertrophy, which is characterized by three sequential pathophysiological events (Table 1) [31–33]. The initial response is characterized by acute heart failure, accompanied by LV dilation, pulmonary congestion, and low cardiac output. Osler called this phase development [32]; Meerson called it transient breakdown [33]. With the development of adaptive hypertrophy, the patient or the experimental animal enters a second phase of compensation or stable hyperfunction in which the increased LV mass restores cardiac output and alleviates the pulmonary congestion along with normalizations of LV wall stress [34]. Then, chronic overloading causes the heart to reach the third and final stage of disrupted compensation. The final stage is associated with progressive myocardial cell death and cardiac fibrosis. Deterioration of the myocytes is normally caused by a long-lasting exposure to cytotoxic agents such as ionizing radiation and free radicals.

Recent recognition of the mechanisms responsible for programmed cell death indicates that the growth response that initiates short-term adaptive hypertrophy becomes maladaptive and may lead to apoptosis [35]. The growth factors that accelerate protein synthesis and stimulate apoptosis include the proto-oncogene c-*myc* and the peptide growth factor TGFb, which are called an immediate-early gene response [36,37]. Katz [28] has proposed the concept that many aspects of the heart's response to overload are of short-term benefit but cause long-term harm when the overload is sustained. Rigorous insights into mechanisms responsible for early alterations in contractile function and for transition from compensated to congestive states might be important for designing therapeutic strategies to reverse the process.

TABLE 1. Three stages in heart response to chronic hemodynamic overload. (From [10], with permission).

Transient breakdown ("development")
Circulatory: acute heart failure; pulmonary congestion, low output
Cardiac: acute left-ventricular dilatation
Myocardial: increased content of mitochondria relative to myofibrils
Stable hyperfunction ("full compensation")
Circulatory: improved pulmonary congestion; increased cardiac output
Cardiac: adaptive hypertrophy
Myocardial: increased content of myofibrils relative to mitochondria
Exhaustion and progressive cardiosclerosis ("broken compensation")
Circulatory: progressive left-ventricular failure
Cardiac: Maladaptive hypertrophy, progressive fibrosis
Myocardial: loss of mitochondria, cell death

5 Remodeling After Myocardial Infarction

In the infarcted region, infarct expansion with regional dilation and thinning of the infarct zone occurs within 24h after infarction [38]. Infarct expansion progresses gradually [39] and provides the basis for aneurysm formation or cardiac rupture [40,41]. Myocardium remote from the area of infarction is subjected to increased diastolic wall stress, resulting in myocardial slippage [42] as well as myocyte hypertrophy [43], which was noted 7–10 days after infarction. These adaptive processes, noted at both infarcted and noninfarcted segments, have the beneficial effects of improved systolic ejection and restoration of cardiac output to normal [44]. Hypertrophy tends to return elevated diastolic and systolic wall stresses toward normal. The reactive hypertrophic response of viable myocytes after infarction has been found to be accompanied by an enhanced norepinephrine-stimulated phosphoinositol turnover and upregulation of α-skeletal actin [45], which support the role of α_1-adrenoceptors and effector pathways associated with these receptors in myocyte hypertrophy.

During the chronic phase of myocardial infarction, there is increasing evidence that prolonged exposure to high wall stress leads to a transition from hypertrophy to failure with decreased contractility, impaired relaxation, and diminished diastolic distensibility [46]. Activation of the renin-angiotensin system occurs acutely and it remains in an active state for months. Vaughan et al. [47] evaluated the role of neurohumoral activation in 36 patients recovering from acute myocardial infarction of less than 45% of ejection fraction. In these patients, plasma angiotensin II and renin but not norepinephrine levels were substantially increased independently of whether the patients were receiving diuretic therapy. Recent studies suggest an important role for the renin-angiotensin system in the development of cardiac hypertrophy [48] and for the renin-angiotensin-aldosterone system in the production of myocardial fibrosis [49]. It was also reported that LV hypertrophy is associated with the induction of gene expression for angiotensin-converting enzyme and increased local synthesis of angiotensin II within the ventricular myocardium [50].

6 Therapeutic Goal

The aim of therapy in patients with chronic congestive heart failure has recently been focused on not only the amelioration of symptoms and signs of heart failure but also the prevention of morbidity and hospitalization, improvement in quality of life, and prolongation of survival. To obtain reliable answers regarding the effect of therapy, minimization of all sources of errors caused by systematic biases and random errors was guaranteed by randomized, double-blind, controlled trials and by increasing the number of events in the study [51]. Five trials have now succeeded in demonstrating a therapeutic effect on life expectancy in these patients (Table 2) [4,52–55]. In the Co-Operative North Scandinavian

TABLE 2. Four major trials of angiotensin-converting enzyme inhibitors versus placebo with congestive heart failure or left-ventricular dysfunction.

Trial	CONSENSUS	SOLVD treatment	SOLVD prevention	SAVE
Design				
Agent	Enalapril	Enalapril	Enalapril	Captopril
Maximum daily dose (mg)	40	20	20	150
Number of patients	240	2559	4228	2231
Entry criteria	NYHA class IV despite 2 weeks of therapy with drugs other than angiotensin-converting enzyme I	EF ≤ 0.35 and on treatment for CHF	EF ≤ 0.35 and not on treatment for CHF	EF ≤ 0.40 and after myocardial infarction
Percent after myocardial infarction	47.5	65.6	80	100
Result (relative risks in):				
Mortality	0.56	0.84	0.82	0.81
Cardiac mortality	0.52	0.81	0.87	0.77
Pump failure deaths	0.40	0.78	0.79	0.64
Presumed arrthythmic deaths	1.05	0.90	0.93	0.82
Hospitalizations for CHF	0.95	0.61	0.64	0.78
Myocardial infarction	0.74	0.77	0.76	0.75

NYHA, New York Heart Association; EF, ejection fraction; CHF, congestive heart failure; CONSENSUS, Co-Operative North Scandinavian Enalapril Study; SOLVD, Study of Left Ventricular Dysfunction; SAVE, Survival and Ventricular Enlargement.

Enalapril Study (CONSENSUS) including patients with severe unstable congestive heart failure, an annual mortality rate of about 60% in the control group was reduced by 31% in 1-year mortality with the addition of enalapril to their regimen [53]. In the Study of Left Ventricular Dysfunction (SOLVD), patients with mild-to-moderate heart failure were given placebo or enalapril in addition to standard therapy for their syndrome. The enalapril-treated patients exhibited a small but statistically significant reduction in overall mortality compared with the placebo group [54]. It is likely, from the data in CONSENSUS, Vasodilator—Heart Failure Trial I (V-HeFT I), and SOLVD that vasodilator-induced relief of pump failure is an important contributor to the improved natural history of the disease.

Clinical trials have provided data that heart failure involves much more than abnormal physiology at organ and cellular levels. For example, only two classes of vasodilators, the angiotensin-converting enzyme inhibitors and a drug combination that includes nitrates, have so far been shown to improve survival in patients with severe heart failure [31]. Other vasodilators as well as most inotropic drugs showed adverse effects on prognosis, which indicates that the underlying problem in chronic heart failure involves more than energy starvation and depressed contractility [6].

7 Perspectives

With the sophistication of molecular biology, cardiac hypertrophy and failure are likely to come under intense scrutiny via mouse models of human cardiovascular disease that are created via transgenic and gene-targeting strategies [56]. The transgenic murine model will offer the potential of creating lines of animals in which expression or overexpression of a particular transgene can be restricted to a specific tissue such as the atrium or ventricle [57]. It will then become necessary to integrate molecular biological techniques with classical cardiovascular physiology.

Acknowledgments. This work was partly supported by Grants for Scientific Research from the Ministry of Education, Science and Culture, Japan. The authors gratefully acknowledge the technical assistance of Eiji Tsuchida and the secretarial assistance of Aiko Funaki and Michiyo Arai.

References

1. Grossman W (1980) Cardiac hypertrophy: adaptation or pathologic process. Am J Med 69:576–583
2. Braunwald E, Ross J Jr (1963) The ventricular end-diastolic pressure: appraisal of its value in the recognition of ventricular failure in man. Am J Med 34:147–150

3. Ross J Jr, Sonnenblick EH, Taylor RR, Spotnitz HM, Covell JW (1971) Diastolic geometry and sarcomere lengths in the chronically dilated canine left ventricle. Circ Res 28:49–61
4. Pfeffer MA, Braunwald E, Moye LA, Basta L, Brown EJ, Cuddy TE, Davis BR, Geltman EM, Goldman S, Flaker GC, Klein M, Lamas GA, Packer M, Rouleau J, Rouleau JL, Rutherford J, Wertheimer JH, Hawkins CM, on behalf of the SAVE investigators (1992) The effects of captopril on mortality and morbidity in patients with left ventricular dysfunction following myocardial infarction: results of survival and ventricular enlargement (SAVE) trial. N Engl J Med 327:669–677
5. Urabe Y, Tomoike H, Ohzono K, Koyanagi S, Nakamura M (1985) Role of afterload in determining regional right ventricular performance during coronary underperfusion in dogs. Circ Res 57:96–104
6. Henquell L, Honig CR (1976) O_2 extraction of right and left ventricles. Proc Soc Exp Biol Med 152:52–53
7. Ellis AK, Klocke FJ (1979) Effects of preload on the transmural distribution of perfusion and pressure-flow relationships in the canine coronary vascular bed. Circ Res 46:68–77
8. Waters DD, da Luz P, Wyatt HL, Swan HJC, Forrester JS (1977) Early changes in regional and global left ventricular function induced by graded reductions in regional coronary perfusion. Am J Cardiol 39:537–543
9. Wyatt HL, Forrester JS, Tyberg JV, Goldner S, Ogan SE, Parmley WW, Swan HJC (1975) Effect of graded reductions in regional coronary perfusion on regional and total cardiac function. Am J Cardiol 69:576–583
10. Banka VS, Bodenheimer MM, Helfant RH (1977) Relation between progressive decreases in regional coronary perfusion and contractile abnormalities. Am J Cardiol 40:200–205
11. Downey JM (1976) Myocardial contractile force as a function of coronary blood flow. Am J Physiol 230:1–6
12. Sakai K, Tomoike H, Ootsubo H, Kikuchi Y, Nakamura M (1982) Preocclusive perfusion area as a determinant of infarct size in a canine model. Cardiovasc Res 16:408–416
13. Ohzono K, Koyanagi S, Urabe Y, Harasawa Y, Tomoike H, Nakamura M (1986) Transmural distribution of myocardial infarction: difference between the right and left ventricles in a canine model. Circulation 59:67–73
14. Kloner RA, Ganote CE, Jennings RB (1974) The "no-reflow" phenomenon after temporary coronary occlusion in the dog. J Clin Invest 54:1496–1508
15. Hirzel HO, Nelson GR, Sonnenblick EH, Kirk ES (1976) Redistribution of collateral blood flow from necrotic to surviving myocardium following coronary occlusion in the dog. Circ Res 39:214–222
16. Grayson J, Irvine M, Parrat JR, Cunningham J (1968) Vasospastic elements in myocardial infarction following coronary occlusion in the dog. Cardiovasc Res 2:54–62
17. Weiss HR, Neubauer JA, Lipp JA, Sinha AK (1978) Quantitative determination of regional oxygen consumption in the dog heart. Circ Res 42:394–401
18. Downey JB, Kirk ES (1975) Inhibition of coronary blood flow by a vascular waterfall mechanism. Circ Res 1975:753–760
19. Sonnenblick EH, Strobeck JE (1977) Current concepts in cardiology: derived indexes of ventricular and myocardial function. N Engl J Med 296:978–982
20. McDonald KM, Garr M, Carlyle PF, Francis GS, Hauer K, Hunter DW, Parish T, Stillman A, Cohn JN (1994) Relative effects of alpha 1-adrenoceptor blockade, con-

verting enzyme inhibitor therapy, and angiotensin II subtype 1 receptor blackade on ventricular remodeling in the dog. Circulation 90(6):3034–46
21. Cohn JN (1995) Structural basis for heart failure; ventricular remodeling and its pharmacological inhibition. Circulation 91:2504–2507
22. Hatt PY (1990) 2. Experimental models. In: Swynghedauw B (ed) Cardiac Hypertrophy and Failure. Libbey Eurotext/INSERM, Paris, pp 13–22
23. Dodge HT, Baxley W (1969) Left ventricular volume and mass and their significance in heart disease. Am J Cardiol 23:528–537
24. Jouannot P, Hatt PY (1975) Rat myocardial mechanics during pressure-induced hypertrophy development and reversal. Am J Physiol 229:355–364
25. Sasayama S, Ross J Jr, Franklin D (1976) Adaptations of the left ventricle to chronic pressure overload. Circ Res 38:172–178
26. Izumo S, Lompre AM, Matsuoka R, Koren G, Schwartz K, Nadal-Ginard B, Mahdavi V (1987) Myosin heavy chain messenger RNA and protein isoform transitions during cardiac hypertrophy. J Clin Invest 79:970–977
27. Schwartz K (1990) Phenoconversion and mechanogenic transduction of the mammalian heart. Med Sci 6:664–673
28. Katz AM (1990) Cardiomyopathy of overload: a major determination of prognosis in congestive heart failure. N Engl J Med 322:100–110
29. Kiuchi K, Shannon RP, Komamura K, Cohen DJ, Bianchi C, Homcy CJ, Vatner SF, Vatner DE (1993) Myocardial beta-adrenergic receptor function during the development of pacing-induced heart failure. J Clin Invest 91(3):907–914
30. Vatner DE, Sato N, Kiuchi K, Shannon RP, Vatner SF (1994) Decrease in myocardial ryanodine receptors and altered excitation-contraction coupling early in the development of heart failure. Circulation 90(3):1423–1430
31. Katz AM (1995) Scientific insights from clinical studies of converting-enzyme inhibitors in the failing heart. Trends Cardiovasc Med 5:37–44
32. Osler W (1892) The principles and practice of medicine. D Appleton, New York, p 634
33. Meerson FZ (1961) On the mechanism of compensatory hyperfunction and insufficiency of the heart. Cor Vasa 3:161–177
34. Grossman W, Jones D, McLaurin LP (1975) Wall stress and pattern of hypertrophy in the human left ventricle. J Clin Invest 56:56–64
35. Cohen JJ (1993) Overview: mechanisms of apoptosis. Immunol Today 14:126–130
36. Izumo S, Nadal-Ginard B, Mahdavi V (1988) Protooncogene induction and reprogramming of cardiac gene expression produced by pressure overload. Proc Natl Acad Sci USA 85:339–343
37. Morgan HE, Baker KM (1991) Cardiac hypertrophy. Mechanical, neural, and endocrine dependence. Circulation 83:13–25
38. Weisman HF, Bush DE, Mannisi JA, Bulkley BH (1985) Global cardiac remodeling after acute myocardial infarction: a study in the rat model. J Am Coll Cardiol 5:1355–1362
39. Crozatier B, Ross J Jr, Franklin D, Bloor CM, White FC, Tomoike H, McKown DP (1978) Myocardial infarction in the baboon: regional function and the collateral circulation. Am J Physiol 235:H413–H421
40. Schuster EH, Bulkley BH (1979) Expansion of transmural myocardial infarction: a pathophysiologic factor in cardiac rupture. Circulation 60:1532–1538
41. Erlebacher JA, Weiss JL, Weisfeldt ML, Bulkley BH (1984) Early dilation of the infarcted segment in acute transmural myocardial infarction: role of infarct expansion in acute left ventricular enlargement. J Am Coll Cardiol 4:201–208

42. Olivetti G, Capasso JM, Sonnenblick EH, Anversa P (1990) Side-to-side slippage of myocytes participates in ventricular wall remodeling acutely after myocardial infarction in rats. Circ Res 67:23–34
43. Anversa P, Loud AV, Levicky V, Guideri G (1985) Left ventricular failure induced by myocardial infarction: I. Myocyte hypertrophy. Am J Physiol 248:H876–H882
44. McKay RG, Pfeffer MA, Pasternak RC, Markis JE, Come GC, Nakao C, Alderman JD, Ferguson JJ, Safian RD, Grossman W (1986) Left ventricular remodeling after myocardial infarction: a corollary to infarct expansion. Circulation 74:693–702
45. Meggs LG, Tillotson J, Huang H, Sonnenblick EH, Capasso JM, Anversa P (1990) Noncoordinate regulation of alpha-1 adrenoreceptor coupling and reexpression of alpha skeletal actin in myocardial infarction-induced left ventricular failure in rats. J Clin Invest 86:1451–1458
46. Pfeffer MA, Braunwald E (1990) Ventricular remodeling after myocardial infarction: experimental observations and clinical implications. Circulation 81:1161–1172
47. Vaughan DE, Lamas GA, Pfeffer MA (1990) Role of left ventricular dysfunction in selective neurohumoral activation in the recovery phase of anterior wall acute myocardial infarction. Am J Cardiol 66:529–532
48. Baker KM, Chernin MI, Wixson SK, Aceto JF (1990) Renin-angiotensin system involvement in pressure-overload cardiac hypertrophy in rats. Am J Physiol 259:H324–H332
49. Brilla CG, Pick R, Tan LB, Janicki JS, Weber KT (1990) Remodeling of the rat right and left ventricles in experimental hypertension. Circ Res 67:1355–1364
50. Schunkert H, Dzau VJ, Tahn SS, Hirsch AT, Apstein CS, Lorell BH (1990) Increased rat cardiac angiotensin converting enzyme activity and mRNA expression in pressure overload left ventricular hypertrophy. J Clin Invest 86:1913–1920
51. Yusuf S, Collins R, Peto R (1984) Why do we need some large, simple randomized trials? Stat Med 3:409–420
52. Cohn JN, Archibald DG, Ziesche S, Franciosa JA, Harston WE, Tristani FE, Dunkman WB, Jacobs W, Francis GS, Flohr KH, Goldman S, Cobb FR, Shah PM, Saunders R, Fletcher RD, Loeb HS, Hughes VC, Baker B (1986) Effect of vasodilator therapy on mortality in chronic congestive heart failure; results of a Veterans Administration Cooperative Study (V-HeFT). N Engl J Med 314:1547–1552
53. The CONSENSUS Trial Study Group (1987) Effects of enalapril on mortality in severe congestive heart failure. N Engl J Med 316:1429–1435
54. The SOLVD Investigators (1991) Effect of enalapril on survival in patients with reduced left ventricular ejection fractions and congestive heart failure. N Engl J Med 325:293–302
55. The SOLVD Investigators (1992) Effect of enalapril on mortality and the development of heart failure in asymptomatic patients with reduced left ventricular ejection fraction. N Engl J Med 327:685–691
56. Mullins JJ, Peters J, Ganten D (1990) Fulminant hypertension in transgenic rats harbouring the mouse Ren-2 gene. Nature 344:541–544
57. Koch WJ, Rockman HA, Samama P, Hamilton R, Bond RA, Milano CA, Lefkowitz RJ (1995) Cardiac function in mice overexpressing the β-adrenergic receptor kinase or a βAPK inhibitor. Science 268:1350–1353

Residual Stress in the Left Ventricle

Hiroyuki Abé[1], Satoru Goto[2], Teruo Kimura[3], Hidetsugu Kushibiki[4], and Shigeru Arai[5]

Summary. The left ventricle is a highly deformable, thick-walled structure that is subjected to intraventricular pressure and myocardial contractile force. Stress in the left-ventricular wall may not be zero even if the intraventricular pressure is not present; such a stress is called the residual stress. The values of stresses must be obtained from the appropriate mathematial model, which should be as simple as possible. In the mathematical models proposed so far, the residual stress has not been considered. If the effect of residual stress is ignored, the extreme concentration of the stress occurs near the endocardium, so that there exists a contradiction between the oxygen consumption and the local mechanical work because of stress and strain along the wall thickness. First, a mathematical model was proposed that was valid for isovolumic and isobaric contractions in which the ventricle was subjected to both intraventricular pressure and myocardial contractile force. Expressions for the stress components were derived without assuming the functional form of the stress–strain relation, thus differing from those assumed previously by many researchers. The residual strain was obtained from experimental work using canine ventricles. Then, the residual stress was introduced to improve the model to avoid the contradiction just mentioned. As a result, the extreme concentration of stress at the endocardium was largely reduced in the improved ventricular model.

Key words: Left ventricle—Isovolumic contraction—Isobaric contraction—Residual strain—Residual stress

---

[1] Department of Mechanical Engineering, Graduate School of Engineering, Tohoku University, Aoba, Aramaki, Aoba-ku, Sendai, 980-77 Japan
[2] Michelin Research Japan, Fujimi Building SF, 1-6-1 Fujimi, Chiyoda-ku, Tokyo, 102 Japan
[3] Institute of Development, Aging and Cancer, Tohoku University, 4-1 Seiryocho, Aoba-ku, Sendai, 980-77 Japan
[4] Department of Internal Medicine, Hirosaki University School of Medicine, 5 Zaifucho, Hirosaki, Aomori, 036 Japan
[5] Department of Pathology, Yamagata University School of Medicine, 2-2-2 Iidanishi, Yamagata, 990-23 Japan

1 Introduction

Quantitative knowledge of stress and strain within the ventricular wall is of great significance for a clear understanding of left-ventricular functional properties. The strain components based on a reference state can be evaluated in principle, although some difficulties still exist depending on shape. On the other hand, there are no means for measuring the wall stress of the intact ventricle, so it must be evaluated by using appropriate mathematical models.

Because the ventricle is a largely deformable solid, many attempts based on finite deformation theory have been made to predict the wall stress by using simplified ventricular geometries when considering the ventricular pressure–time or pressure–volume relationship.

The left ventricle is deformed by the following forces: intraventricular pressure (a), myocardial contractile force (b), and myocardial stress in the absence of intraventricular pressure (residual stress) (c). The introduction of b is different from the usual engineering problems in solid mechanics. Actually, many publications on left-ventricular mechanics have not included the effects of b and c. The myocardial contractile force introduced as b can be ignored at diastole. However, it plays a significant role at systole [1].

In the mathematical left-ventricular models proposed so far, all the stresses vanish when the intraventricular pressure approaches zero. This state has been treated as the reference. Thus, the circumferential stress becomes extremely large near the endocardium compared with that near the epicardium. The work expressed, such as oxygen consumption, is considered to be almost uniform throughout the wall thickness, or to be somewhat larger at the endocardium than at the epicardium. The models may be improved by taking account of the residual stress introduced as c. Fortunately, the residual strain was found experimentally by Omens and Fung [2] in rat left ventricles. The residual stress is therefore expected to exist.

The constitutive equation that relates the stress and strain components is generally necessary for formulation of the mathematical model. For this purpose, the functional form of the strain energy density function has often been assumed. With such an assumption, it is occasionally difficult to give the accurate constitutive relationship of the solid, which has a different dimension from that used for the determination of the stress–strain relationship [3]. Therefore, the thick spherical shell model is used as a first approximation in which the form of the strain energy density function is not assumed in advance. Furthermore, the reference state should carefully be selected for evaluating stress and strain.

This chapter is concerned with proposing a mathematical model that takes account of a, b, and possibly c. Two typical cases were treated: isovolumic and isobaric contractions. The myocardial contractile forces as well as the intraventricular pressure were considered. Next, the residual strains in canine left ventricles for isovolumic contractions were measured experimentally, and an attempt was made to formulate a mathematical model that included the effect of residual stress and which was compatible with the distribution of oxygen consumption through the left-ventricular wall thickness.

The symmetrical deformation of a thick spherical ventricle was assumed as a first approximation. Also, the ventricle was assumed to be incompressible and transversely isotropic with respect to the direction through the wall thickness.

2 Isovolumic Contraction

The pressure–time (p–t) relation is given at different left-ventricular volumes of isovolumic beat. The pressure–volume (p–V) relation can be obtained from it as follows:

$$p(t) = a(t)\left[V - V_0(t)\right] \tag{1}$$

where $V_0(t)$ is the conceptual volume obtained by extending the p–V line at time t to intersect the horizontal axis ($V = 0$).

Let (r^0, θ, ϕ) and (R, θ, ϕ) be systems of spherical coordinates introduced in ventricles for which $V = V_0(t)$ and $V = V(t)$ respectively. The deformation may be characterized by the strain energy density function denoted by W, defined per unit volume of $V_0(t)$. The function W is simply given by

$$W = \frac{1}{2}k\left(\bar{\lambda}_\theta^3 - 1\right)^2 \tag{2}$$

where $\bar{\lambda}_\theta = R/r^0$ [1] and

$$k = aV_0\left(1 + \frac{V_0}{V_w}\right) \tag{3}$$

with V_w being the left-ventricular wall volume. It is important that k is independent of t [1].

The strain components are easily obtained from the stretch ratio $\bar{\lambda}_\theta$. The corresponding radial and circumferential stress components are given from [1] by

$$\bar{\sigma}_R = -k\left(\bar{\lambda}_\theta^3 - \bar{\lambda}_{\theta_2}^3\right) \tag{4a}$$

$$\bar{\sigma}_\theta = \bar{\sigma}_R + \frac{3}{2}k\bar{\lambda}_\theta^3\left(\bar{\lambda}_\theta^3 - 1\right) \tag{4b}$$

where the subscript 2 denotes the value at the epicardium. The stress components $\bar{\sigma}_R$ and $\bar{\sigma}_\theta$ vary as a function of the myocardial contractile force.

So far, functional forms of W together with some unknown coefficients have been assumed and the coefficients have been determined to agree with the experimental results, such as the myocardial stress–strain relation obtained by the uniaxial tensile test or the ventricular p–V relation. Such an assumption however often gives an inaccurate constitutive relation, particularly for a configuration that has a different dimension from that used for the determination of the coefficient [3].

It is noted that the stress–strain relationship (Eq. 4) has been obtained without assuming the functional form of W.

3 Isobaric Contraction

3.1 Pressure–Volume Relation

In Fig. 1, the range $t < t^*$ is for the isovolumic contraction ($V = V^* =$ constant), and $t > t^*$ is for the isobaric contraction ($p = p^* =$ constant), where V^* is the end-diastolic volume, p^* is the ejection pressure, and p_v is the pressure for isovolumic contractions.

If the p–V relation during isobaric contraction is expressed in the form of Eq. 1, volume V should decrease (as the broken line as shown in Fig. 1a) and take the minimum value at $t = t_n$ when p_v attains the maximum. However test results of canine left ventricles show V as the solid line (Fig. 1a).

Let

$$\Delta V_A(t) = V^* - V(t) \tag{5}$$

According to Starling's law of the heart, the linear relation is assumed during ejection:

$$\Delta V_A(t) = b\left[V^* - V_b(t)\right] \tag{6}$$

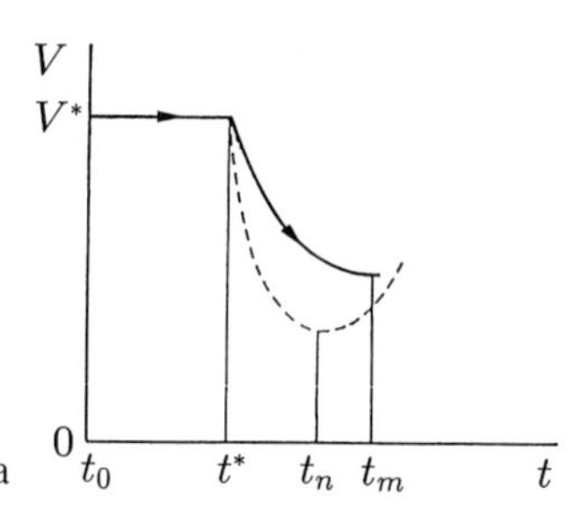

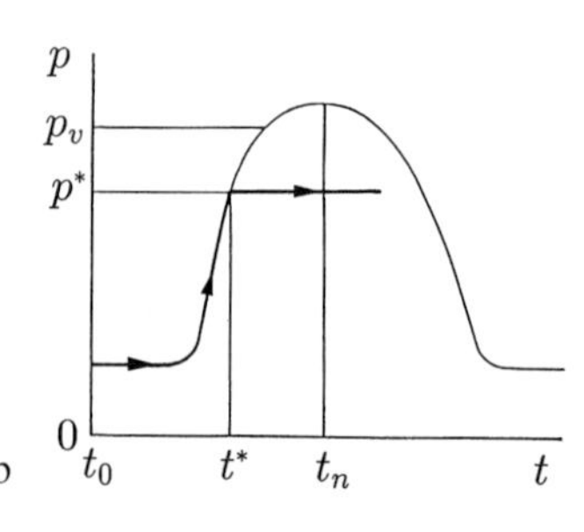

Fig. 1a,b. Schematic illustration of left-ventricular volume V and intraventricular pressure p as function of time t

where b is independent of t and V_b is the volume obtained by extending the ΔV_A–V^* line to the horizontal axis ($\Delta V_A = 0$).

The p–V relation can therefore be expressed by the following simple equation:

$$p = \frac{V'}{bg} + \frac{a}{b}\left[V - \left\{(1-b)V^* + bV_0\right\}\right] \tag{7}$$

where ( )′ denotes differentiation with respect to t.

Equation 7 is reduced to the form of Eq. 1 during isovolumic contractions. For isobaric contractions, Eq. 7 is written as

$$\left(\Delta V_A\right)' + ag\Delta V_A = bg\left(p_v - p^*\right) \tag{8}$$

The experimental work was carried out by using four isolated canine hearts, denoted by D_1–D_4. They were caused to beat 120/min by pacemaker stimulation. Data were collected every 2.5 ms; collection time was denoted by t_i ($i = 0, 1, \ldots, 200$), where t_0 is the time when the pacemaker signal was given and the contractile force can be neglected at $t = t_0$.

Figure 2 shows the ΔV_A–V^* relations of D4 as an example in which the test results are shown as a set of dots. The slope b of Eq. 6 was determined from the

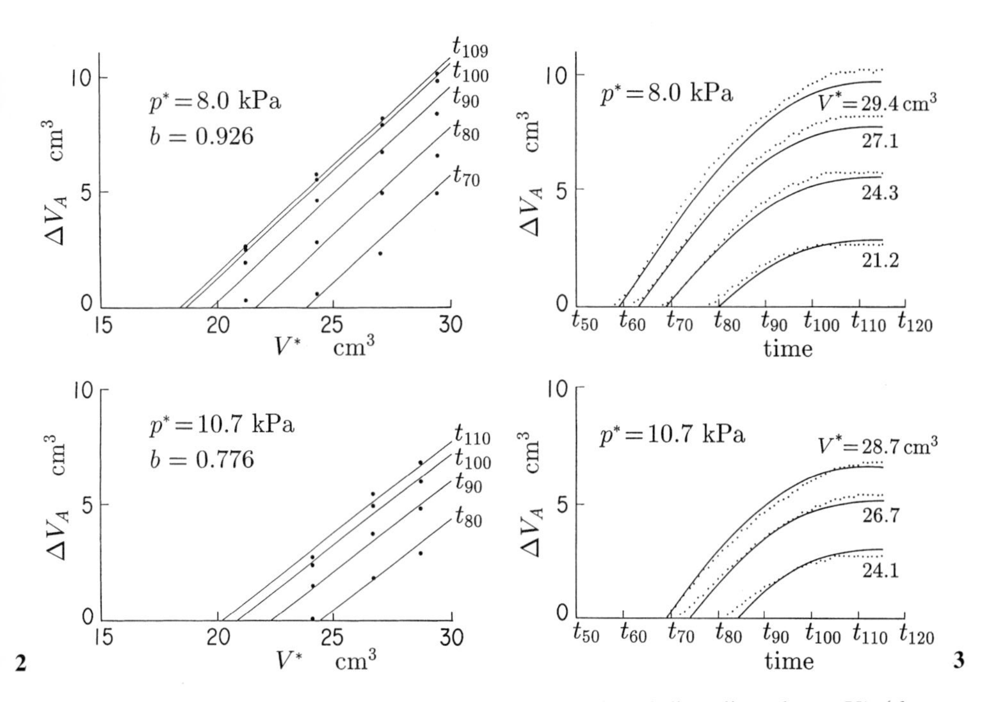

FIG. 2. Relationship between stroke volume ΔV_A and end-diastolic volume V^* (*dots*, experimental data) at times t

FIG. 3. Stroke volumes: numerical, *solid lines*; experimental, *dotted lines*

TABLE 1. Slope b of ΔV_A–V^* relationship.

Dog	V_w (ml)	p^* (kPa)	b
D1	61.36	8.0	0.491
		10.7	0.503
D2	47.28	8.0	0.919
		10.7	0.776
D3	61.94	8.0	0.926
		10.7	0.701
D4	41.17	8.0	0.926
		10.7	0.776

experimental data and Eq. 6 by means of the least-squares method [4]. The results are shown in Fig. 2 by the straight lines. The values of b thus obtained for the four dogs are shown in Table 1. It was found from the test results of canine hearts that b appeared to be in the range $0 < b < 1$.

Equation 8 was solved by using the Runge–Kutta method; the results obtained are shown as solid lines in Fig. 3 and the experimental results by dotted lines. Both sets of data were found to agree well. Thus, Eq. 8 is the appropriate left-ventricular model during isobaric constractions and Eq. 7 is valid both for isovolumic and for isobaric contractions.

Equation 8 is an improved expression of the work by Shroff et al. [5] where $b = 1$ and $V_0 \neq V_0(t)$ [4].

3.2 Expressions for Stress Components

Let

$$\bar{\lambda}_\theta^* = \frac{R^*}{r^0}, \qquad \Delta\bar{\lambda}_\theta^3 = 1 + \bar{\lambda}_\theta^{*3} - \bar{\lambda}_\theta^3 \tag{9}$$

where * denotes the quantities at $V = V^*$. The stress components during isobaric contractions are easily obtained as in [6]. They are as follow:

$$\bar{\sigma}_R = -k\left[\left(\bar{\lambda}_\theta^{*3} - \bar{\lambda}_{\theta 2}^{*3}\right) - \frac{1}{b}\left(\Delta\bar{\lambda}_\theta^3 - \Delta\bar{\lambda}_{\theta 2}^3\right) - \frac{1}{abg}\left(\frac{1}{r^{03}} - \frac{1}{r_2^{03}}\right)\left(R^{*3} - R^3\right)'\right] \tag{10a}$$

$$\bar{\sigma}_\theta = \bar{\sigma}_R + \frac{3}{2}k\bar{\lambda}_\theta^3\left[\bar{\lambda}_\theta^3 - 1 - \frac{1}{b}\left(\Delta\bar{\lambda}_\theta^3 - 1\right) - \frac{1}{abg}\left(\Delta\bar{\lambda}_\theta^3 - 1\right)'\right] \tag{10b}$$

where ( )′ denotes the differentiation with respect to t as before. It is shown that Eq. 10 is reduced to Eq. 4 for $p \leqq p^*$.

Let ξ be the parameter defined by

$$\frac{4}{3}\pi r^{0^3} = \frac{4}{3}\pi r_1^{0^3} + \xi V_w = V_0 + \xi V_w \tag{11}$$

where $\xi(r_1^0) = 0$ and $\xi(r_2^0) = 1$. The subscript 1 denotes the value at the endocardium and 2 denotes the value at the epicardium. Equation 10 can be written as

$$\overline{\sigma}_R = -\frac{(1-\xi)V_0}{V_0 + \xi V_w} p \tag{12a}$$

$$\overline{\sigma}_\theta = \overline{\sigma}_R + \frac{3}{2}\frac{V_0(V_0 + V_w)(V + \xi V_w)}{V_w(V_0 + \xi V_w)^2}\left[p + \frac{\Delta V_A}{bg}\frac{(V_0)'}{V_0 + \xi V_w}\right] \tag{12b}$$

If $\Delta V_A = 0$, then Eq. 12 is reduced to that valid for isovolumic contractions.

4 Residual Stress and Strain

4.1 Background

In the left-ventricular models proposed so far by many researchers, there was a large stress concentration near the endocardium because the models were constructed under the assumption that the stresses vanish when the intraventricular pressure is zero. These models ignored the so-called residual stress. In sections 2 and 3 of this chapter, residual stress was still ignored so that $\overline{\sigma}_{\theta_1}$ is extremely large compared with $\overline{\sigma}_{\theta_2}$.

On the other hand, direct measurement of regional oxygen consumption showed little difference between endocardial and epicardial sites [7–9]. Assuming that myocardial mechanical work is closely related to myocardial oxygen consumption, the extreme stress concentration near the endocardium makes it difficult to understand regional oxygen consumption. Omens and Fung [2] reported the existence of residual strain by using potassium-arrested rat left ventricles.

The models may be improved by taking account of residual stress. For this purpose one must measure residual strains in the state of zero intraventricular pressure. Residual strains were calculated by using sliced canine left ventricles obtained during isovolumic contractions. The method was essentially based on the work developed by Omens and Fung [2] at diastole. The residual strains determined were applied to the new model. As a result, the stress concentration near the left-ventricular endocardium was found to be largely reduced.

4.2 Mathematical Model

Let the functional properties of the muscle fibers be essentially the same throughout the wall thickness. The residual stress may then be generated by the geometrical configuration of muscle fibers, such as their mismatches.

The left ventricle, being a thick-walled shell, is therefore assumed to be fabricated by a set of thin spherical shells in which the change in circumferential length of two adjacent shells produces the fundamental part of the residual strain:

$$\left(r^0 + dr^0\right)^3 - r^{0^3} = \left(r^F + \frac{1}{\alpha} dr^F\right)^3 - r^{F^3} \tag{13}$$

where r^F is the radial coordinate of a point in the shell before the fabrication, i.e., in the stress-free state. The coordinate r^0 is that of the set of shells after the fabrication. Equation 13 is written as

$$\alpha = \left(\frac{r^F}{r^0}\right)^2 \frac{dr^F}{dr^0} \tag{14}$$

which expresses the mismatch before the fabrication. The radial and circumferential stretch ratios are defined as $\lambda_r^0 = dr^0/dr^F$ and $\lambda_\theta^0 = r^0/r^F$, respectively.

The p–V relation obtained during isovolumic contractions is shown schematically in Fig. 4, where Eq. 1 is assumed to hold. The coordinate r^0 is introduced in V_0 as before. Let V_s be the volume under consideration and R_s be the coordinate in V_s at the point corresponding to r^0. The radial and circumferential residual stress components in V_0 are denoted by σ_r^0 and σ_θ^0, respectively. They must satisfy the equilibrium equations:

$$\frac{d\sigma_r^0}{dr^0} + \frac{2}{r^0}\left(\sigma_r^0 - \sigma_\theta^0\right) = 0 \tag{15}$$

and

$$\sigma_r^0\left(\alpha_1, \lambda_{\theta 1}^0\right) = \sigma_r^0\left(\alpha_2, \lambda_{\theta 2}^0\right) = 0 \tag{16}$$

The model has to express both the tensile and compressive states by means of α and λ_θ^0. It is noted that $\bar{\sigma}_\theta$–$\bar{\sigma}_R$ in Eq. 4b is the fundamental term of the series of $\bar{\lambda}_\theta^3$–1. The fundamental terms of the present model corresponding to Eq. 4 may therefore be written as follow:

$$\begin{aligned}
\sigma_r^0 = & -2A_0\left(\alpha\lambda_\theta^{0^3} - \alpha_1\lambda_{\theta 1}^{0^3}\right) \\
& - b_0\left[\left(\alpha^{-1}\lambda_\theta^{0^{-3}} - 1\right)^2 - \left(\alpha_1^{-1}\lambda_{\theta 1}^{0^{-3}} - 1\right)^2\right] \\
& + 2C_0\left(\alpha^{-1}\lambda_\theta^{0^3} - \alpha_1^{-1}\lambda_{\theta 1}^{0^3}\right) \\
& - d_0\left[\left(\alpha\lambda_\theta^{0^{-3}} - 1\right)^2 - \left(\alpha_1\lambda_{\theta 1}^{0^{-3}} - 1\right)^2\right]
\end{aligned} \tag{17a}$$

FIG. 4. Pressure–volume (p–V) relation during isovolumic contraction

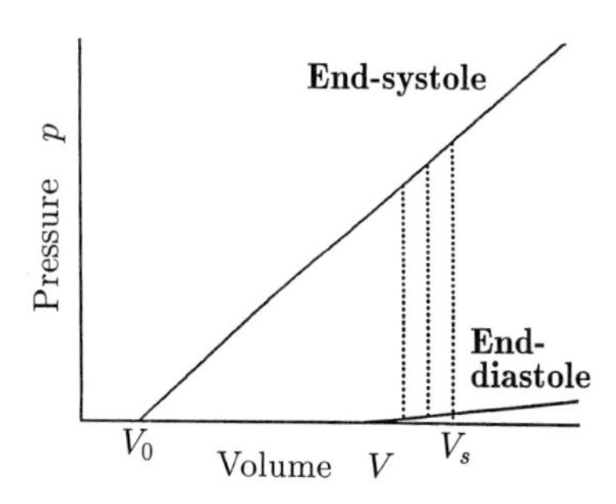

$$\sigma_\theta^0 = \sigma_r^0 + A_0\left(3+\frac{\alpha'}{\alpha}\lambda_\theta^0\right)\alpha\lambda_\theta^{0^3}\left(\alpha\lambda_\theta^{0^3}-1\right)$$

$$+ b_0\left(3+\frac{\alpha'}{\alpha}\lambda_\theta^0\right)\left(\alpha^{-1}\lambda_\theta^{0^{-3}}-1\right)^2$$

$$+ C_0\left(3-\frac{\alpha'}{\alpha}\lambda_\theta^0\right)\left(1-\alpha\lambda_\theta^{0^3}\right)\alpha^{-1}\lambda_\theta^{0^3}$$

$$+ d_0\left(3-\frac{\alpha'}{\alpha}\lambda_\theta^0\right)\left(1-\alpha\lambda_\theta^{0^3}\right)\alpha\lambda_\theta^{0^{-3}}\left(\alpha\lambda_\theta^{0^{-3}}-1\right) \tag{17b}$$

where $\alpha' = d\alpha/d\lambda_\theta^0$. The coefficients A_0, b_0, C_0, and d_0 are material constants that satisfy the following relations:

$$C_0 = \frac{\alpha_2\lambda_{\theta 2}^{0^3} - \alpha_1\lambda_{\theta 1}^{0^3}}{\alpha_2^{-1}\lambda_{\theta 2}^{0^3} - \alpha_1^{-1}\lambda_{\theta 1}^{0^3}} A_0$$

$$d_0 = -\frac{\left(\alpha_2^{-1}\lambda_{\theta 2}^{0^{-3}}-1\right)^2 - \left(\alpha_1^{-1}\lambda_{\theta 1}^{0^{-3}}-1\right)^2}{\left(\alpha_2\lambda_{\theta 2}^{0^{-3}}-1\right)^2 - \left(\alpha_1\lambda_{\theta 1}^{0^{-3}}-1\right)^2} b_0 \tag{18}$$

Because A_0 relates to the tensile stress, it may be written that $A_0 = k/2$. The constant b_0 is to be determined from β, which is defined in Section 4.3.

The total values of the stress components in V_s are given by

$$\sigma_R^T = \frac{1}{\overline{\lambda}_\theta^2}\sigma_r^0 + \overline{\sigma}_R, \quad \sigma_\theta^T = \overline{\lambda}_\theta\sigma_\theta^0 + \overline{\sigma}_\theta \tag{19}$$

They are the results of the intraventricular pressure, the myocardial contractile force, and the residual stress. It is easily shown that they satisfy the following equilibrium equation:

$$\frac{d\sigma_R^T}{dR_s} + \frac{2}{R_s}\left(\sigma_R^T - \sigma_\theta^T\right) = 0 \tag{20}$$

The total stretch ratio from the stress-free state to V_s is given by

$$\lambda_\theta^T = \frac{R_s}{r^F} = \lambda_\theta^0 \overline{\lambda}_\theta \tag{21}$$

The strain energy density function W may be expressed as $W = W(\lambda_\theta^T)$. The function W is closely related to the local mechanical work done by muscle fibers. The local work is considered to depend essentially on myocardial oxygen consumption.

4.3 Experiments and Results

Two adult mongrel dogs were used; the body weights and left-ventricular wall volumes are shown in Table 2. The dogs were anesthetized and their hearts were rapidly excised. The left main coronary was cannulated and the heart was immediately perfused via the cannula with the arterial blood from a donor dog. The *p–t* relationship was obtained at different ventricular volumes of isovolumic beat. Then, *p–V* relations were determined.

The perfusate was changed from blood to Ca^{2+}-free Tyrode solution (see Table 3). Munch et al. [10] have reported that the barium contracture in a whole heart is a reasonable model for sustained systolic contracture. After the contraction disappeared, therefore, the perfusate was again changed to 1 m*M*/l Ba^{2+}-added Tyrode solution. During perfusion, the coronary artery was maximally dilated with adenosine and carbocromen, and blood coagulation was prevented with heparin-sodium (10^4 unit/l). Perfusion with Ba^{2+}-added Tyrode solution was maintained for 5 min to obtain the expected systolic state.

The heart was then sliced perpendicular to its long axis near the equatorial region into slices 5–6 mm thick using a 5-mm-thick cutting guide. After removing the solution, brass markers (150-μmϕ wire, cut to 100–200 μm in length) were scattered and pressed onto the surface of the slice. The slice was submerged again in the Ba^{2+}-containing perfusate in a shallow dish to reduce the morphological influence caused by gravity and friction and to resemble the configuration at the no-load state. After photographs were taken, the left-ventricular wall opposite to the right ventricle was cut radially and photographs were taken again. Our definition for the center of the left ventricle before cutting was simply the centroid of the outer configuration of the left ventricle. This is different from the definition employed by Omens and Fung [2].

The positions of three markers that form a triangle on the surface of the slice were decided. By assuming that the strain distribution is uniform within the triangle, the strain components and therefore the principal stretch ratios were determined. Because the stretch ratio just determined was referred to the no-load state before cutting, the stretch ratio referred to the stress-free state was obtained as the inverse of the former.

TABLE 2. Dogs used for measurements of residual strain.

Dog	B_w (kg)	V_w (ml)
A	14.5	118.4
B	15.0	71.7

TABLE 3. Ca^{2+}-free Tyrode solution.

Contents	Concentration mol/l (× 10^{-3})
NaCl	130.00
KCl	4.00
$MgCl_2$	1.00
$NaHCO_3$	10.00
NaH_2PO_4	0.44
$C_6H_{12}O_6$ (glucose)	5.56

The distributions of the principal residual stretch ratios measured for dog A and dog B are given in Figs. 5 and 6, respectively. The dot in these figures denotes the centroid of the triangle.

The p–V relations for dog A and dog B are given in Fig. 7. The stretch ratios along R2 taking account of the mathematical relation between λ_r^0 and λ_θ^0 were obtained as shown in Figs. 8 and 9, together with $\bar{\lambda}_\theta$ and λ_θ^T. The stretch ratios obtained were two dimensional. Although they do not express the stretch ratios of the intact ventricle, we use them as an approximation in evaluating the stresses as a first step.

The distributions of the total circumferential stress, including the effect of residual stress, were obtained for dog A and dog B and are shown in Figs. 10 and 11, respectively. The parameter β, which is a function of the constant b_0, was introduced as shown in Figs. 10 and 11, where it varies from 0 at end-diastole to 1 at end-systole. The total circumferential stress $\sigma_{\theta_1}^T$ ($\equiv \sigma_\theta^T(R_{s1})$) attains the maximum compressive stress when β is equal to zero, while it is the maximum tensile stress when β is unity. The parameter β (Figs. 10 and 11) denotes the value at which $\sigma_{\theta 1}^T$ becomes zero between the end-diastole and end-systole; σ_θ^T shows the maximum tensile stress along the wall thickness. It was found that the stress concentration near the endocardium was largely reduced. The residual stress σ_θ^0 was found to be compressive at the inner part of the left-ventricular wall and tensile at its outer part.

4.4 Discussion

Our model, taking account of the residual stress, was an attempt, although the results were expected to be in agreement with the distribution of oxygen con-

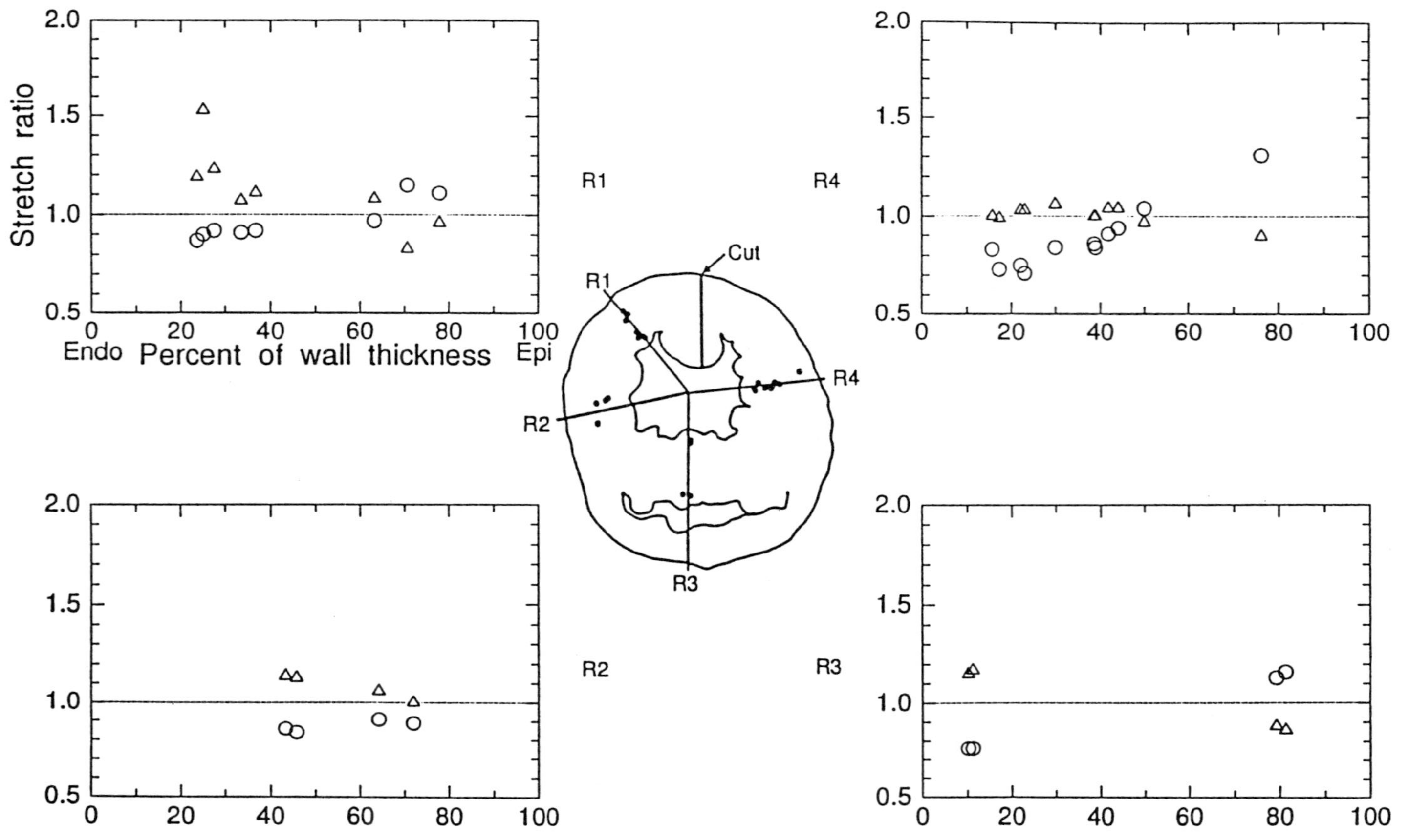

Fig. 5. Distributions of principal residual stretch ratios for dog A: *circles*, circumferential; *triangles*, radial

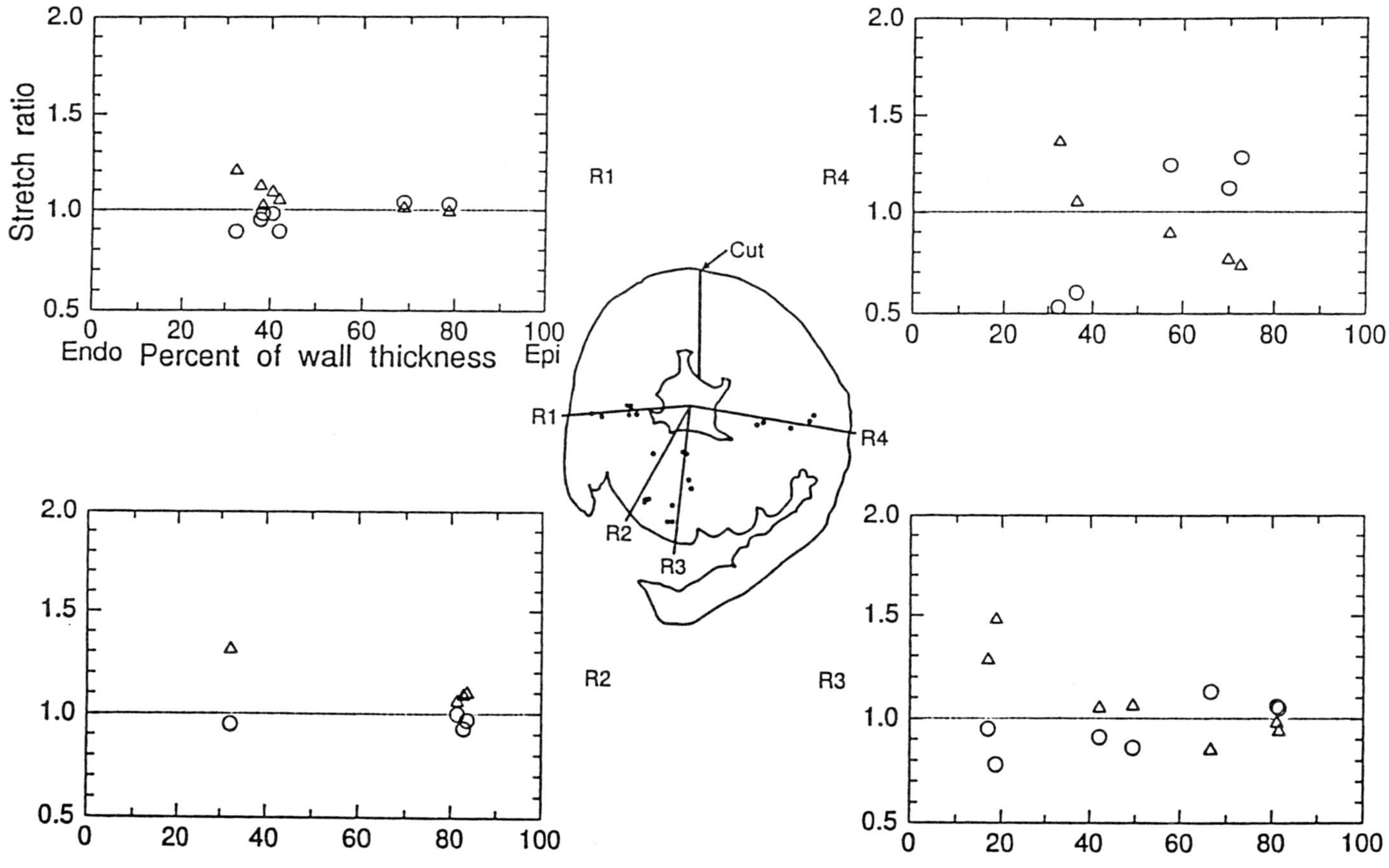

FIG. 6. Distributions of principal residual stretch ratios for dog B: *circles*, circumferential; *triangles*, radial

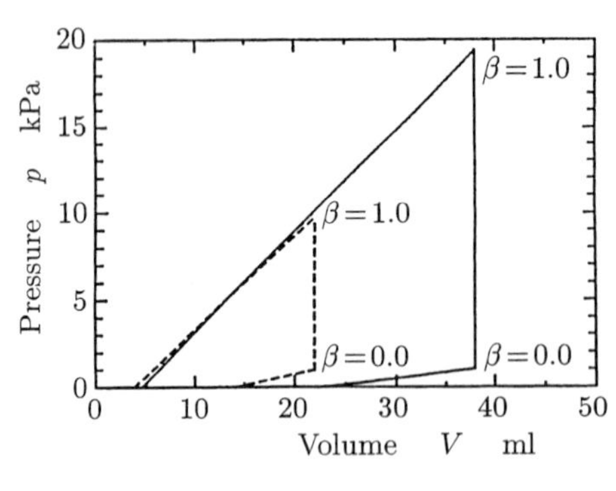

FIG. 7. Pressure–volume (p–V) relationships obtained from experiments for dog A (*solid lines*) and dog B (*dashed lines*)

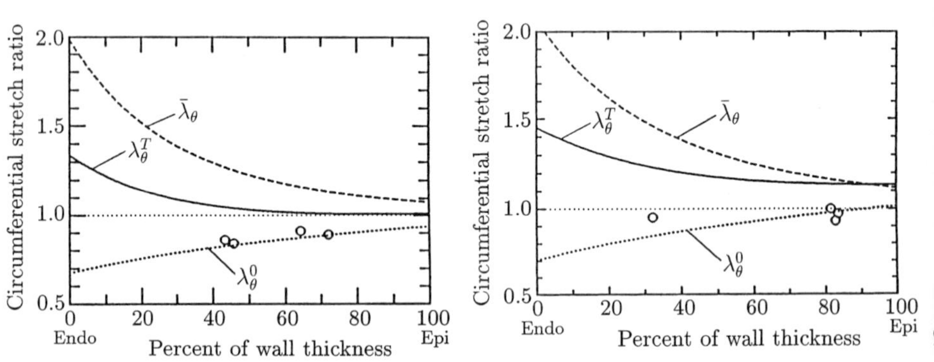

FIG. 8. Stretch ratios along cut R2 for dog A (*circles*, experimental data)

FIG. 9. Stretch ratios along cut R2 for dog B (*circles*, experimental data)

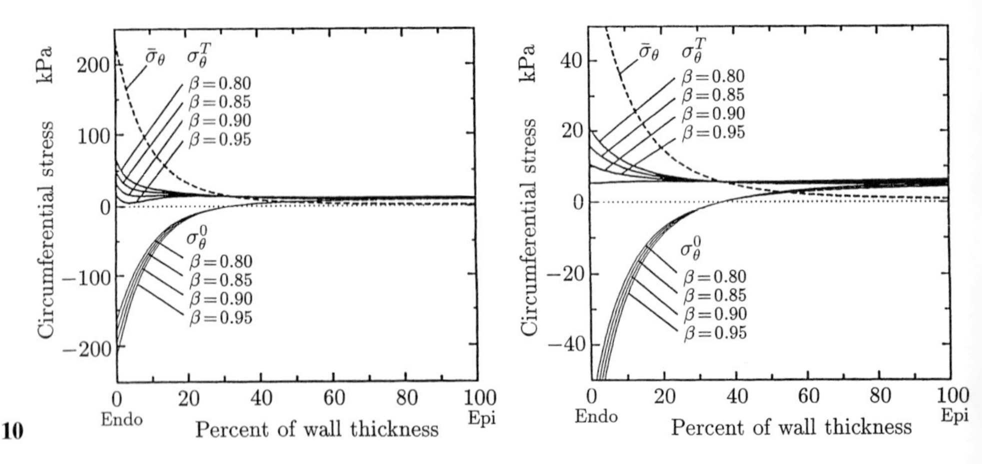

FIG. 10. Distribution of circumferential stress (in *KPa*) for dog A. β is a function of the constant b_0

FIG. 11. Distribution of circumferential stress (in *KPa*) for dog B

sumption for the values of β shown in Figs. 10 and 11. However, the most suitable value of β has not yet been definitely determined. More accurate measurement of the distribution of oxygen consumption throughout ventricular wall thickness and a more refined mathematical model would be necessary for the determination of β. A more refined approach would be to use the mechanical work of one cycle of the left-ventricular motion instead of the distribution of the maximum tensile stress and stretch ratio as was used in the current model. Also, a left-ventricular model with residual stress for isobaric contractions is still something to be tackled in the future.

Acknowledgments. The authors acknowledge Mr. T. Shōji for preparation of the manuscript. This work was supported by the Ministry of Education, Science and Culture, Grant-in-Aid for Scientific Research on Priority Areas, No. 04237104.

References

1. Abé H, Nakamura T, Kimura T, Motomiya M, Konno K, Arai S, Suzuki N (1981) Stress-strain relation of cardiac muscle determined from ventricular pressure-time relationships during isovolumic contractions. J Biomech 14:357–360
2. Omens JH, Fung YC (1990) Residual strain in rat left ventricle. Circ Res 66:37–45
3. Abé H, Nakamura T, Motomiya M, Konno K, Arai S (1978) Stresses in left ventricular wall and biaxial stress-strain relation of the cardiac muscle fiber for the potassium-arrested heart. Trans ASME J Biomech Eng 100:116–121
4. Abé H, Miyano I, Nakamura T (1985) The pressure-volume relation of left ventricles during isobaric contractions (in Japanese). Trans Jpn Soc Mech Eng (A) 51:2239–2242
5. Shroff SG, Janicki JS, Weber KT (1983) Left ventricular systolic dynamics in terms of its chamber mechanical properties. Am J Physiol 245:110–124
6. Abé H, Miyano I, Nakamura T (1985) Stress in the left ventricular wall during isobaric contractions and biaxial stress-strain relation of the unit cardiac muscle fiber (in Japanese). Trans Jpn Soc Mech Eng (A) 51:2243–2246
7. Weiss HR, Neubauer JA, Lipp JA, Sinha AK (1978) Quantitative determination of regional oxygen consumption in the dog heart. Circ Res 42:394–401
8. Scholz PM, Kedem J, Sideman S, Beyar R, Weiss HR (1990) Effect of hyper- and hypovolaemia on regional myocardial oxygen consumption. Cardiovasc Res 24: 81–86
9. Wright C, Weiss HR, Kedem J, Scholz PM (1991) Effect of dopamine on regional myocardial function and oxygen consumption in experimental left ventricular hypertrophy. Basic Res Cardiol 86:449–460
10. Munch DF, Comer HT, Downey JM (1980) Barium contracture: a model for systole. Am J Physiol 239:H438–H442

Response of Arterial Wall to Hypertension and Residual Stress

Takeo Matsumoto[1] and Kozaburo Hayashi[2]

Summary. The mechanical response of the arterial wall to hypertension is reviewed here with special reference to the rat thoracic aorta. In response to hypertension, the aortic wall thickens rapidly, as if the thickening maintain the mean circumferential stress and the luminal area at constant levels. The inherent elastic modulus of wall material at in vivo pressure, which has increased once because of pressure elevation, also seems to be restored to the normotensive value after a relatively long period of time. In addition, the opening angle and therefore the residual strain increase with elevation of blood pressure; by this means, the stress distribution and stress concentration factor seem to be kept similar to those in the normotensive aorta. These changes are accompanied by thickening of the lamellar unit in the media, especially near the inner wall where increase in mechanical stress caused by the pressure elevation is dominant. These phenomena are considered to be a mechanical adaptation process of the vascular wall. The smooth muscle cells in the aortic media seem to sense the local mechanical stress. On the other hand, the aortic wall does not seem to maintain mechanical stress in the axial direction at a constant level: mean axial stress and strain are significantly lower in the hypertensive aorta than in the normotensive one. The aortic wall might not respond to the change in stress in this direction as it responds to that in the circumferential direction. Recent findings on residual stress in the arterial wall are also reviewed.

Key words: Arterial mechanics—Hypertension—Residual stress—Opening angle—Stress distribution

---

[1] Biomechanics Laboratory, Graduate School of Mechanical Engineering, Tohoku University, Aoba, Aramaki, Aoba-ku, Sendai, 980-77 Japan
[2] Department of Mechanical Engineering, Faculty of Engineering Science, Osaka University, 1-3 Machikaneyama-cho, Toyonaka, Osaka, 560 Japan

1 Introduction

Hypertension causes various changes in the arterial wall, both morphologically and mechanically. General mechanical changes include artery wall thickening, altered passive stiffness of vessel walls, altered active vascular smooth muscle contractility, increased sensitivity to pharmacological stimulation, and increased viscoelasticity [1,2].

From a biomechanical point of view, hypertension is a state in which blood pressure, i.e., circumferential tension in the arterial wall, has been elevated. Many of the arterial wall responses to hypertension can be viewed as an adaptation process of soft biomechanical tissues to increased load. For example, aortic walls in hypertensive rats increase their thickness as if the hypertrophy maintained the circumferential stress at a level similar to that in normotensive animals [3–6]. As for mechanical properties, it has been reported that hypertensive arteries have a lower incremental elastic modulus than normotensive counterparts when compared at the same pressure, suggesting the restoration of elastic properties under physiological conditions [3,7,8].

The compatibility of mean stress values between hypertensive and normotensive aortas suggests that there may be a normal level of stress in the artery wall. If such a "normal level" is maintained not only in a particular tissue from different subjects but also among individual cells in the tissue, i.e., if stress distributions inside these are uniform under physiological conditions, residual stresses must appear after all external loads are removed [9,10]. In fact, an increasing number of papers report that unloaded blood vessel walls opened up when they were cut radially [9–17], implying the presence of compressive and tensile circumferential residual stresses in the inner and outer walls, respectively, before they were cut. Because residual stress is closely correlated with local tissue response to the mechanical environment, study of the change of residual stresses in response to hypertension may give us much information on tissue adaptation.

One of our major interests is to know how vascular tissues adapt to their mechanical environment. To reveal the adaptation mechanism quantitatively, we need to know intramural stress and strain fields under physiological condition. However, it is very difficult to measure stress and strain distributions in vivo. We have to employ reliable methods for their estimation by combining residual stresses and constitutive equations.

This chapter reviews the effects of hypertension on the bulk mechanical properties of the arterial wall and recent studies of residual stress in the arteries, introduces a method for estimating transmural distributions of stress and strain, and discusses the in vivo mechanical environment in normotensive and hypertensive aortas. (See other reviews such as Dobrin [2] and Cox [1] for the effects of hypertension on smooth muscle contractility and viscoelasticity.)

2 Effect of Hypertension on the Bulk Mechanical Properties of the Arterial Wall

There have been a number of studies on the mechanical properties of arterial walls in hypertensive rats [5–7,18–22], rabbits [23,24], dogs [4,25,26], and humans [27,28]. Many of these reported wall thickening [5–7,18–21,23] and decreased arterial distensibility [7,18,19,23–25,27]. In this section, these changes are viewed from the standpoint of mechanical adaptation.

2.1 Wall Dimensions and Circumferential Stress

Hypertension causes blood pressure elevation. Elevated pressure increases circumferential tension in the arterial wall. Wall thickening reduces the increase in circumferential stress that would be caused by this increased circumferential tension. Interestingly, it has been reported that circumferential wall stress of hypertensive subjects calculated for their physiological condition is almost the same as that of normotensive subjects [3–6].

Wolinsky [5] reported that circumferential stress in the thoracic aorta in systole did not differ between hypertensive rats and age-matched control animals at 20 weeks after the initiation of hypertension. He induced renal hypertension in rats by constricting their renal arteries (Goldblatt hypertension). Aortic wall dimensions were measured on specimens pressure-fixed at their in vivo blood pressure. We obtained recently similar results in more extensive experiments. We measured the pressure–diameter relation of hypertensive and normotensive rat aortas to obtain circumferential stress over a wide range of pressure.

Figure 1 shows the relations between in vivo systolic blood pressure of each rat, *Psys*, and circumferential stresses calculated for three pressure levels [3]. Goldblatt hypertension was induced by constricting the renal artery 16 weeks before the measurement. The circumferential stress calculated for the systolic blood pressure was almost independent of this pressure. On the other hand, the stresses calculated for 100 and 200 mmHg varied with *Psys*, having a negative correlation. These results indicated that the aortic wall thickens with the progress of hypertension so as to maintain the circumferential stress developed by the in situ blood pressure at a constant level and that this is not the case for other pressure conditions.

Diameter change accompanies wall thickening. How do the inner and outer diameters change with hypertension? There appears to be a lack of research on this subject. Our recent result may provide a key: Fig. 2 shows an example of the relation between in vivo systolic pressure (*Psys*) and the wall dimensions of unloaded aortas used in Fig. 1 [3]. Because body weight was significantly different between the hypertensive and control rats in this case, wall thickness and diameters were normalized by the cube root of body weight; the open circles show the mean values in age-matched controls. There was a good correlation between the wall thickness and *Psys*. The outer diameter of unloaded vessels correlated

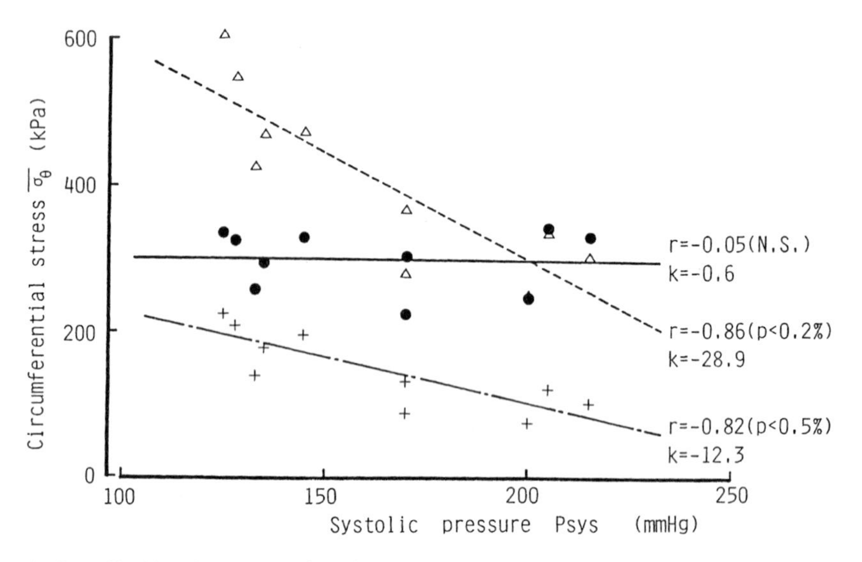

FIG. 1. Systolic blood pressure (*Psys*) versus circumferential stresses at three different pressure levels for each of 10 rats rendered hypertensive for 16 weeks. Circumferential stress ($\overline{\sigma_\theta}$, kPa) was calculated from aortic wall dimensions and blood pressure by applying the law of Laplace (see Eq. 7). *r*, Correlation coefficient; *P*, its significance level; *k*, gradient of regression line (kPa/mmHg); *triangles*, 200 mmHg; *circles*, *Psys*; *crosses*, 100 mmHg. (From [3], with permission)

significantly with *Psys*, although the inner diameter did not. Similar relations were observed for the data before normalization.

These data suggest that the aortic wall thickens as a result of hypertension, keeping the inner diameter almost constant. Kamiya and Togawa [29] observed in arteriovenous shunt experiments that the inner diameter of the canine carotid artery changed so as to maintain the wall shear rate constant. According to Ledingham and Pelling [30], changes in the cardiac output of rats following the induction of Goldblatt hypertension were within 10% and were not significant. Assuming laminar flow in the aorta, the relation between the cardiac output, Q, and the wall shear rate, $\dot{\gamma}$, can be expressed by

$$Q = \left(\pi r^3/4\right)\dot{\gamma} \tag{1}$$

where r is the inner radius. The cardiac output is roughly proportional to the body weight, *BW*. If we assume this and that the wall shear rate is maintained at a constant value, then the inner radius, or the inner diameter, is proportional to the cube root of *BW*. Our results (Fig. 2) indicate that the inner diameter normalized by the cube root of body weight is independent of *Psys*, i.e., the inner diameter is proportional to the cube root of *BW*. Stacy and Prewitt [20] observed that the inner radius of a hypertensive rat aorta was almost equal to that of a normotensive rat having almost the same body weight. These results might indicate that

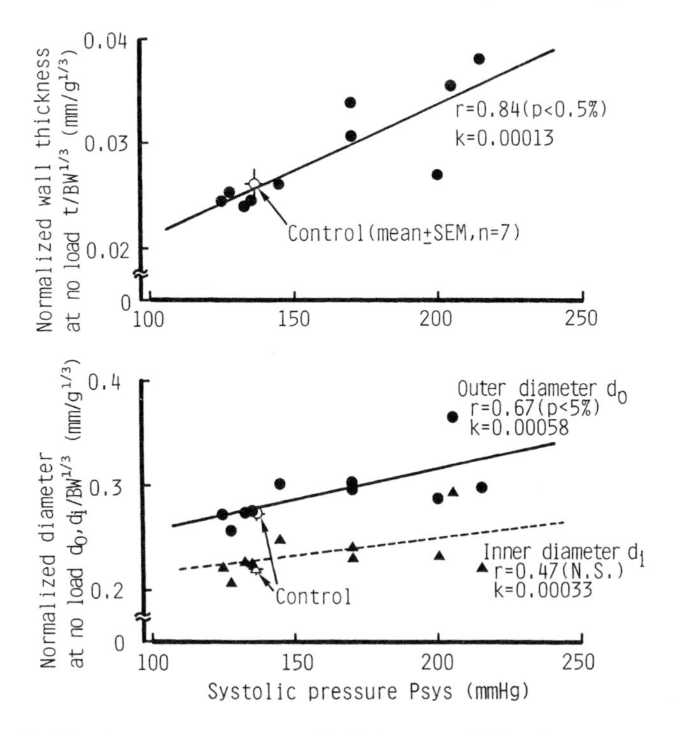

FIG. 2. Systolic blood pressure versus wall thickness and blood pressure vs. outer and inner diameters of unloaded aortas of Fig. 1. The wall thickness t and diameters d_o and d_i were normalized by the body weight BW of each animal. *Open circles*, mean values in the control groups; correlation coefficient, r, its significance level, P, and gradient of regression line, k (mm/g$^{1/3}$/mmHg), were calculated for the 10 operated rats. (From [3], with permission)

wall thickening occurs toward the outside (adventitial) direction to maintain the wall shear rate, i.e., wall shear stress, developed by blood flow at a constant level.

Diametrical change in response to hypertension is an interesting phenomenon because it may reflect mechanical response of the smooth muscle cell and that of the endothelial cell. These responses may interact. Detailed studies are necessary.

2.2 Distensibility

It is commonly noted that hypertension causes decreased arterial distensibility. This wording, however, should be used with caution. Figure 3 shows examples of the pressure–outer diameter curves of the midthoracic aortas of hypertensive and control rats observed at 2, 4, 8, and 16 weeks after induction of hypertension [3]. The diameter is expressed by the distension ratio. The distension ratio was smaller in the hypertensive rats than in the control when compared at the same pressure. In a low-pressure region where the aorta had high distensibility, the

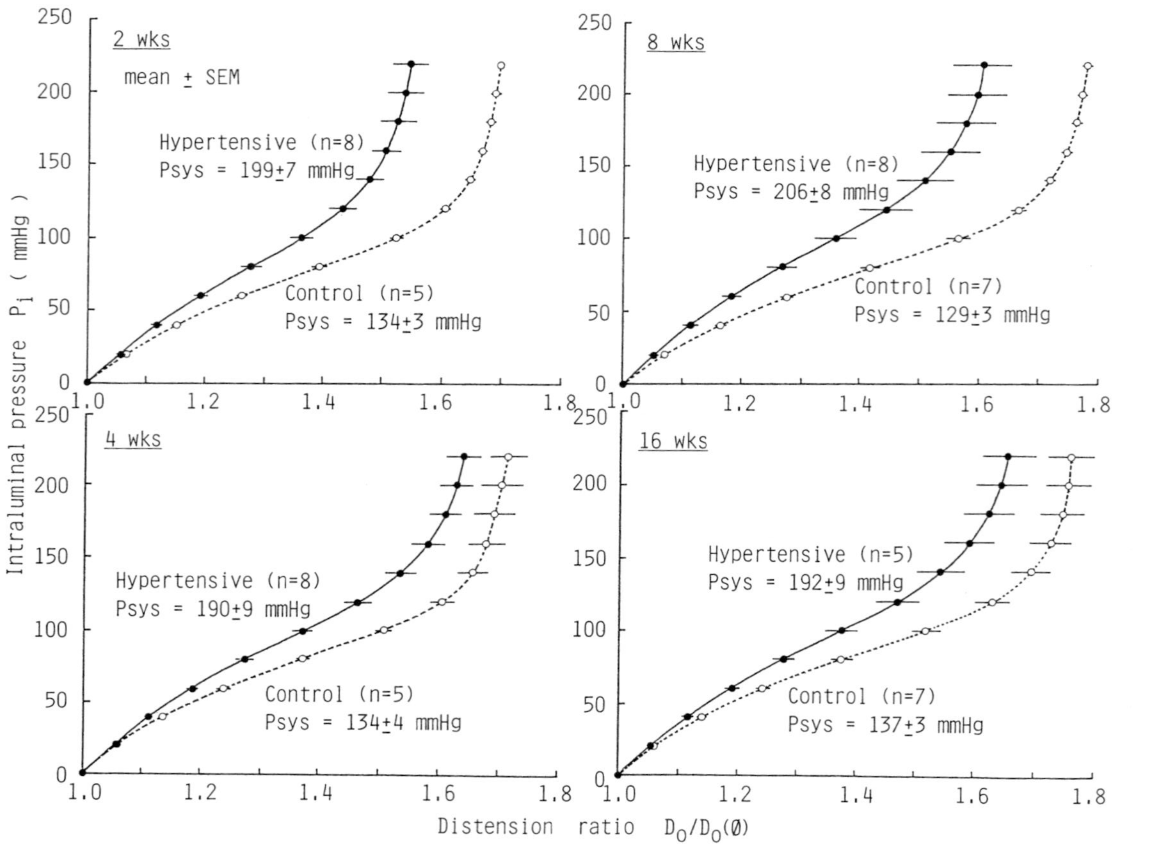

FIG. 3. Pressure–diameter curves of the aorta in hypertensive and control rats. *Psys*, Systolic blood pressure before sacrifice; D_o, outer diameter; $D_o(0)$, outer diameter at zero pressure. (From [3], with permission)

hypertensive aorta was less distensible than the control. At higher pressures above 140 mmHg, however, the slopes of the curves of the hypertensive aortas were lower than those of the controls. This means that although hypertensive aortas are less distensible than normotensive aortas in a lower pressure region, they are more distensible in a higher pressure region.

Elastic properties of the artery were expressed in two parameters: an apparent stiffness as a distensible tube (which corresponds to the inverse of distensibility) and an inherent elastic modulus of the wall material. Pressure–strain elastic modulus [31] and incremental elastic modulus [32,33] are frequently used as the apparent stiffness and the elastic modulus, respectively. These two parameters are not independent of each other: the apparent stiffness depends on both the elastic modulus and the dimensions of the wall material. Thus, change in arterial distensibility with hypertension should be analyzed by changes in both wall dimensions and elastic modulus.

An example of this analysis is shown in Tables 1 and 2 for rat aortas used in Fig. 3 [3]. The pressure-strain elastic moduli *Ep* calculated for three different blood pressure levels are shown in Table 1. At a lower pressure (100 mmHg), the modulus was significantly larger in the hypertensive animals than in the control except at 4 weeks. On the other hand, the hypertensive animals had smaller moduli than the controls at a higher pressure (200 mmHg), although differences were significant only at 4 weeks. Under in vivo condition (*Psys*), the modulus was significantly larger in the hypertensive animals except at 4 weeks. The structural stiffness of blood vessels is determined by the elastic modulus of the wall mate-

TABLE 1. Pressure–strain elastic modulus (*Ep*) of hypertensive and control rat midthoracic aortas at three blood pressure levels. (From [3], with permission).

Parameter	Weeks after operation			
	2	4	8	16
Control				
Number of animals	5	5	7	7
Psys (mmHg)	134 ± 3	134 ± 4	129 ± 3	137 ± 3
Ep (mmHg):				
At 100 mmHg	3.8 ± 0.2	3.4 ± 0.3	3.3 ± 0.1	3.2 ± 0.1
At *Psys*	14.0 ± 2.3	12.2 ± 2.4	9.4 ± 1.2	9.7 ± 1.7
At 200 mmHg	53.4 ± 5.8	52.4 ± 5.2	55.4 ± 6.1	50.8 ± 6.8
Hypertensive (*Psys* ≥ 160 mmHg)				
Number of animals	8	8	8	5
Psys (mmHg)	199 ± 7*6	190 ± 9*6	206 ± 8*6	192 ± 9*6
Ep (mmHg):				
At 100 mmHg	4.8 ± 0.3*1	3.9 ± 0.2	4.3 ± 0.4*2	4.0 ± 0.3*1
At *Psys*	41.1 ± 6.3*3	26.7 ± 5.3	35.3 ± 4.8*6	22.4 ± 2.8*5
At 200 mmHg	40.0 ± 4.3	31.0 ± 2.6*5	36.9 ± 6.6	31.3 ± 8.0

Values are expressed as mean ± SEM. *Ep*, Pressure–strain elastic modulus [31] ($= \Delta P_i/\Delta D_o/D_o$), where ΔD_o is the increment of outer diameter D_o induced by pressure increment ΔP_i); *Psys*, systolic blood pressure before sacrifice. *1,2,3,5,6 indicate significant differences from the control data ($P < 5, 2, 1, 0.2, 0.1\%$, respectively).

rial, $H_{\theta\theta}$, and the wall thickness/inner radius ratio, *H/Ri*; i.e., *Ep* increases with the increase of $H_{\theta\theta}$ and also with the increase of *H/Ri*. To know the cause of the change in wall stiffness in the hypertensive rats, *H/Ri* and $H_{\theta\theta}$ were calculated for the operated rats for the intraluminal pressures of 100 and 200 mmHg, and their correlations with *Psys* were examined.

As shown in Table 2, wall thickness at both pressures had significant correlation with *Psys* except at 8 weeks. The incremental elastic modulus at 100 mmHg had negative and nonsignificant correlation with in vivo blood pressure except at 4 weeks, while the modulus at 200 mmHg had significant negative correlation with in vivo pressure at all periods. These results indicate that the increased stiffness of the hypertensive aorta in the low-pressure range was primarily caused by wall thickening, and that the increased distensibility in the high-pressure range was caused by a decrease in the elastic modulus from the control value. The softening of wall material that appeared in the high-pressure range may be regarded as an adaptation process of the aortic wall, although the pressure–strain elastic modulus calculated for *Psys* was different between the hypertensive and control animals.

2.3 Difference Observed in the Circumferential and Axial Directions

Studies on the mechanical adaptation of vascular tissue have been mainly concerned with changes of the mechanical properties in the circumferential direction [2]. Information obtained from these studies, however, cannot be extrapolated to other directions, because the vascular wall is anisotropic.

An example of differences between the circumferential and axial directions is shown in Table 3 [34], in which the in vivo stresses and strains in the midthoracic aorta of hypertensive and normotensive rats have been calculated from pressure–diameter–axial force relations obtained in vitro. Axial force measured under the

TABLE 2. Correlation coefficients between in vivo systolic blood pressure and factors affecting aortic stiffness. (From [3], with permission).

Parameter	Weeks after operation			
	2	4	8	16
At 100 mmHg				
H/Ri	0.639	0.842	(0.626)	0.787
$H_{\theta\theta}$	(–0.181)	–0.711	(–0.261)	(–0.574)
At 200 mmHg				
H/Ri	0.682	0.775	(0.588)	0.843
$H_{\theta\theta}$	–0.706	–0.859	–0.673	–0.896

H, Aortic wall thickness; *Ri*, inner radius; $H_{\theta\theta}$, incremental elastic modulus (see Fig. 6). Parentheses, nonsignificant correlation ($P >$ 5%).

TABLE 3. Mean stresses and strains in circumferential and axial directions and axial force in midthoracic aortas of hypotensive and normotensive rats. (From [34], with permission).

Group	n	$Psys$	$\overline{\sigma_\theta}^{(p)}$ (kPa)	$\overline{\varepsilon_\theta}^{(p)}$ (%)	$\overline{\sigma_z}^{(p)}$ (kPa)	$\overline{\varepsilon_z}^{(p)}$ (%)	$F_z^{(p)}$ (10^{-4}N)
Hypertensive	5	215 ± 9	339 ± 28	72.1 ± 8.4	99 ± 12*2	17.3 ± 1.0*4	−945 ± 165*3
Normotensive	5	128 ± 3	317 ± 13	85.6 ± 3.2	159 ± 11	37.2 ± 1.7	25 ± 99

Hypertensive, rats rendered hypertensive by constricting their renal artery for 8 weeks; normotensive, age-matched controls; $\overline{\sigma_\theta}$, $\overline{\sigma_z}$, mean circumferential and axial stresses (see Eqs. 7 and 8 for calculation); $\overline{\varepsilon_\theta}$, $\overline{\varepsilon_z}$, mean circumferential and axial strains; F_z, axial force required to maintain the specimen at in vivo length. Superscript (p) denotes the in vivo load condition ($P_i = P_{sys}$ and $\overline{\lambda_z} = \overline{\lambda_z}^{(p)}$). *1 $P < 0.05$; *2 $P < 0.01$; *3 $P < 0.005$; *4 $P < 0.001$; *vs* normotensive. Data are means ± SEM.

in vivo load condition is also included in Table 3. The mean stress and strain in the circumferential direction were not significantly different between hypertensive and control groups. Although the systolic blood pressure in the hypertensive rats was 68% higher than that in the control rats, the circumferential stress and strain calculated for systolic blood pressure were only 7% higher and 15% lower on average than the control values, respectively, and the differences were not significant. In contrast, those in the axial direction were significantly different between the two groups: the axial stress and strain were 38% and 47% lower, respectively, than the control values. Furthermore, the axial force, which is equivalent to the in situ tethering force, was significantly lower in the hypertensive animals. The decrease in the axial strain or the axial extension ratio that appeared in response to hypertension was also observed by Wolinsky [21] and Vaishnav et al. [4]. These results indicate that stress and strain in the circumferential direction are maintained at a constant level irrespective of blood pressure, while those in the axial direction are reduced with the elevation in blood pressure.

The reason for this directional difference is unknown. However, potential explanations can be given by taking the structure of aortic wall into account. The mechanical properties of rat aortas were almost exclusively determined by those of the media. The aortic media is mainly composed of concentric layers of elastin lamellae and intervening smooth muscle cells and collagen and elastin fibers. Spindle-shaped vascular smooth muscle cells are oriented in a helical pattern with a predominantly circumferential orientation, which develops a greater contraction in the circumferential direction than in the axial direction [35]. The wall thickening occurring in response to hypertension is associated with hypertrophy of smooth muscle cells (enlargement of existing cells) and increase in collagen and elastin fibers [36]. Because smooth muscle cells are preferentially oriented in the circumferential direction, their hypertrophy may increase the volume as the result of dimensional changes in both the radial and axial directions. If this is the case, the volume increase in the radial direction causes the wall thickening.

However, the volume change does not increase the axial length of the aorta because the vessel is tethered by its branches and surrounding tissues, which results in reduction of the axial strain, concomitantly with decrease in the axial stress as well as the axial force. If there are no mechanisms in the vascular wall to return the decreased axial stress to the normal level, then the changes that occurred in the axial direction remain. The result indicating that circumferential stress did not differ between the hypertensive and normotensive aortas suggests that vascular smooth muscle cells change not only themselves but the surrounding environment to maintain the circumferential stress at a constant level. The smooth muscle cells, however, may have less response to stresses in directions other than the circumferential direction. There may be other possible explanations: for example, that the axial direction takes a longer time to adapt than does the circumferential direction, or that tissue responds more slowly to stress reduction than to stress increase. Detailed studies on the mechanical response of smooth muscle cells to applied load are necessary.

2.4 Histological Change Associated with Hypertension

Figure 4 shows micrographs of the aortic sections of control, hypertensive, and severely hypertensive rats [34]. It is apparent that the total wall thickness was increased by the elevation in blood pressure. The lamellar units had almost the same thickness throughout the wall thickness in the normotensive rat (Fig. 4a), while in the severely hypertensive rat (Fig. 4c) the units were much thicker near the inner surface than near the outer surface. The thickening of the lamellar units

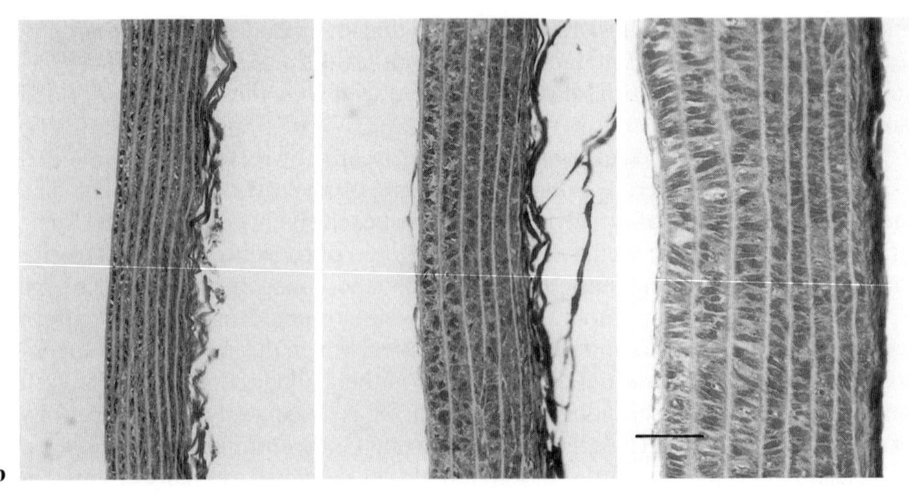

Fig. 4a–c. Micrographs of the thoracic aorta in three operated rats fixed under in vivo loading condition and stained with Azan. **a** Normotensive rat ($Psys$ = 145 mmHg). **b** Hypertensive rat ($Psys$ = 200 mmHg). **c** Severely hypertensive rat ($Psys$ = 240 mmHg). Sections are parallel to the longitudinal axis of the vessel; intimal surface faces *left*. Length marker (50 μm) in **c** applies to all parts of the figure. (From [34], with permission)

was associated mainly with the increase of ground substance and partly with the hypertrophy of smooth muscle cells.

Using micrographs of each specimen as shown in Fig. 4, total medial thickness, number of lamellar units, and thickness of each lamellar unit were measured at five randomly selected locations. The results are summarized in Table 4; the subintimal and subadventitial units are the lamellar units closest to the intima and the adventitia, respectively. The number of the lamellar units and the relative thickness of the subadventitial lamellar units did not differ between the two groups. The medial thickness and the relative thickness of the subintimal lamellar units were significantly larger in the hypertensive rats than in the controls. These results suggest that hypertension caused the wall thickening, especially in the subintimal region.

It is interesting to note that the increase in the stresses and strains caused by the elevation of intraluminal pressure is greater in the inner wall than in the outer wall. The hypertrophy was accompanied by the increase of the ground substance as well as hypertrophy of smooth muscle cells. Vascular smooth muscle cells produce the ground substance such as elastin, collagen, and glycosaminoglycan. The hypertrophy of smooth muscle cells and their production of the ground substance seem to depend on the magnitude of mechanical stress applied to the cells. As is discussed later, nonuniform hypertrophy in the radial direction might affect the opening angle; the increase in the opening angle in the hypertensive aortas might be caused by the hypertrophy of smooth muscle cells and the increase of ground substance in the subintimal region (see Effect of Hypertension on Arterial Opening Angle, in the next section).

2.5 *Chronological Course of Adaptation to Hypertension*

Knowledge about the temporal change of blood vessels in response to hypertension is important to obtain a more quantitative and detailed analysis of mechanical adaptation. An example of this analysis is shown in Figs. 5 and 6 [3]. The

TABLE 4. Summary of morphological observation of hypertensive and normotensive rat aortas. (From [34], with permission).

Group	n	Total medial thickness (μm)	Number of lamellar units	Relative thickness of lamellar units[a] (%)	
				Subintimal	Subadventitial
Hypertensive	5	109.7 ± 3.8*4	7.38 ± 0.25	17.0 ± 0.9*1	9.7 ± 0.6
Normotensive	5	57.3 ± 2.4	7.52 ± 0.53	12.9 ± 1.2	11.0 ± 0.8

Hypertensive, rats rendered hypertensive by constricting their renal artery for 8 weeks; normotensive, age-matched controls. Data are means ± SEM.

*1 $P < 0.05$; *4 $P < 0.001$; *vs* normotensive.

[a] Mean thickness of the subintimal and subadventitial lamellar units were normalized by the total medial thickness.

temporal change in the relationship between *Psys* in each rat and the circumferential wall stress at this pressure is shown in Fig. 5. The circumferential stress calculated for in vivo blood pressure was almost constant in all animals at all periods, even though the systolic blood pressure continued to rise until 6 weeks after the operation. These results indicate that the dimensional adaptation associated with wall thickening occurs very rapidly, as if the aortic wall responds to the elevation of blood pressure instantly.

Relations between *Psys* and the incremental elastic modulus at this pressure are shown in Fig. 6. Unlike circumferential stress, the incremental elastic modulus had a significant correlation with blood pressure until 8 weeks after the operation. Generally speaking, the arterial wall has nonlinear elasticity; i.e., the higher the pressure is, the stiffer the wall becomes. If the blood pressure–wall elasticity relation is similar among all rats, then the elastic modulus at the in vivo pressure should be larger in rats having higher blood pressure. However, the elastic modulus was independent of the blood pressure at 16 weeks, which suggests that the aortic wall in hypertensive animals restores the in vivo elastic properties to a normal level in 16 weeks.

The rapid thickening of the aortic wall and the relatively slow change in the elastic properties indicate that dimensional adaptation precedes material adapta-

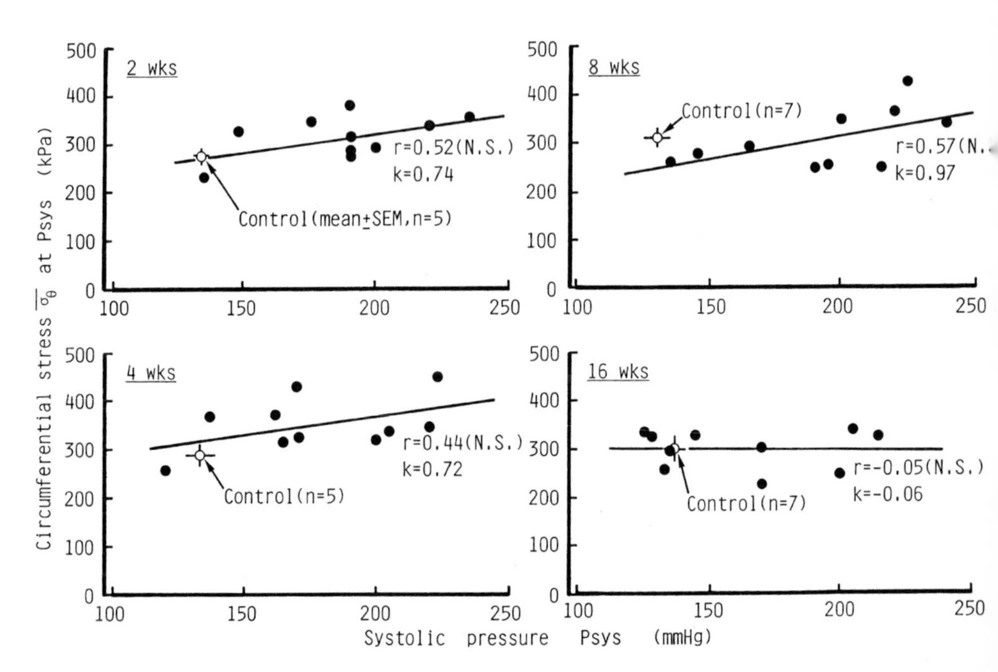

Fig. 5. Temporal change in relation between systolic blood pressure (*Psys*) before sacrifice in each rat and circumferential wall stress at this pressure. *Open circles*, mean values in control groups; correlation coefficient, *r*, its significance level, *P*, and gradient of regression line, *k* (kPa/mmHg), were calculated for the operated rats. (From [3], with permission)

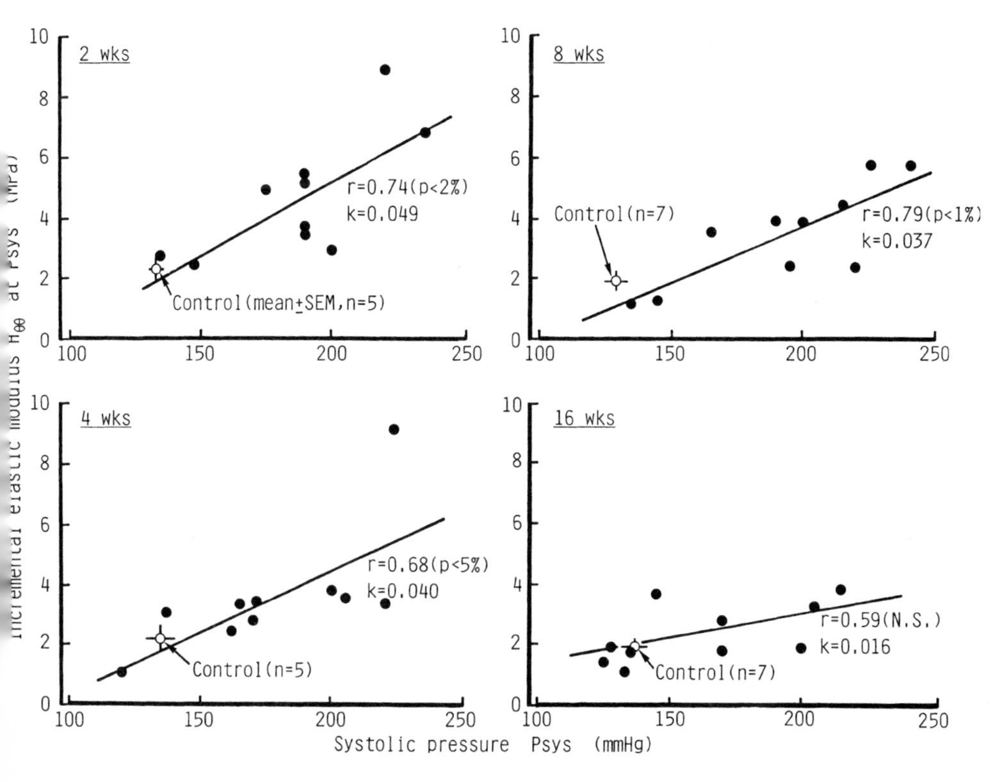

FIG. 6. Relation between systolic blood pressure before sacrifice and incremental elastic modulus at this pressure. According to Hudetz [33], incremental elastic modulus is defined as $H_{\theta\theta} = 2D_o/(D_o^2 - D_i^2)(D_i^2 \Delta P_i/\Delta D_o + P_i D_o)$ for orthotropic material where ΔD_o is the increment of D_o induced by the increment of P_i, ΔP_i. *Open circles*, mean values in the control groups; correlation coefficient, r, its significant level, P, and gradient of regression line, k (MPa/mmHg), were calculated for the operated rats. (From [3], with permission)

tion in the arterial wall. The aortic wall responds to the change in the mechanical load in two processes: first, the wall thickness changes to keep the stress in an optimal range, and second, the wall elasticity changes to maintain wall function in an optimal fashion.

3 Effect of Hypertension on the Zero-Stress State of Arteries

3.1 *Residual Stress in the Arterial Wall*

According to Vossoughi [37], residual stress/strain in arteries was first noted by Bergel [38]. However, it was not until recently that the existence of residual stress in soft biological tissues was recognized and its physiological significance has been discussed widely. Vaishnav and Vossoughi [11] found that ringlike specimens of the bovine and porcine aortas sprung open after being cut in the radial direction and pointed out the advantageous residual stress by virtue of which the actual maximum in vivo circumferential stress is lower than that computed assuming the natural configuration to be stress free.

In 1984, Fung [10] observed a similar phenomenon in the rabbit thoracic aorta, and ascribed the residual stress to the "principle of optimal operation" in living tissues and organs. He introduced a simple parameter called an opening angle to represent the amount of residual stress/strain in the arterial wall. This parameter is defined as the angle subtended by the two radii drawn from the midpoint to the tips of the arc of the artery inner wall (see Fig. 8 later in this chapter). In connection with this phenomenon, Chuong and Fung [39] calculated stress and strain distributions in the rabbit thoracic aorta, considering the shape change of the ring specimen following radial cutting. They showed that residual stress reduced the stress concentration factor in the arterial wall from 6.5 to 1.42.

Takamizawa and Hayashi [9] proposed that the vascular wall has a constant circumferential strain through the wall thickness under the physiological loading condition, and called this the "uniform strain hypothesis". On the basis of this hypothesis, they calculated the distributions of residual stress and showed that residual stress cancels the high stress concentration near the intimal surface at the in situ blood pressure, which is obtained assuming that a material is stress free when all the external loads are removed (zero initial stress assumption). Since then, many researchers have contributed to this field using various tissues including the arteries [9–17,34,39–44], veins [45], heart [46–48], and trachea [49]. (See Vossoughi [37] for an extensive review of the literature on biological residual stress and strain during the past decade.)

3.2 Effect of Hypertension on Arterial Opening Angle

Residual stress and strain are associated with the adaptation of arterial wall tissue to the mechanical environment (stress and strain fields). Therefore, studies on the effects of growth and disease on residual stress and strain should give us much information on how biological tissues adapt themselves to their mechanical environment and how mechanical factors affect the initiation and progression of disease.

Liu and Fung [14] and Fung and Liu [16] reported that residual stress in the rat aortic wall represented by the opening angle changed rapidly in response to the induction of hypertension. When they induced hypertension by constricting the abdominal aorta, they found a marked increase in the opening angle, from 171° to 214° in 4 days after constriction, followed by gradual decrease to an asymptotic value of 126° in 40 days in the ascending aorta. Fung and Liu [50] also observed similar and faster change in the opening angle in rat pulmonary arteries subjected to hypoxic hypertension: the angle increased from 294° to 385° in 12h and then decreased gradually to 193° in 240h. By comparing these changes in opening angle with histological observation, they explained the swing of opening angle by the nonuniform remodeling of the vessel wall.

The foregoing studies were mainly concerned with short-term changes of the opening angle following induction of hypertension and did not show the long-term effect of hypertension on opening angle. Also, the relation between the amount of mechanical stimulation and change in the zero-stress state was not clear. To answer these questions, we measured the opening angle of the rat midthoracic aorta 8 and 16 weeks after constriction of the renal artery [34]. Figure 7 shows the relation between the opening angle and systolic blood pressure. There was a significant correlation between the opening angle and systolic blood pressure in both groups. The linear regression line between angle and blood pressure appeared to shift upward with increase in duration of hypertension. These results suggest that the opening angle increases in accordance with the amount and duration of mechanical stimulation. As shown in Fig. 4 and Table 4, hypertension appeared to cause the hypertrophy of smooth muscle cells and the increase of ground substance, especially in the subintimal region. This nonuniform hypertrophy might increase the opening angle.

The increase in the opening angle in the hypertensive artery has also been shown theoretically. Employing a model in which the artery was considered to be a thick-walled, two-layered tube made of nonlinear elastic incompressible materials [51], Rachev [44] simulated the adaptive change of the arterial wall in response to hypertension. He introduced growth parameters into the model, which had the dimensions of a normotensive artery, and determined their values to meet a basic hypothesis that the remodeling lasts until the distribution of strain

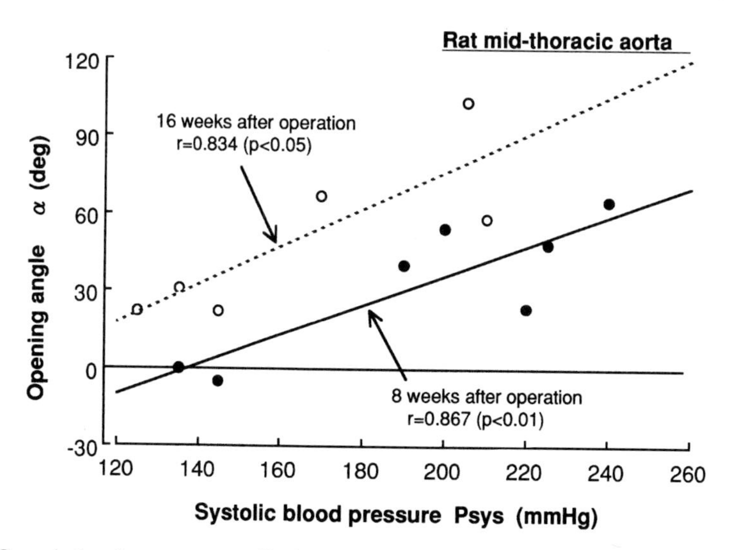

FIG. 7. Correlation between systolic blood pressure and opening angle (α). *Closed* and *open circles* show data obtained from the midthoracic aorta of rats whose renal arteries were constricted for 8 and 16 weeks, respectively

and stress in the wall of the hypertensive vessel becomes equivalent to that of the normotensive vessel. The remodeled artery showed wall thickening in both unloaded and working conditions, an increase of the inner diameter in the deformed state, and an increase of the opening angle.

3.3 Other Factors Affecting the Opening Angle

Because residual stress and strain are expected to be sensitive to minute changes in the arterial wall, various factors other than hypertension affecting the opening angle have also been studied, including diabetes [15], atherosclerosis [41], and aging [43]. Liu and Fung [15] studied the zero-stress state of pulmonary and systemic arteries of diabetic rats to find an increase in the opening angle during diabetes development. Matsumoto et al. [41] observed that intimal hyperplasia increased the opening angle of the rabbit atherosclerotic aorta and that tissue calcification decreased it. They also found marked nonaxisymmetry in the zero-stress state of the atherosclerotic aorta and proposed a novel method to calculate its local strain distribution [41]. The effect of aging on the opening angle has been studied by Greenwald et al. [42], who found that ring opening increased steadily and significantly with age both in rat and in human aortas. Based on their data, Rachev et al. [43] studied temporal change in intramural stress distribution and concluded that changes in the geometrical configuration of the zero-stress state of the rat aorta during the period of development could be considered to be "growth modulated" by the stresses existing in the arterial wall.

3.4 Limitation of the Opening Angle as an Index of Residual Stress/Strain

The opening angle has been frequently used for its simplicity as a parameter expressing the amount of residual strain in the arteries. However, this parameter should be used with caution. It has been reported that the parameter is dependent on the longitudinal position of the artery [13–16,40,50], species [17], age [42,43], gender [42], and so on. The use of this parameter requires axisymmetry of the opened-up ring, although this is not always the case in the aorta. As Fung and Liu [16] pointed out, the aortic ring sometimes shows nonaxisymmetry. When the aorta has a local atherosclerotic lesion, its opened-up shape can be far from an arc [41]. Thus, it should be noted that residual strain indexed by the opening angle is an averaged or an effective value through the aortic circumference. For a more correct estimation of residual strain, local variation in the shape of the opened-up ring should be taken into consideration [41]. Furthermore, the opening angle is affected by several factors such as contraction of smooth muscle and the structural complexity of the aortic wall.

Matsumoto et al. [52] reported that smooth muscle contraction/relaxation had a significant effect on the opening angle of the rat midthoracic aorta. They showed that smooth muscle contraction increased the opening angle significantly by about 50° and the following relaxation decreased it significantly by about 80°, and suggested that this contraction/relaxation may play a role in controlling the intramural strain distribution. Vossoughi et al. [53] found that the configuration of an opened-up ring specimen of the bovine aorta changed when it was cut into two inner and outer layers. The opening angle of the inner layer was significantly larger than that of the outer, indicating that one radial cut is not enough to obtain a stress-free state in some cases. Based on change in the opening angle of the rat aorta following digestion of various wall components, Greenwald et al. [40] suggested that residual strains and stresses exist mainly in the elastic layers and are practically absent in collagen and smooth muscle fibers. Although the opening angle has been frequently used for its simplicity as a parameter expressing the amount of residual strain, we should use this parameter cautiously until the factors that affect opening angle have been studied more rigorously.

4 Stress/Strain Analysis in the Wall of the Hypertensive Aorta

In the previous section, it was shown that hypertension may increase the opening angle. Change in the opening angle, however, does not provide enough information on adaptational change of the arterial wall; we need to know the strain and stress field in the arterial wall. Few studies have been devoted to the calculation of intramural stress/strain distribution considering residual stress/strain [39,44]. In this section, we introduce our approach to obtain intramural strain/stress distribution [34] in which the opening angle is measured as an index of residual strain in the circumferential direction. The mechanical properties of the aorta were determined by a static pressure–diameter–axial force test. The distribution of strain was then calculated from the pressure–diameter data, aortic configuration in the state at no load, and opening angle. The distribution of stress was calculated using a strain energy density function proposed by Takamizawa and Hayashi [9].

4.1 Mechanical Test

To know the in vivo dimensions and to obtain parameters for the constitutive equation used, the pressure–diameter–axial force relation was obtained from tubular segments of arteries. Each segment was attached to a special jig in a bath filled with physiological saline solution at 37°C. After being stretched to the in vivo length, the segment was inflated and deflated between 0 and 250 mmHg at a rate of about 3 mmHg/s. The outer diameter, intraluminal pressure, and the axial force of the specimen were measured with a video dimension analyzer, a pressure transducer, and a cantilever-type load cell, respectively. After a stable pressure–diameter–axial force relation was obtained, the axial length of the specimen was changed to 110% and 120% of the in vivo length, and the same procedures were repeated for each specimen.

4.2 Aortic Wall Morphometry

Figure 8 shows the configurations of a vascular wall in three states. After the mechanical test, a thin ring specimen was sliced from each vessel to measure the internal and external radii at no load. The internal radius at the loaded state, R_i, was calculated assuming the incompressibility of the wall material:

$$R_i = \sqrt{R_o^2 - \frac{1}{\overline{\lambda_z}}\left(r_o^2 - r_i^2\right)} \tag{2}$$

where R_o is the external radius in the loaded state, and $\overline{\lambda_z}$ is the mean axial extension ratio, which is the ratio of the vascular length under load, L, to that in the state at no load, l.

The ring was then cut radially to measure its opening angle, α. If we assume that the aorta is a homogeneous and cylindrically orthotropic tube, the opened-up ring forms an arc of constant curvature and thickness. The effective external radius of the opened-up ring specimen, r_o^*, was calculated assuming that the mean circumference and the cross-sectional area of the specimen are not changed by cutting:

$$r_o^* = \frac{1}{2}\left\{\frac{\pi}{\theta}\left(r_o + r_i\right) + \left(r_o - r_i\right)\right\} \tag{3}$$

where θ is one-half of the central angle of the arc and is equal to the angle of circumference of the arc, i.e., $\theta = \pi - \alpha$. The angle θ is negative when the opening angle exceeds 180°. In this case, r_o^* is also negative.

Extension ratios at the radius R in a loaded wall are

$$\lambda_r(R) = \frac{\partial R}{\partial r*} = \frac{\theta}{\pi\lambda_z}\frac{r*}{R}, \quad \lambda_\theta(R) = \frac{\pi}{\theta}\frac{R}{r*}, \quad \lambda_z = \frac{L}{l} \tag{4}$$

where

$$\begin{aligned} r* &= \sqrt{r_o^{*2} - \frac{\pi\overline{\lambda_z}}{\theta}\left(R_o^2 - R^2\right)}, \quad (0 < \theta) \\ &= -\sqrt{r_o^{*2} - \frac{\pi\overline{\lambda_z}}{\theta}\left(R_o^2 - R^2\right)}, \quad (\theta < 0) \end{aligned} \tag{5}$$

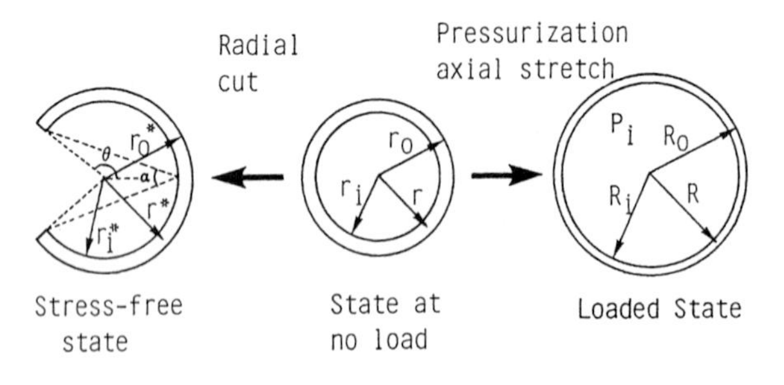

Fig. 8. Configurations of vascular wall in three states. From *left* to *right* stress-free state, state at no load, and loaded state. (From [34], with permission)

or

$$\theta r* = \frac{\pi}{2}(r_o + r_i), \quad (\theta = 0)$$

The subscripts, r, θ, and z indicate the radial, circumferential, and axial directions, respectively. The axial extension ratio, λ_z, was assumed to be constant through the wall thickness, i.e., $\lambda_z = \overline{\lambda}_z$. The in vivo axial extension ratio is denoted as $\overline{\lambda}_z^{(p)}$. Green's strains referred to the locally stress-free configuration are given by

$$\varepsilon_n = \frac{1}{2}(\lambda_n^2 - 1), \quad (n = r, \theta, z) \tag{6}$$

4.3 Stress Analysis

The mean circumferential stress, $\overline{\sigma}_\theta$, was calculated by the law of Laplace:

$$\overline{\sigma}_\theta = \frac{P_i R_i}{R_o - R_i} \tag{7}$$

The mean axial stress, $\overline{\sigma}_z$, was calculated as follows:

$$\overline{\sigma}_z = \frac{F_z + \pi R_i^2 P_i}{\pi(R_o^2 - R_i^2)} \tag{8}$$

The mean circumferential extension ratio, $\overline{\lambda}_\theta$, was calculated as

$$\overline{\lambda}_z = \frac{R_o + R_i}{r_o + r_i} \tag{9}$$

To calculate the stress distributions inside the wall, the authors used the strain energy density function [9] given by

$$W = -C\ln(1 - \psi)$$
$$\psi = \frac{1}{2}(a_{\theta\theta}\varepsilon_\theta^2 + 2a_{\theta z}\varepsilon_\theta\varepsilon_z + a_{zz}\varepsilon_z^2) \tag{10}$$

where the coefficients C, $a_{\theta\theta}$, $a_{\theta z}$, and a_{zz} characterize the mechanical properties of the material.

The stress–strain relation is derived as follows:

$$\sigma_n - \sigma_r = (1 + 2\varepsilon_n)\frac{\partial W}{\partial \varepsilon_n} \equiv Q_n, \quad (n = \theta, z) \tag{11}$$

where σ_r and σ_n represent Cauchy's stresses in the radial and n-directions, respectively.

From the equilibrium of forces, the following equation is obtained:

$$\frac{d\sigma_r}{dR} + \frac{\sigma_r - \sigma_\theta}{R} = 0 \tag{12}$$

The stress in the radial direction at the radius R, $\sigma_r(R)$, is obtained from this equation with the boundary condition of $\sigma_r(R_o) = -P_o = 0$:

$$\sigma_r(R) = -\int_R^{R_o} Q_\theta \frac{dR}{R}, \quad (R_i \le R \le R_o) \tag{13}$$

The stresses in the circumferential and axial directions at the radius R, $\sigma_\theta(R)$ and $\sigma_z(R)$, respectively, are calculated from Eqs. 10, 11, and 13. In this study, the strains obtained from Eqs. 3–6 were applied to the strain energy density function (Eq. 10) to take residual strains into consideration.

Intraluminal pressure and axial force are expressed, respectively, by

$$P_i = -\sigma_r(R_i)$$

and

$$F_z = -\pi R_i^2 P_i + 2\pi \int_{R_i}^{R_o} \sigma_z R dR = 2\pi \int_{R_i}^{R_o} \left(Q_z - \frac{Q_\theta}{2} \right) R dR \tag{14}$$

The material constants included in the strain energy density function were determined for each vessel by a least squares fit with the experimental data for the pressure and axial force.

5 Effect of Hypertension on Intramural Distribution of Stress and Strain

In this section, intramural distributions of stress and strain are calculated [34] for the midthoracic aortas of rats rendered hypertensive by renal artery constriction for 8 weeks and their age-matched controls by the method described in the previous section. The effect of hypertension on the intramural mechanical field is discussed.

5.1 *Residual Stress and Strain*

Table 5 summarizes the residual strains and stresses at the inner wall of hypertensive and normotensive rat aorta along with their opening angles. Be-

TABLE 5. Opening angle and residual strains and stressess at the inner wall in hypertensive and normotensive rat aortas. (From [34], with permission).

Group	n	α (degree)	ε_θ (%)	ε_r (%)	σ_θ (kPa)	σ_z (kPa)
Hypertensive	5	46 ± 7	−3.1 ± 0.5	3.3 ± 0.5	−3.16 ± 0.50	−0.78 ± 0.04
Normotensive	5	5 ± 8	−0.3	0.3	−0.21	−0.06

Hypertensive, rats rendered hypertensive by constricting their renal artery for 8 weeks; normotensive, age-matched controls; α, opening angle; ε_θ and ε_r, circumferential and radial strains, respectively; σ_θ and σ_z, circumferential and axial stresses, respectively. For normotensive rats, strains and stresses were calculated from the averaged pressure–diameter–axial force curves of five rats. Data are means ± SEM.

cause of the large opening angle, the residual strains and stresses were large in the hypertensive rats: compressive strain and stress in the circumferential direction were particularly great. As it was assumed that the axial strain is constant across the width of the wall and that the wall material is incompressible, the compressive strain in the circumferential direction caused a large tensile strain and compressive stress in the radial and axial directions, respectively. The strain ε_z was very small and was of the order of 0.1% for hypertensive rats and 0.01% for normotensive animals. The radial stress caused by the radial strain was less than 0.05 kPa.

5.2 Strain and Stress Distribution Under In Vivo Conditions

Figure 9 shows the effect of the opening angle on the strain and stress distribution in a hypertensive aorta ($Psys$ = 240 mmHg) under the in vivo load condition ($Pi = Psys$, $\varepsilon_z = 0.150$). The calculations were made for the opening angles of 65° and

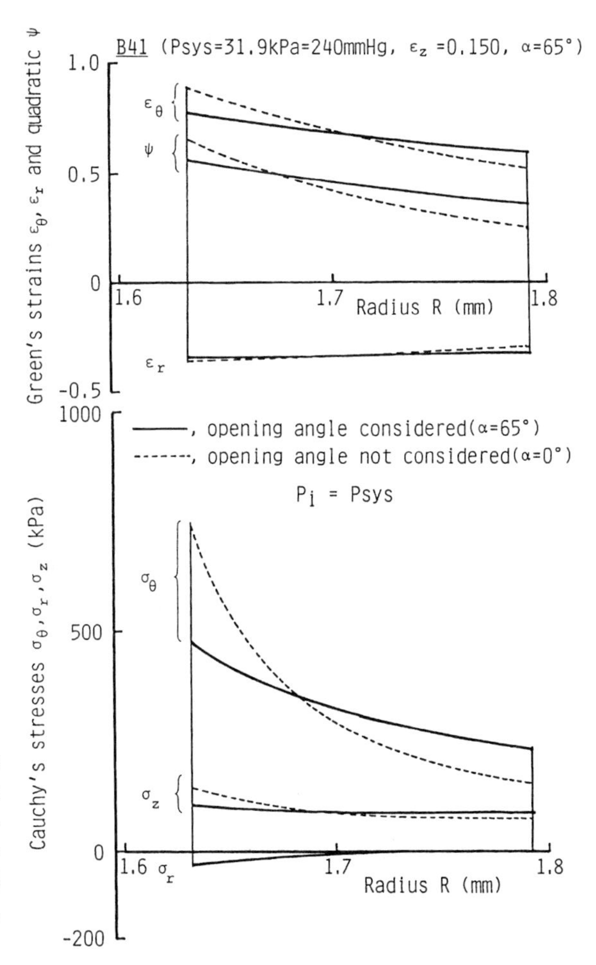

FIG. 9. Effect of opening angle on strain and stress distribution in a hypertensive aorta. Distributions were calculated for experimentally obtained radii (R) at Pi = 240 mmHg. (From [34], with permission)

0°. The magnitudes of strains and stresses were larger at the inner wall than at the outer wall. The stress gradient was especially large for the circumferential stress because of the large strain in this direction and the stiffening of wall material with increasing strain, that is commonly observed in soft tissue. The strain and stress were not constant through the wall thickness even though the opening angle was considered. However, if we did not consider the opening angle, that is, if we assumed no residual strain, the concentration of strain and stress in the inner layer increased: the increase was marked in the circumferential stress.

Examples of the distribution of strain and stress through the wall thickness in the control aorta ($Psys$ = 128 mmHg) and in a hypertensive aorta ($Psys$ = 240 mmHg) are shown in Figs. 10 and 11, respectively. Although the opening angles were considered in the calculations, the strains and stresses were not constant throughout the wall thickness. The large opening angle of the hypertensive aorta resulted from large strains at $P_i = 0$, which moderated the

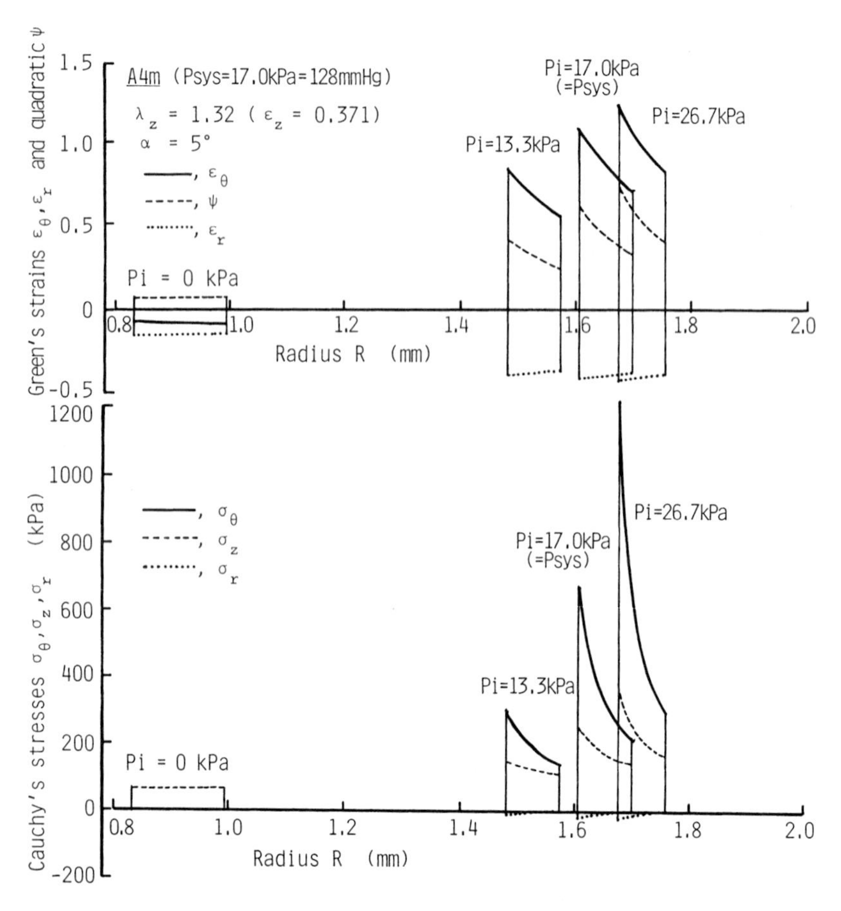

FIG. 10. Distribution of strain and stress through the wall thickness in the control aorta (opening angle considered). (From [34], with permission)

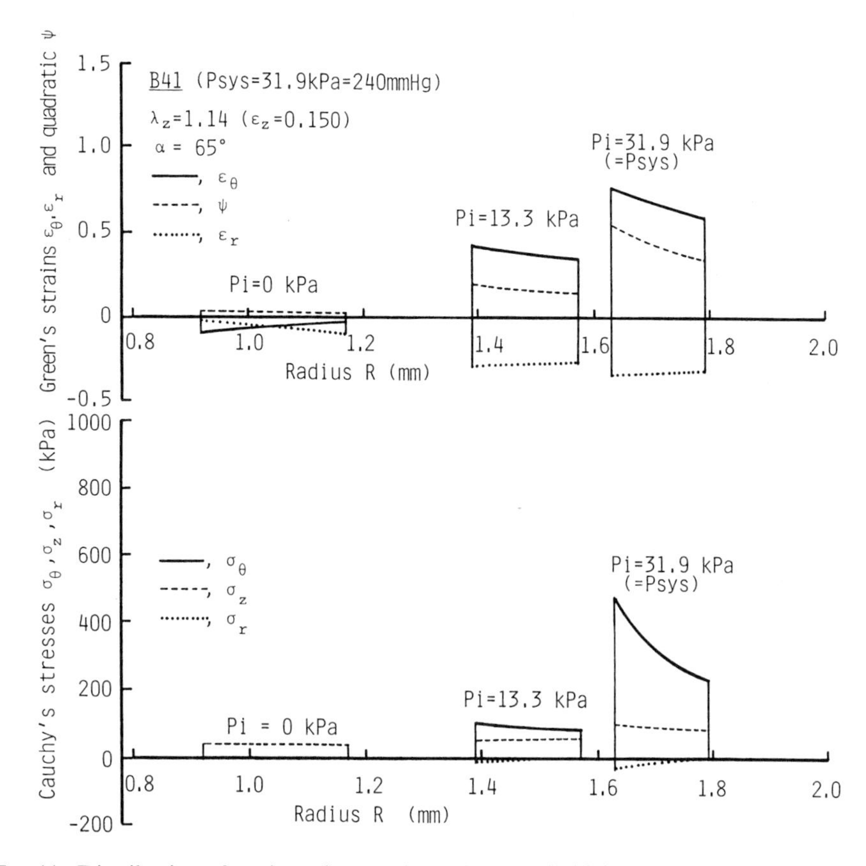

FIG. 11. Distribution of strain and stress through the wall thickness in a hypertensive aorta (opening angle considered). (From [34], with permission)

stress and strain concentrations at *Psys* although *Psys* was very high (Fig. 11). If the intraluminal pressure in the control aorta was this high, a markedly high stress concentration should appear at the inner wall.

5.3 Effect of Hypertension on the Intramural Mechanical Environment

As shown in Fig. 7 and Table 5, hypertension increased the opening angle and circumferential residual stress. This increase was, however, not large enough to attain a uniform stress/strain environment throughout the wall thickness (Fig. 9). Furthermore, stress/strain distributions were not uniform even in the normotensive aorta (Fig. 10). Does the opening angle not change so as to maintain the uniform distributions of stress and strain? Are there any other possible ways to explain the change in the opening angle that occurred in response to hypertension? To answer these questions, the significance of the increase in the opening angle was studied with a stress concentration factor, *Ks*, which is defined by the ratio of the peak value to the mean value in the stress distribution at systolic blood pressure.

Figure 12 shows the relation between systolic blood pressure and *Ks* in the circumferential direction. The stress concentration factor calculated considering the opening angle was almost independent of blood pressure. On the other hand, if the opening angle was not taken into consideration, *Ks* had a significant correlation with blood pressure. These results suggest that the residual stress changes with the progression of hypertension so as to keep the stress concentration factor at the level in the normotensive rats.

Stress distribution accompanied by a slight stress concentration in the inner layer might happen if the stress developed in the aortic wall is controlled within a certain allowable range. The calculated circumferential stress at the inner wall was greater than that at the outer wall by a factor of 1.8 in the rat B41. Although the actual stress range has never been measured in the aortic wall because of technical difficulties, the allowable range of stress could be one of the reasons why the calculated distribution of stress was not uniform.

6 Concluding Remarks

As an example of mechanical adaptation in soft biological tissues, the mechanical response of arterial wall to hypertension has been reviewed. In response to change in its mechanical environment, arterial tissue remodels itself very quickly and actively in some cases. Wall thickening in rat thoracic aorta occurred as if the wall responded to the pressure elevation instantly. A pulmonary artery subjected to hypoxic hypertension remodeled in several hours. Although many of these responses such as wall thickening and increase in the opening angle could be

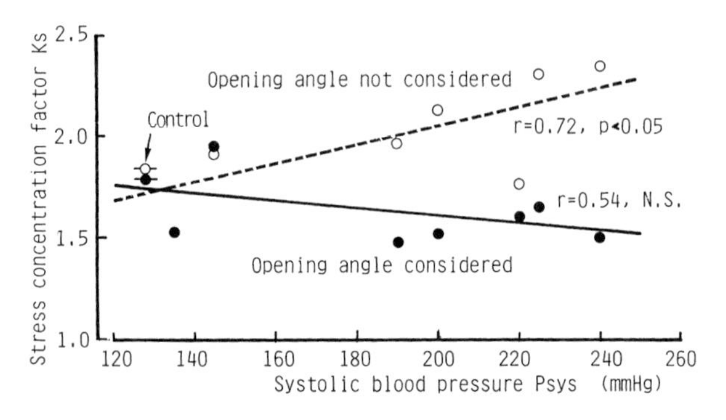

FIG. 12. Relation between systolic blood pressure and stress concentration factor of circumferential stress. Data obtained from the control aorta are also shown with the standard error of *Psys* and were included for the statistical calculation. *Solid line*, opening angle considered; *dashed line*, opening angle not considered. (From [34], with permission)

attributed to adaptive phenomena, not all of them can be. Reduction in axial stress and strain seemed not to be an adaptive process. However, all these changes are driven by the responses of individual cells to the mechanical environment. To obtain a unified view of the mechanical adaptation phenomenon, we need to accumulate detailed knowledge on vascular cell response to the mechanical field and also on the microscopic mechanical field in the three-dimensional structure of the vascular wall, although these are very difficult questions to address.

Acknowledgment. This work was supported in part by the Grant-in-Aid for Scientific Research from the Ministry of Education, Science and Culture, Japan (05221204, 06213206, and 06855012).

References

1. Cox RH (1989) Mechanical properties of arteries in hypertension. In: Lee RMKW (ed) Blood vessel changes in hypertension: structure and function. CRC Press, Boca Raton, FL, pp 62–98
2. Dobrin PB (1983) Vascular mechanics. In: Shepherd JT, Abboud FM, Geiger SR (eds) Handbook of physiology, sect 2, The cardiovascular system, vol III, Peripheral circulation, part 1. American Physiological Society, Bethesda, MD, pp 65–102
3. Matsumoto T, Hayashi K (1994) Mechanical and dimensional adaptation of rat aorta to hypertension. J Biomech Eng 116:278–283
4. Vaishnav RN, Vossoughi J, Patel DJ, Cothran LN, Coleman BR, Ison-Franklin EL (1990) Effect of hypertension on elasticity and geometry of aortic tissue from dogs. J Biomech Eng 112:70–74
5. Wolinsky H (1971) Effects of hypertension and its reversal on the thoracic aorta of male and female rats. Circ Res 28:622–637
6. Wolinsky H (1972) Long-term effects of hypertension on the rat aortic wall and their relation to concurrent aging changes. Circ Res 30:301–309
7. Berry C, Greenwald S (1976) Effects of hypertension on the static mechanical properties and chemical composition of the rat aorta. Cardiovasc Res 10:437–451
8. Cox RH (1981) Basis for the altered arterial wall mechanics in the spontaneously hypertensive rat. Hypertension (Dallas) 3:485
9. Takamizawa K, Hayashi K (1987) Strain energy density function and uniform strain hypothesis for arterial mechanics. J Biomech 20:7–17
10. Fung YC (1984) Biodynamics. Springer Berlin Heidelberg, New York, pp 54–66
11. Vaishnav RN, Vossoughi J (1983) Estimation of residual strains in aortic segments. In: Hall CW (ed) Biomedical engineering, II, recent developments. Pergamon Press, New York, pp 330–333
12. Vaishnav RN, Vossoughi J (1987) Residual stress and strain in aortic segments. J Biomech 20:235–239
13. Liu SQ, Fung YC (1988) Zero-stress state of arteries. ASME J Biomech Eng 110:82–85
14. Liu SQ, Fung YC (1989) Relationship between hypertension, hypertrophy, and opening angle of zero-stress state of arteries following aortic constriction. ASME J Biomech Eng 111:325–335

15. Liu SQ, Fung YC (1992) Influence of STZ-induced diabetes on zero-stress states of rat pulmonary and systemic arteries. Diabetes 41:136–46
16. Fung YC, Liu SQ (1989) Change of residual strains in arteries due to hypertrophy caused by aortic constriction. Circ Res 65:1340–1349
17. Han HC, Fung YC (1991) Species dependence of the zero-stress state of aorta: pig versus rat. ASME J Biomech Eng 113:446–451
18. Bandick N, Sparks H (1970) Viscoelastic properties of the aorta of hypertensive rats. Proc Soc Exp Biol Med 34:56–60
19. Cox R (1979) Comparison of arterial wall mechanics in normotensive and spontaneously hypertensive rats. Am J Physiol 237:H159–H167
20. Stacy D, Prewitt R (1989) Effects of chronic hypertension and its reversal on arteries and arterioles. Circ Res 65:869–879
21. Wolinsky H (1970) Response of the rat aortic media to hypertension. Circ Res 26:507–522
22. Haidu MA, Baumbach GL (1994) Mechanics of large and small cerebral arteries in chronic hypertension. Am J Physiol 266:H1027–H1039
23. Aars H (1968) Static load-length characteristics of aortic strips from hypertensive rabbits. Acta Physiol Scand 73:101–110
24. Sharma M, Hollis T (1974) Rheological properties of arteries under normal and experimental hypertension conditions. J Biomech 4:293–300
25. Feigl E, Peterson L, Jones A (1963) Mechanical and chemical properties of arteries in experimental hypertension. J Clin Invest 42:1640–1647
26. Cox RH, Bagshaw RJ (1988) Effects of hypertension and its reversal on canine arterial wall properties. Hypertension 12:301–309
27. Greene M, Friedlander R, Boltax A (1966) Distensibility of arteries in human hypertension. Proc Soc Exp Biol Med 121:580–585
28. Laurent S, Girerd X, Mourad JJ, Lacolley P, Beck L, Boutouyrie P, Mignot JP, Safar M (1994) Elastic modulus of the radial artery wall material is not increased in patients with essential hypertension. Arterioscler Thromb 14:1223–1231
29. Kamiya A, Togawa T (1980) Adaptive regulation of wall shear stress to flow change in the canine carotid artery. Am J Physiol 239:H14–H21
30. Ledingham J, Pelling D (1967) Cardiac output and peripheral resistance in experimental renal hypertension. Circ Res (suppl II) 20,21:187–199
31. Peterson L, Jensen R, Parnell R (1960) Mechanical properties of arteries in vivo. Circ Res 8:622–639
32. Bergel DH (1961) The static elastic properties of the arterial wall. J Physiol 156:445–457
33. Hudetz A (1979) Incremental elastic modulus for orthotropic incompressible arteries. J Biomech 12:651–655
34. Matsumoto T, Hayashi K (in press) Analysis of stress and strain distributions in hypertensive and normotensive rat aorta considering residual strain. J Biomech Eng
35. Takamizawa K, Hayashi K, Matsuda T (1992) Isometric biaxial tension of smooth muscle in isolated cylindrical segments of rabbit arteries. Am J Physiol 263:H30–H34
36. Ito H (1989) Vascular connective tissue changes in hypertension. In: Lee RMKW (ed) Blood vessel changes in hypertension: structure and function. CRC Press, Boca Raton, FL, pp 99–122
37. Vossoughi J (1994) Biological residual stress and strain. In: Vossoughi J (ed) Biomedical engineering: recent developments. University of the District of Columbia, Washington, DC, pp 200–206

38. Bergel DH (1960) The viscoelastic properties of the arterial wall. PhD thesis, University of London
39. Chuong CJ, Fung YC (1986) On residual stresses in arteries. J Biomech Eng 108:189–192
40. Greenwald S, Rachev A, Moore JE Jr, Meister J-J (1994) The contribution of the structural components of the arterial wall to residual strains. In: Vossoughi J (ed) Biomedical engineering: recent developments. University of the District of Columbia, Washington, DC, pp 215–218
41. Matsumoto T, Hayashi K, Ide K (1995) Residual strain and local strain distributions in the rabbit atherosclerotic aorta. J Biomech 28:1207–1217
42. Greenwald SE, Saini A, Badrek Amoudi A (1994) Age-related changes of residual strain in the rat and human aorta. In: Vossoughi J (ed) Biomedical engineering: recent developments. University of the District of Columbia, Washington, DC, pp 1015–1018
43. Rachev A, Greenwald SE, Kane T, Moore JE Jr, Meister J-J (1994) Effect of age-related changes in the residual strains on the stress distribution in the arterial wall. In: Vossoughi J (ed) Biomedical engineering: recent developments. University of the District of Columbia, Washington, DC, pp 409–412
44. Rachev A (1994) Theoretical study on the effect of stress-dependent remodeling on arterial geometry under hypertensive conditions. In: Vossoughi J (ed) Biomedical engineering: recent developments. University of the District of Columbia, Washington, DC, pp 799–802
45. Xie JP, Liu SQ, Yang RF, Fung YC (1991) The zero-stress state of rat veins and vena cava. ASME J Biomech Eng 113:36–41
46. Omens JH, Fung YC (1990) Residual strain in the rat left ventricle. Circ Res 66:37–45
47. Rodriguez EK, Omens JH, Waldman LK, McCulloch AD (1993) Effect of residual stress on transmural sarcomere length distributions in rat left ventricle. Am J Physiol 264:H1048–H1056
48. Taber LA, Hu N, Pexieder T, Clark EB, Keller BB (1993) Residual strain in the ventricle of the stage 16–24 chick embryo. Circ Res 72:455–462
49. Han HC, Fung YC (1991) Residual strains in porcine and canine trachea. J Biomech 24:307–315
50. Fung YC, Liu SQ (1991) Changes of zero-stress state of rat pulmonary arteries in hypoxic hypertension. J Appl Physiol 70:2455–2470
51. Berry JL, Rachev A, Moore JE Jr, Meister J-J (1992) Analysis of the effects of a non-circular two layer stress-free state on arterial wall stresses. In: Proceedings of the 14th annual international conference of IEEE-EMBS, Paris, pp 65–66
52. Matsumoto T, Tuchida M, Sato M (1995) Effect of smooth muscle contraction/relaxation on strain distribution in aortic walls. In: Proceedings of the 4th China-Japan-USA-Singapore conference on biomechanics, Taiyuan, Shanxi, China, 21–27 May 1995 pp 107–110
53. Vossoughi J, Hedjazi Z, Borris FS II (1993) Intimal residual stress and strain in large arteries. In: Langrana NA, Friedman MH, Grood ES (eds) Proceedings of 1993 bioengineering conference, Breckenridge, CO. American Society of Mechanical Engineers, New York, pp 434–437

Tissue Remodeling and Biomechanical Response in Orthopedics and Orthodontics

Mechanical Stresses and Bone Formation

JUHACHI ODA[1], JIRO SAKAMOTO[1], KAZUHIRO AOYAMA[2], YASUNOBU SUEYOSHI[2], KATSURO TOMITA[2], and TAKESHI SAWAGUCHI[3]

Summary. Although it is well known that bone responds to mechanical stimuli, details of this phenomenon are still uncertain. To study the quantitative relation between bone formation and mechanical stimuli, experiments using rabbits and three-dimensional finite-element analysis were conducted. In the experiments, an intermittent compressive or tensile load was applied for 1 h per day to the rabbit's tibia in which a circular defect was made at the center along the axial direction. This process continued for 1, 2, or 4 weeks, and then bone formation at the circular bone defect was evaluated histologically. The histology of bone formation and the strain energy density at the defect, which is calculated by the finite-element method, were compared. It is confirmed, by the experimental results, that bone formation is promoted by mechanical stimuli. It was necessary to establish a theoretical model of bone formation caused by mechanical stress stimuli to apply the experimental results to a clinical situation. We attempted to apply the theory of time-dependent bone remodeling proposed by Beaupré et al. to bone formation during repair of defects. Computer simulation of bone formation around the circular defect corresponding to the experiment was carried out using an algorithm based on the theory; the effectiveness of this method is discussed in comparison with the experimental results.

Key words: Bone formation—Mechanical stimulation—Theoretical model—Finite-element method—Computer simulation

[1]Department of Mechanical Systems Engineering, Kanazawa University, 2-40-20 Kodatsuno, Kanazawa, Ishikawa, 920 Japan
[2]Department of Orthopaedic Surgery, Kanazawa University School of Medicine, 13-1 Takara-machi, Kanazawa, Ishikawa, 920 Japan
[3]Department of Orthopaedic Surgery, Toyama Municipal Hospital, 292 Imaizumi, Toyama, Toyama, 939 Japan

1 Introduction

The functional adaptation of bone tissue is known to follow Wolff's law [1]. The concept was proposed about 100 years ago, but mechanisms of the functional adaptation, or how the bone tissue can adapt to a mechanical environment, are not yet completely evident. Investigation of bone tissue metabolism has been progressing rapidly in this century with the progress of bone cell biology. Some materials that promote bone cell activity and induce bone formation have been found; for example, bone morphogenic protein (BMP) [2]. The influence of some hormones on bone remodeling, such as parathyroid hormone (PTH) and calcitonin (CT) [3], also has become apparent. Although knowledge obtained by studies of bone tissue metabolism and cell biology is very useful when considering adaptive bone remodeling, it cannot give an answer to a question such as "How can the bone tissue adapt to mechanical stress?"

Determining the relation between mechanical stimulation and bone remodeling is needed to answer this question. Investigation of the influence of mechanical stress on bone remodeling has not progressed greatly because chemical action receive much of the attention in the field of bone metabolism. Furthermore, some serious complexities are related to mechanical stress on bone remodeling; that is, analysis and measurement of stress occurring in bone tissue are intricate, and estimates of stress on the bone cells is also difficult. We cannot discuss stress concerning bone remodeling directly even if the stresses have been determined because stress seems to affect the bone cells indirectly. Chemical and electrical conditions around the bone cells undergo change because the stress condition changes through nonstress factors such as interstitial fluid flow [4] or the piezoelectric effect [5]. The bone cells seem to sense the chemical or electrical signals and transform their activity. To show the mechanism of adaptive bone remodeling in detail, it is necessary to elucidate mechanical stress in bone tissues and cells, interactions between stress and the chemical or electrical environment, and signals that promote activity of bone cells.

The approach described here is required to clarify the adaptive bone remodeling phenomenon in general, but it is too complicated to apply to clinical problems in practical use. In the clinical situation, determination of the quantitative relationship between bone remodeling and the mechanical stress stimulus is most important, because it is necessary to know how much stress stimulus is needed for promoting bone formation and constructing firm bone tissue. There is no need to consider mediating factors between stress stimulus and bone remodeling such as the chemical or electrical condition around the bone cells in the clinical situation. In clinical studies, change of thickness of cortical bone or apparent density of cancellous bone is evaluated when the mechanical stress is changed. Immobilization studies that evaluate bone resorption with no stress are frequently carried out [6–8]. On the other hand, stimulation studies evaluating the bone apposition and resorption after providing stress in controls are relatively few [9,10], because it is difficult to control bone stress in vivo. If the quantitative relation between bone remodeling and the stress stimulus is once established, it will be applicable for many clinical problems. For example, bone resorption around the hip component is a serious problem in clinics. If we could determine a stress limit at which bone resorption begins, this knowledge would be very useful for hip component design preventing bone resorption. It is also applicable to clinical problems such as bone healing and remodeling around bone grafts or grafting structures.

In this section, we describe an experiment for determining the relation between mechanical stress stimulus and bone formation in rabbit tibia in vivo plus our computer simulation of the experiment based on the adaptive bone remodeling theory. We focused on bone formation in such abnormal conditions as bone fracture and the effect of mechanical stress on bone, which is more meaningful than bone remodeling in normal conditions in clinical problems. We evaluated bone formation in an abnormal condition at a defect, which was drilled into the tibia, when a mechanical stimulus was applied. It is apparent from this experiment that bone formation in such an abnormal condition is promoted by stress stimulus, the same as normal adaptive remodeling. Moreover, we confirmed by means of the computer simulation of this experiment that the adaptive bone remodeling theory is applicable to bone recovery. The experiment and computer simulation are described in the following two sections.

2 Experimental Study of Adaptive Bone Formation

Many researchers have reported experiments to estimate the quantitative relation between mechanical stress stimuli and bone formation and remodeling [9–15]. For example, the influence of static force on bone formation of callus was studied by Yasuda et al. [9], and Rubin and Lanyon [10] have provided good studies on adaptive bone remodeling in response to mechanical stimuli. Mabuchi et al. [13,14] have experimented on bone fracture healing in various mechanical environments. The authors have also experimented on bone formation in the rabbit tibia during intermittent mechanical stress stimuli in vivo [16,17]. In these experiments, a compressive or tensile load along the axial direction is applied for a few weeks to a rabbit tibia in which a circular defect has been made at the center. The quantity of bone formation around the circular defect was estimated relative to the degree of stress stimuli. The basic relationship between bone formation and mechanical stimuli has been clarified gradually from the studies just described. It is considered that a general theoretical model of bone formation based on experimental results is necessary if the experimental data are to be applied to more complicated clinical problems such as healing of bone fractures.

2.1 Materials and Methods

Forty-five female Japanese white rabbits were used in this investigation. For each rabbit, two pins each 2 mm in diameter were inserted into the proximal and distal parts of the right tibia 70 mm apart, and a 3-mm circular bone defect was made halfway between the two pins (Fig. 1a). The right leg was fixed with external fixators through the two pins and a cast post-operatively (Fig. 1b). In the experimental groups, an intermittent compressive or tensile axial load of 49 N at a frequency of 0.5 Hz was applied for 1 h per day to the rabbit tibia through the two pins by the self-monitoring loading apparatus devised by us [17] (Fig. 1c). In the control group, no load was applied and the rabbit tibia remained fixed by the external fixator and the cast. The tibia was resected at 1, 2, or 4 weeks after the operation, and bone formation at the circular bone defect was evaluated histologically. Compression, tension, and control groups were established, and each group was divided into three subgroups: a 1-week group, a 2-week group, and a 4-week group; each subgroup contained five rabbits. Image-analyzing software ("ULTIMAGE" of GRAFTEK, Yokohama, Japan) was used for quantitative analysis of bone formation. The distribution of the mechanical stimulus was analyzed by the three-dimensional finite-element method (3D-FEM). Strain energy density was used as a representative quantity of mechanical stimuli [18].

2.2 Histology Results

No bone formation at the circular bone defect was observed in the compression group, tension group, or control group after 1 week. After 2 weeks of stimulation, comparatively greater immature bone formation was observed at the area parallel to the loading axis in both the compression and tension groups. In contrast, there was no or only slight bone formation at the area perpendicular to the loading axis. In the control group, there were two cases in which bone formation was observed and three cases in which bone formation was not observed. In the former two cases, however, the amount of bone formation was slight and the site of bone formation was random. After 4 weeks of stimulation, the woven bone that had been observed in both the compression group and tension group at 2 weeks had become mature, and bone formation extended to the center of the circular bone defect. In the control group at 4 weeks, bone formation was observed, but it was immature and of lesser quantity than that of the compression and tension groups (Fig. 2).

2.3 Quantitative Analysis of Bone Formation by Image-Analyzing Software

The amount of bone formation within the circular bone defect was compared in the compression, tension, and control groups quantitatively by means of image-analyzing software. The results showed that the quantity of bone formation in both compression and tension groups was larger than that in the control group. The quantity of bone formation in the tension group tended to be larger at 2 weeks; at 4 weeks, however, that in the compression group tended to be larger.

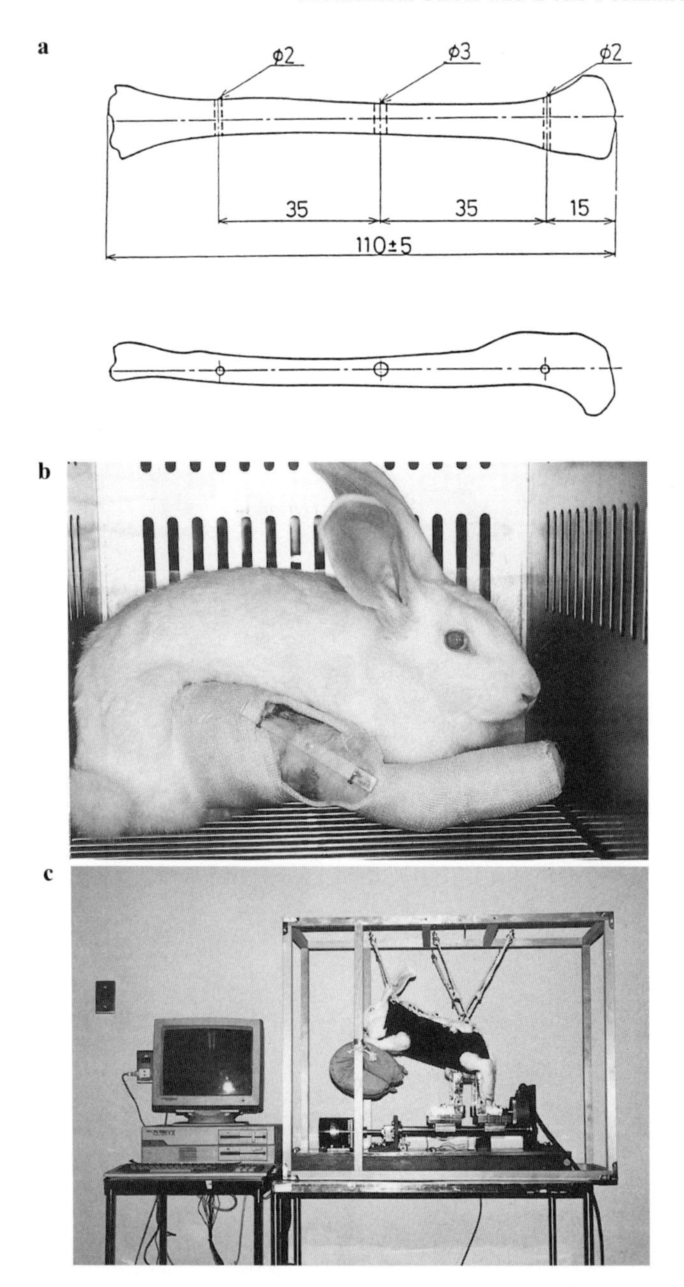

Fig. 1. **a** Schema of the rabbit tibia (the fibula was excised). Two pins (Ø, 2 mm) were inserted 70 mm apart and a hole 3 mm in diameter was drilled halfway between the two pins. **b** Non-stimulated status. The tibia was fixed with an external fixator and cast. **c** Setup of experiment. The rabbit tibia was fixed to the stimulator. Intermittent compressive or tensile axial loads of 49 N at 0.5 Hz were applied for 1 h per day

FIG. 2a,b. Photographs of the sections perpendicular to the axis. **a** Compression (*top*), tension (*middle*), and control (*bottom*) at 2 weeks. **b** Compression (*top*), tension (*middle*), and control (*bottom*) at 4 weeks. Loading direction is horizontal

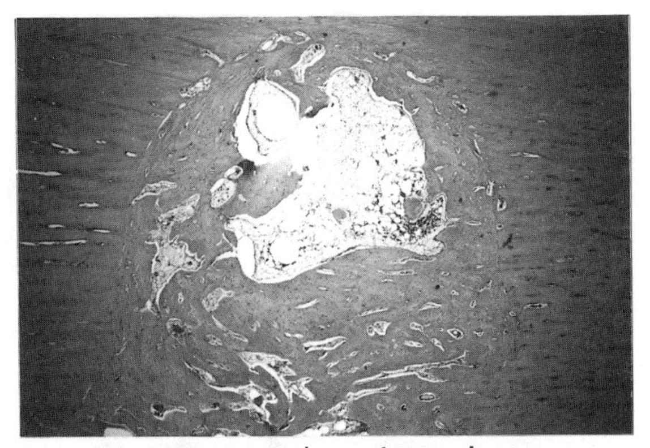

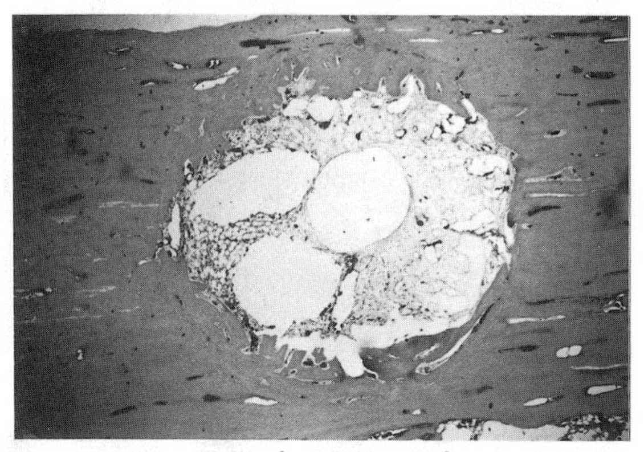

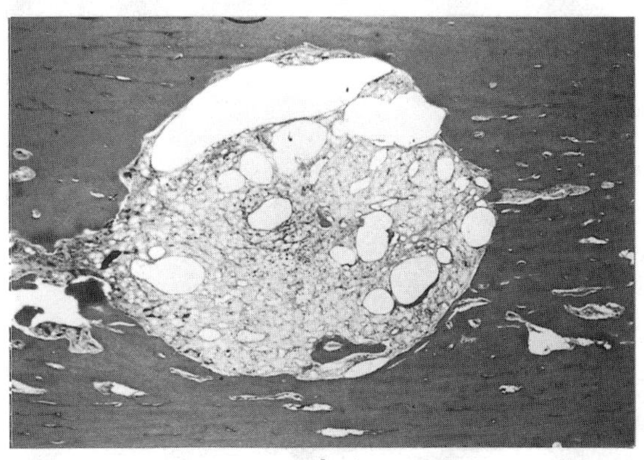

FIG. 2a,b. *Continued*

2.4 Three-Dimensional Finite-Element Method

Strain energy density around the circular bone defect was analyzed by means of 3D-FEM to relate the level of mechanical stimulus to the quantity of bone formation. Figure 3 shows the schema of rabbit tibia in the mechanical stimulation experiment. Analysis was carried out by approximating a part of the rabbit tibia near the circular bone defect to an elliptical cylinder model (Fig. 4). A quarter of the model was used for analysis because the model was symmetrical (Fig. 5). We took the axial direction to be 0°. The distribution of strain energy density at a border of the circular bone defect that was generated by the compressive or tensile load of 49 N along the axial direction is shown in Fig. 6. The quarter of the circular bone defect was divided into two areas: one was a high strain energy density area (45°–90°) and the other a low strain energy density area (0°–45°). A strain energy density between 0.3 and 4.9 ($\times 10^{-3}$) MPa was generated in the higher area and strain energy density between 0.0 and 0.7 ($\times 10^{-3}$) MPa was

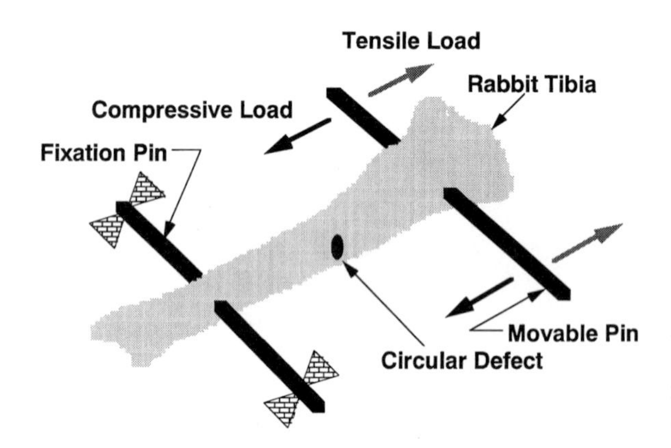

FIG. 3. Schema of rabbit tibia with circular defect in mechanical stimulation experiment

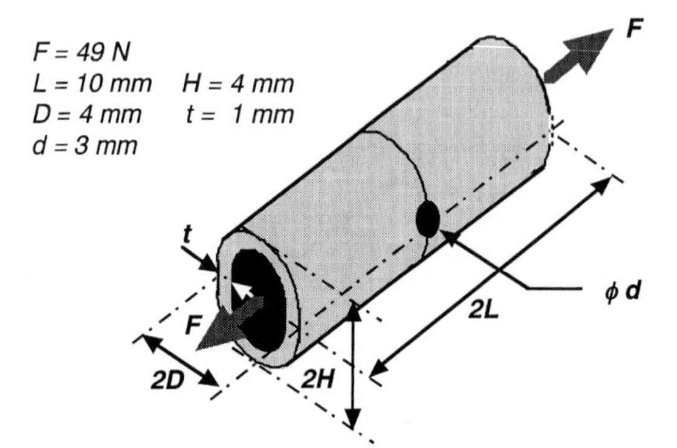

FIG. 4. Analytical model for three-dimensional finite-element method

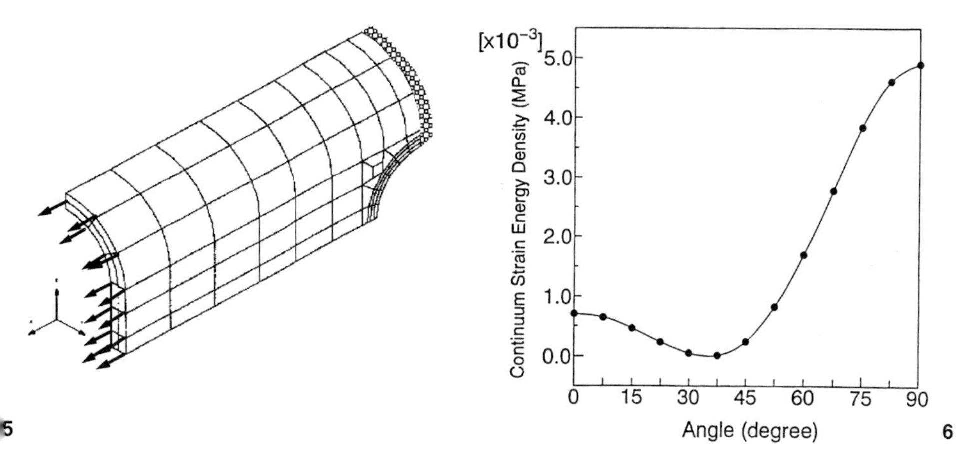

FIG. 5. Finite-element subdivision of computer simulation model of rabbit tibia

FIG. 6. Strain energy density around circular bone defect. The loading direction is taken as 0°

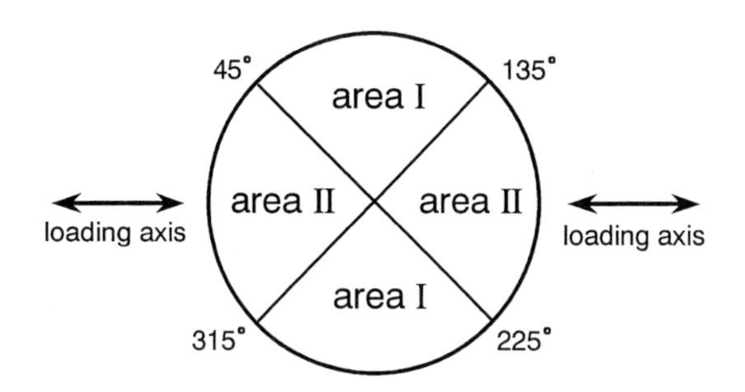

FIG. 7. Division of circular bone defect into two areas. Relatively higher strain energy density was generated in area I than in area II

generated in the lower area. Quantitative evaluation of bone formation was carried out separately, and the two areas were compared.

2.5 Relation Between Level of Mechanical Stimuli and Bone Formation

The circular bone defect was divided by lines at 45°, 135°, 225°, and 315° (Fig. 7), and we assumed the higher strain energy density area to be area I and the lower

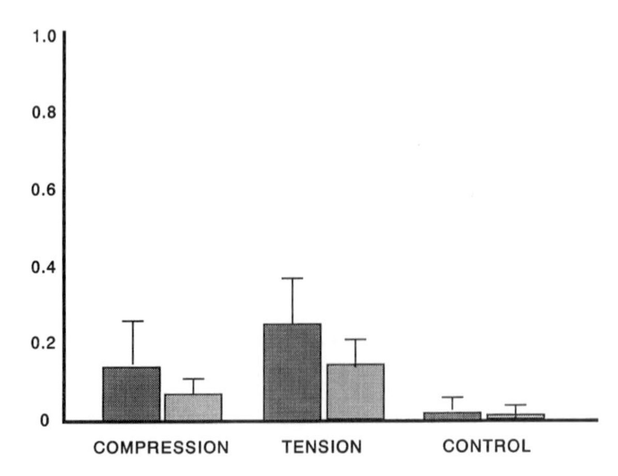

FIG. 8. Quantitative analysis of bone formation (ratio of bone formation area to initial bone defect area) compared between area I (*darker shading, left bars*) and area II (*lighter shading, right bars*)

strain energy density area to be area II. The control group was divided into the same two areas to be compared. The strain energy density at the border of the bone defect changes as the bone is generated and the shape of the bone defect changes. Thus, the quantitative evaluation of bone formation was performed in the 2-week specimen, which had the least change of shape. The quantity of bone formation was statistically tested between area I and area II in each group. There was no significant difference in the control group, but the quantity of bone formation in area I was larger than that in area II in both the compression group and the tension group ($P < 0.05$) (Fig. 8).

2.6 Discussion

Since the formulation of Wolff's law, it has been recognized that the structure of bone adapts itself to the mechanical environment modifying its morphology and density. Factors involved in bone formation include electrical [19], bone morphological [20], circulatory, and systemic factors such as hormones and metabolism in addition to mechanical stimuli, with these factors thought to interact. Because many varied interactions occur among these factors, not all of which are understood, the mechanisms involved in bone formation are extremely complex and require much additional study. We have aimed at clarifying the quantitative relation between mechanical stimuli and bone formation in this research. As regards the type of mechanical stimulus, it has been clarified by past research [21], including ours [16], that compressive stress promotes bone formation. It is widely considered that tensile stress acts as a bone absorption effect, as it has hitherto been explained by the example of self-correcting of malunion in long bone, but Yasuda et al. [22], Okada [23], and Matsushita and Kurokawa [24] concluded that tensile stress promotes bone formation. However, this issue is still controversial. In this study, we performed an experiment in which tensile load was applied to the rabbit tibia to clarify the influence of tensile stress on bone formation. Also, the loading apparatus used in the former experiment was devised for the purpose of comparing tension with compression under the same conditions [17].

Strain energy density on the border of the circular bone defect analyzed by 3D-FEM was compared with the histological results and the results of image analysis after 2 weeks. More bone formation was observed in area I, in which 0.3–4.9 $\times 10^{-3}$ MPa of strain energy density was produced during the initial period than in area II, in which 0.0–0.7 $\times 10^{-3}$ MPa of strain energy density was produced. Although bone formation was observed in two of the control rabbits at 2 weeks, the site of bone formation was irregular and the formation was less than that of the experimental group. This bone formation is not thought to be caused by mechanical stimuli but by the natural repair mechanism, because a bone defect, constituting damage to the bone, was caused in this experiment. More bone, and also more mature bone, was generated after 4 weeks than after 2 weeks in both the compression group and the tension group, while less bone that was immature was generated in the control group. Our experiment demonstrated that tensile stress as well as compressive stress promotes both the amount and maturation of bone, and that the promotion of bone formation depends on the quantity of the stress. After 2 weeks the quantity of bone formation was greater in the tension group than in the compression group, but after 4 weeks the result was reversed. It is unclear which kind of stress exerts a greater promoting action on bone formation, but it is clear that these two types of stress differ in some way in their bone growth-promoting properties.

3 Theoretical Study of Adaptive Bone Formation

Various theoretical models have been proposed for bone remodeling, because various factors mediate between bone remodeling and mechanical stimuli. For example, the piezoelastic model of Gjelsvik [25], the electromechanical model of Guzelsu and Saha [26], the cell biological model of Hart et al. [27], and the bone fluid model of Takakuda [20] have been proposed. These models are useful for examination of the hypotheses, but not for application to clinical problems. The phenomenological model called the adaptive elasticity model in which the relation of bone formation and stress stimuli are determined directly is adequate for clinical problems. In fact, Weinans et al. [28] applied their phenomenological model to bone remodeling around femoral hip components, and discussed the effect of material properties of the components. The most effective theory of the phenomenological models is the time-dependent model proposed by Beaupré et al. [29,30]. In this remodeling theory, not only the influence of stimulus magnitude but also the number of stimulation cycles is considered.

It is known that the bone formation at injury sites occurs by adaptation to mechanical stimuli, just as bone remodeling occurs in the absence of an injury. Investigation of the theoretical model of adaptive bone formation at an injury site compared with that of the bone remodeling model as described here, is rare in spite of its importance to clinical problems. Thus, we investigated the theoretical model of adaptive bone formation with repair of injured bone applying the phenomenological remodeling theory. The time-dependent theory of Beaupré is used in the model because of its efficiency. Computer simulation corresponding to our experiments on rabbit tibia was carried out, and the effectiveness of the theoretical model is discussed by comparing the results.

3.1 Bone Remodeling Theory

Mechanical stress stimulus per day, ψ_b, is defined by Beaupré as follows:

$$\psi_b = \left(\sum_{day} n_i \sigma_i^m \right)^{1/m} \qquad \sigma_i = \sqrt{2EU} \tag{1}$$

where n and σ_i are the number of stimuli per day and the effective stress that occurred at the bone, respectively. The subscript i corresponds to various loads applied to the bone, i.e., tensile, compression, bending, and torsional load; m is an empirical constant that can be thought of as a weighting factor for the relative importance of stress magnitude and the number of stimuli. The effective stress is defined as Eq. 1, where E is an elastic modulus and U is the strain energy density. By using the equations, we can treat the mechanical stress stimuli per day numerically in the bone remodeling theory.

It is considered that bone apposition or resorption occur according to the level of the stress stimuli at each point. Bone apposition and resorption are in equilibrium with each other when values of the stress stimuli are normal, but bone apposition will occur when the stimuli are more than the daily stress stimulus values. For values of stress stimuli less than normal, bone resorption will occur. The normal stress stimulus is called the attracter state stimulus, ψ_{bAS}, which is constant depending on the genotype and metabolic status. Considering the simplest case, it is assumed that bone apposition or resorption occur in proportion to the difference between the applied stress stimulus ψ_b and the attracter state stimulus ψ_{bAS}. The idealized relationship between the stress stimulus and bone apposition or resorption per day assumed here is shown in Fig. 9. The relation is defined as the following equations:

$$\dot{r} = \begin{cases} c\left(\psi_b - \psi_{bAS}\right) + cw & \left(\psi_b - \psi_{bAS} < -w\right) \\ 0 & \left(-w \le \psi_b - \psi_{bAS} \le w\right) \\ c\left(\psi_b - \psi_{bAS}\right) - cw & \left(\psi_b - \psi_{bAS} > w\right) \end{cases} \tag{2}$$

where $\dot{r}$ is bone apposition or resporption per day at the bone surface, c is an empirical rate constant, and w is the half-width of the normal stress stimulus region; c is represented by a gradient of the remodeling line in Fig. 9.

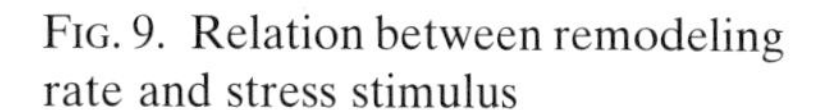
FIG. 9. Relation between remodeling rate and stress stimulus

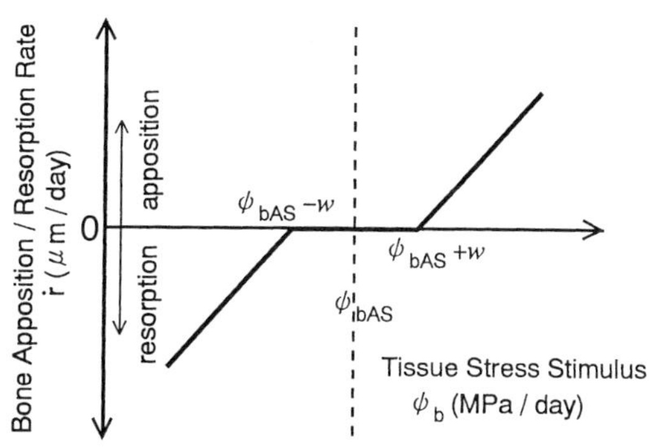

3.2 *Theoretical Model for Experiments*

The adaptive remodeling theory just described was applied to the computer simulation of bone formation with bone repairing. The simulation corresponding to our bone-forming experiment on the rabbit tibia in vivo was carried out using the remodeling theory. Computer simulation of bone formation focused on the circular defect boundary, and results obtained by the calculation were compared with the experimental results.

The process of the computer simulation is described as follows:

1. Calculate the strain energy density around the circular defect using the finite-element method and obtain the effective stresses.
2. Calculate the stress stimuli per day, substituting the effective stress value and the number of stimulations per day into Eq. 1.
3. Obtain the quantity of bone apposition or resorption per day by the relation of Eq. 2.
4. Change the defect shape according to the quantity of the bone apposition or resorption that has been calculated.

The process from step 1 to step 4 corresponds to the bone formation process during a day, so that the process is iterated for the duration of the experimental period in this simulation. This calculation requires great computational effort because the three-dimensional finite-element analyses are needed for each iteration. To improve calculation efficiency, an elliptical cylindrical model that approximates the rabbit's tibia (see Fig. 4) is used here, applying a load equivalent to the experimental load. The finite-element subdivision of this model is shown in Fig. 5. The elastic modulus and Poisson's ratio are defined as 11.8 GPa and 0.5, respectively, which correspond to the material properties of cortical bone.

In applying the remodeling theory of Beaupré to simulation of our experiment, the relation between the bone apposition or resorption and stress stimulus is kept as it is, and only the value of parameters used in Eqs. 1 and 2 are changed. That is, ψ_{bAS}, c, m, and w are the parameters that must be determined in advance before the calculation. Beaupré et al. gave the values as $\psi_{bAS} = 50$ MPa/day, $c =$ 0.02 (μm MPa/day^2), $m = 4$, and $w = 10\%$ of ψ_{bAS}. Because the values of ψ_{bAS} and c were determined for the normal human condition, it is considered that the parameters would take different values for the rabbit in the case in which bone forming occurs with bone repairing at the defect. The values were determined in our calculation as follows. The attracter state stimulus ψ_{bAS} is calculated from average effective stresses that occurred at the rabbit tibia per day and the number of stimulations per day. Assuming that 150% of the average rabbit weight (25 N) is applied to the tibia along the axial direction in daily stimuli that occur by normal movement of the rabbits and that the cross-sectional area of the center of the tibia is 18 mm^2, the average effective stress is approximately 2.1 MPa. Furthermore, assuming that 2000 stimulations occur per day, the value of ψ_{bAS} is given as 14 MPa/day. On the other hand, it is difficult to determine the value of c because it is only an empirical parameter. Thus, we assumed the value of c from one of the experimental results in the 4-week stimulation group, $c = 0.8$ (μm MPa/day^2). Values of the other parameters m and w are the same as given in the analyses by Beaupré.

3.3 Computer Simulations

Changes in the shape of the defect in the rabbit tibia model obtained by the computer simulation are shown in Fig. 10. Only one-quarter of the region is shown here. In this figure, loading direction is along the horizontal axis, and solid, dotted, and broken lines correspond to the shape of the middle, outer, and inner surfaces, respectively. The shapes of the defects were estimated after 1, 2, and 4 weeks of stimulation in the experiment, so that the calculation is iterated 28 times corresponding to 4 weeks. Figure 10 shows the defect shape at intervals of 1 week. As the effective stresses around the upper-side boundary on the initial shape of the defect are larger in comparison with the stresses around the right-side bound-

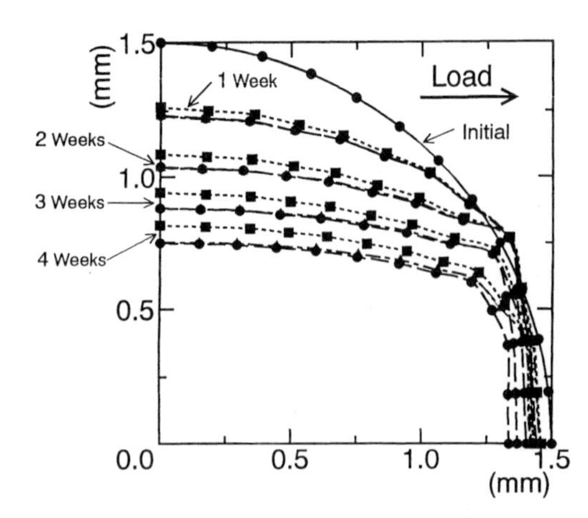

Fig. 10. Changes of defect shape of rabbit tibia model on computer simulation. *Squares*, outer surface; *circles*, middle section; *triangles*, inner surface

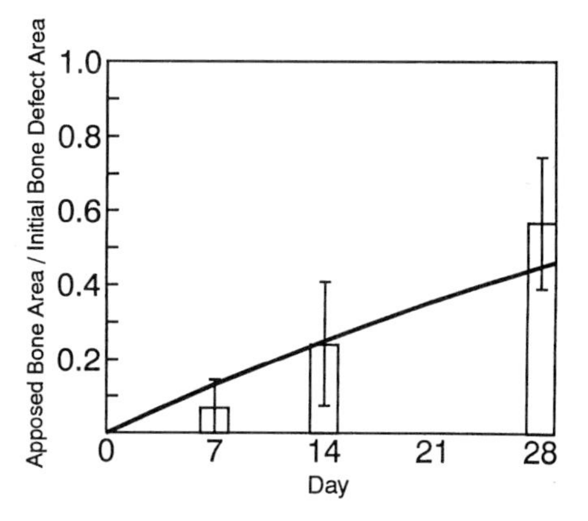

FIG. 11. Time history of bone formation quantities on the computer simulation (*solid line*) and the experiment (*bars*). $n = 30$, tension and compression data combined

ary, bone apposition occurs more around the upper boundary than the right boundary in the bone formation process. At the lowest stimulation point of initial shape, the bone resorption can also be seen temporally. To compare these numerical results, bone formation after 2 weeks and 4 weeks of stimulation in the experiment (see Fig. 2) was evaluated by hematoxylin-eosin stain. From the computational and experimental results, the quantities of bone formation that occurred inside the initial defect are evaluated. Time histories of the bone formation quantities normalizing the initial defect area are shown in Fig. 11.

3.4 Discussion

Comparing Fig. 10 with tension stimulation specimens of 2 and 4 weeks in Fig. 2, the locations at which bone apposition occurred are approximately the same on both, and the defect shapes after 4 weeks of stimulation are similar. The tendency of the bone apposition increase in the calculation is very similar to the tendency of the experiment (Fig. 11). In the results, the bone remodeling theory of Beaupré is applicable to bone formation with the bone repairing by tuning the parameters. Some serious problems, however, can be seen by examining the results in detail. Bone resorption was not found in the experimental results although it occurred temporarily at a low stimulation point in the calculation. In the experiment, quantities of bone apposition after 1 week stimulation are very small and apposition after 4 weeks of stimulation tended to be larger in comparison with the calculation. It is considered that nonlinearity of bone apposition change and the large dispersion of the results in the experiment were influenced by bone repair that occurred from the circumference of the defect. The major cause of the differences between the calculation and the experiment is the effect of the bone-repairing circumference of the defect, which is not taken into account in the calculation. To solve such a problem, an improved theoretical model considering the bone repair proper is necessary.

4 Conclusion

In the experimental study, both compressive and tensile stresses generated bone formation after 2 weeks of stimulation, with more bone formation observed in areas subjected to higher stress. After 4 weeks, woven bone was also observed in the control specimen, but mature bone formation was observed in both experimental groups whether loaded with a compressive or tensile load. The quantity of bone formation in the high-energy-density area was greater than that in the low-energy-density area when both compressive and tensile stresses were applied.

The theoretical model of bone remodeling adapted to mechanical stress stimulus provided by Beaupré was applied to the bone-forming simulation in which bone repair was included. The framework of the theory was not changed, and some parameters related to bone formation were tuned in adapting the theory to our problem. The bone formation process around the defect in the rabbit tibia under an intermittent axial load was estimated using computer simulation based on the theory, and compared with the experimental result obtained previously to evaluate the efficiency of the theoretical model. In the results, the bone formation process obtained by the calculation tended to be approximately similar to those obtained experimentally, and the possibility of the practical use of the theoretical model is confirmed. Furthermore, it is clarified that improvements of the theory considering the bone-repairing effect are necessary to apply to the problem treated here.

Acknowledgments. This study was supported by funds provided by the Ministry of Education, Science and Culture, Japan, Grant-in-Aid for Scientific Research on Priority Area "Biomechanics of Structure and Function of Living Cells, Tissue and Organs," 1993, 1994, and 1995. We would like to thank the authorities concerned. We are grateful to Dr. Koichiro Sakai, Mr. Shigeyuki Tabata, Mr. Noriaki Kobayashi, Mr. Yaoki Nakano, and Mr. Satoru Kenmochi.

References

1. Wolff J (1892) Das gesetz der transformation der knochen. A Hirshwald, Berlin
2. Wozney JM (1989) Bone morphogenetic protains. Prog Growth Factor Res 1:267–280
3. Habener JF, Potts JT Jr (1978) In: Avioli LV, Krane SM (eds) Metabolic bone disease, vol II. Academic Press, New York, p 1
4. Hillsley MV, Frangos JA (1994) Bone tissue engineering: the role of interstitial fluid flow. Biotechnol Bioeng 43:573–581
5. Pollack SR, Salzstein R, Pienkowiski D (1984) The electric double layer in bone and its influence on stress-generated potentials. Calcif Tissue Int 36:S67
6. Uhthoff HK, Jaworski ZFG (1978) Bone loss in response to long-term immobilization. J Bone Joint Surg 60B:420
7. Jaworski ZFG, Liskova-Kair M, Uhthoff HK (1980) Effect of long-term immobilization on the pattern of bone loss in older dogs. J Bone Joint Surg 62B:104

8. Uhthoff HK, Sekaly G, Jaworski ZFG (1985) Effect of long-term nontraumatic immobilization on metaphyseal spongiosa in young adult and old beagle dogs. Clin Orthop Relat Res 192:278
9. Yasuda Y, Noguchi K, Sata T (1955) Dynamic callus and electric callus. J Bone Joint Surg 37A:1292
10. Rubin CT, Lanyon LE (1987) Osteoregulatory nature of mechanical stimuli: function as a determinant for adaptive remodeling in bone. J Orthop Res 5:300–310
11. Pead MJ, Skerry TM, Lanyon LE (1988) Direct transformation from quiescence to bone formation in the adult periosteum following a single brief period of bone loading. J Bone Miner Res 3:647–656
12. Pead MJ, Suswillo R, Skerry TM, Vedi S, Lanyon LE (1988) Increased ^{3}H-uridine levels in osteocytes following a single short period of dynamic bone loading in vivo. Calcif Tissue Int 43:92–96
13. Mabuchi K, Fujie H, Yamatoku Y, Yamamoto M, Sanou Y, Sasada T (1992) Utility of robotics to control the mechanical environment on the fracture healing of bone and to assess the healing (in Japanese). Trans Jpn Soc Mech Eng 58(551A):1030–1035
14. Mabuchi K, Fujie H, Yamatoku Y, Yamamoto M, Sasada T (1992) A new methodology with an application of robotics to control the mechanical environment around experimentally fractured bone. In: Niwa S, Perren SM, Hattori T (eds) Biomechanics in orthopedics. Springer-Verlag, Tokyo, pp 183–193
15. Turner CH, Akhter MP, Raab DM, Kimmel DB, Recker RR (1991) A noninvasive, in vivo model for studying strain adaptive bone modeling. Bone 12:73–79
16. Sakai K, Tomita K, Sawaguchi T, Akagawa S, Oda J (1990) An experimental study on quantitative relation between the mechanical stimuli and the bone formation. Orthop Biomech 12:99–103
17. Oda J, Sakamoto J, Kobayashi K, Tomita K, Sawaguchi T, Sueyoshi Y, Aoyama K (1994) Bone formation in vivo due to mechanical stimulation and its evaluation (in Japanese). Trans Jpn Soc Mech Eng 60(579B):185–190
18. Huiskes R, Weinans H, Grootenboer HJ, Dalstra M, Fudala B, Slooff TJ (1987) Adaptive bone-modeling theory applied to prosthetic-design analysis. J Biomech 20:1135–1150
19. Hirasawa Y, Okada K (1989) Application of electric phenomena to fracture treatment: recent trend and future. Sogo Rehabil 17(1):5–11
20. Takakuda K (1992) A hypothetical mechanism for adoptive bone remodeling (transportation of growth factors by mechanical loads) (in Japanese). Trans Jpn Soc Mech Eng 58(511A):1015–1021
21. Nakabayashi Y, Shiba R (1987) An experimental study of bone remodeling influenced by mechanical stress. J Jpn Orthop Assoc 61:1429–1436
22. Yasuda I, Nagayama H, Kato T, Hara O, Okada K, Noguchi K, Sata T (1953) Fundamental problems in the treatment of fracture. J Kyoto Med Soc 4:395–406
23. Okada K (1956) Dynamic studies on bone fractures. J Jpn Orthop Assoc 30(2):105–133
24. Matsushita T, Kurokawa T (1991) Mechanical condition promoting bone healing (comparison between compression and tension). J Jpn Orthop Assoc 65(7):S1216
25. Gjelsvik A (1973) Bone remodeling and piezoelectricity. J Biomech 6:69
26. Guzelsu N, Saha S (1984) Electro-mechanical behavior of wet bone. J Biomech Eng 106:249–261
27. Hart RT, Davy DT, Heiple KG (1984) Mathematical modeling and numerical solution for functionally dependent bone remodeling. Calcif Tissue Int 36:S104

28. Weinans H, Huiskes R, Grootenboer HJ (1992) Effect of material properties of femoral hip components on bone remodeling. J Orthop Res 10:845–853
29. Beaupré GS, Orr TE, Carter DR (1990) An approach for time-dependent bone modeling and remodeling—theoretical development. J Orthop Res 8:651–661
30. Beaupré GS, Orr TE, Carter DR (1990) An approach for time-dependent bone modeling and remodeling-application: a preliminary remodeling simulation. J Orthop Res 8:662–670

Fatigue Fracture Mechanism of Cancellous Bone

Masafumi Morita[1] and Tadashi Sasada[2]

Summary. Femoral neck fractures occurring in senior osteoporotic patients are closely related to bone fatigue. It is considered that the fatigue strength and durability of bone are decreased by the low metabolic activity in such cases. In this chapter, a theoretical analysis and an experimental study were carried out to estimate the progression of osteoporosis and the fatigue life of bone in living body. According to these studies, it was found that fatigue life is determined by the accumulation rate of mechanical damage and by self-healing activity. From these results, it is clear that remodeling activity varied with mechanical stimulation and that fatigue life is short in a high-frequency-cycle loading condition.

Kew words: Osteoporosis—Cancellous bone—Bone metabolism—Femoral neck fracture—Aging

1 Introduction

Bone is maintained by active and ongoing metabolism in the living body. Bone resorption by osteoclasts and the formation of osteoblasts are continually in operation to regenerate bone tissue. Osteoporosis, which is observed frequently in senior patients, is caused by degenerative change in metabolic activities [1]. That is, as a result of relative decline in bone formation activity, bone volume gradually decreases, cortical bone thins, and the porosity of cancellous bone increases remarkably. These changes thus greatly impair the mechanical properties of bone [2,3]. Thus, osteoporosis may cause bone fractures, such as femoral neck fracture and vertebral collapse, to be induced by a weak external force [4–6].

[1] Department of Biomedical Engineering, Kitasato University School of Medicine, 1-15-1 Kitasato, Sagamihara, Kanagawa, 228 Japan
[2] Department of Precision Mechanical Engineering, Chiba Institute of Technology, 2-17-1 Tsudanuma, Narashino, Chiba, 275 Japan

Cancellous bone, which constitutes a large percentage of the proximal end of the femur, has a structure similar to that of porous functional materials. Therefore, its trabecular structure is not homogeneous at the proximal end, but is distributed so as to adapt to mechanical conditions in each area. Trabecular structures thus have various mechanical properties, some of them being resistant to compression and others being tolerant of bending [7]. To understand the conditions that generate fractures resulting from osteoporosis in senior patients, it is necessary to quantify the progressive stages of osteoporosis in relation to the trabecular structures and then to clarify the fatigue propagation of bone in the living condition.

In this chapter, we first describe the theoretical analysis of bone volume change of cancellous bone, and then report on experimental clarification of the relationship between metabolic activity and bone fatigue strength.

2 Progression of Osteoporosis

2.1 *Theoretical Analysis of Bone Volume Change*

Bone is maintained by the metabolic cycle in the living body. Resorption by osteoclasts and formation by osteoblasts occur on the bone surface. The mechanism of bone remodeling is illustrated in Fig. 1. When the amount of bone formation per unit time, V_f, is obtained from a microscopic specimen of bone double-labeled with tetracycline [8], V_f may be expressed as

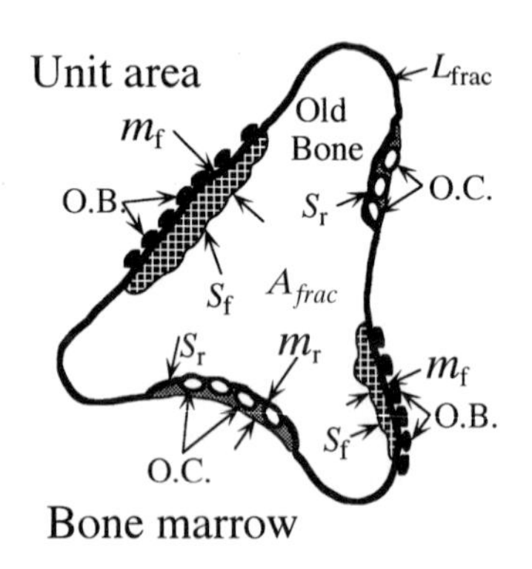

Fig. 1. Diagram of bone formation and resorption in cancellous bone (cut section). Osteoclasts crowd on the bone surface that requires remodeling and resorb the old bone. When resorption is completed, the osteoclasts move to the next target. When osteoblasts appear in the resorbed cavity and form an osteoid layer, calcification thus occurs to form new bone tissue. *Cross-hatching*, bone formation area; *dots*, resorption area; *O.B.*, osteoblast; *O.C.*, osteoclast; m_f, rate of bone formation; L_{frac}, bone total circumference of trabeculae per unit area; A_{frac}, total bone sectional area per unit area enclosed with square; m_r, rate of bone resorption; S_f, mean circumference of osteoid layer; S_r, mean circumference of resorbed cavity

$$V_f = n_f \cdot S_f \cdot m_f \tag{1}$$

where n_f is the number of sites of bone formation per unit area of cut section, S_f is the mean circumferential length of the osteoid layer, and m_f is the rate of bone formation obtained from the width of the image double-labeled with tetracyline. Similarly, the amount of bone resorbed per unit of time, V_r, can be expressed as

$$V_r = n_r \cdot S_r \cdot m_r \tag{2}$$

where n_r is the number of sites in which bone resorption has taken place, S_r is the mean circumferential length of the resorbed cavity, and m_r is the rate of bone resorption. If $V_f > V_r$, bone volume increases, and if $V_f = V_r$, there is equilibrium between the two activities and no change occurs in the bone volume. When $V_f < V_r$, osteoporosis progresses (Fig. 2). Using these parameters, the change in the amount of bone per sectional area fraction, ΔA_{frac}, can be defined by the following equation:

$$\begin{aligned} \Delta A_{frac} &= V_f - V_r \\ &= \left(n_f \cdot b_f \cdot m_f - n_r \cdot b_r \cdot m_r\right) \cdot l_{frac} \\ &= R_{act} \cdot l_{frac} \end{aligned} \tag{3}$$

Here, R_{act} is a parameter of the remodeling activity, the total circumferential length of bone trabeculae involved in a sectional unit area is termed l_{frac}, and $S_{f,r} = b_{f,r} \cdot l_{frac}$.

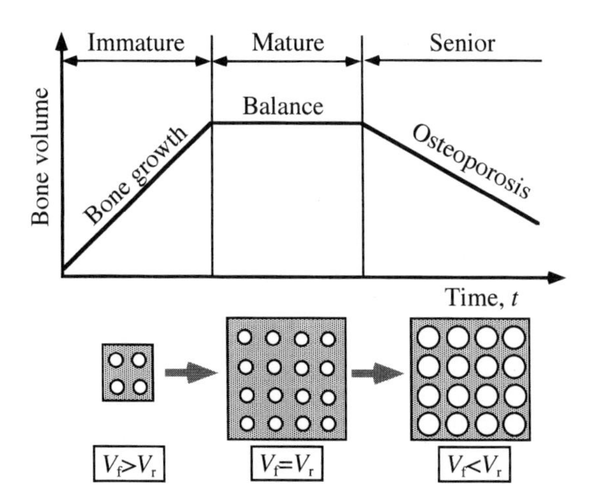

Fig. 2. Relation between bone metabolic activity change and bone volume. If the rate of bone formation is greater than that of resorption, $V_f > V_r$, bone volume increases. If the rate of formation is equal to that of resorption, $V_f = V_r$, the volume does not change. If the rate of formation is less than that of resorption, $V_f < V_r$, osteoporosis progresses. *Circles*, pores; *dotted area*, bone

The total circumferential length of trabeculae per unit area, l_{frac}, and the bone sectional area fraction, A_{frac}, are given by the following equations:

$$l_{frac} = l/A_0 \tag{4}$$

$$A_{frac} = A/A_0 \tag{5}$$

where A_0 is a field of view of a specimen, and l and A are the circumferential length and sectional area of trabecular bone, respectively.

The relation between l_{frac} and A_{frac} is given by the following equation:

$$\begin{aligned} l_{frac} &= (l/A) \cdot A_{frac} \\ &= K_{tr} \cdot A_{frac} \end{aligned} \tag{6}$$

Here the parameter K_{tr} shows the features of the trabecular structure, that is, the K_{tr} is defined as the circumferential length encircled the unit bone area in acute section. Therefore, ΔA_{frac} may be expressed as follows:

$$\Delta A_{frac} / A_{frac} = R_{act} \cdot K_{tr} \tag{7}$$

where A_{frac} is integrated with respect to time. To simplify the problem for the senior stage, it can be assumed that the parameter R_{act} and K_{tr} are independent of time t, because these values are relatively insensitive to time compared with those during the immature or mature stage. The time-course of changes in $A_{frac}(t)$ can be estimated as shown by the following equation:

$$A_{frac}(t) = A_{frac}(0) \cdot \exp(R_{act} \cdot K_{tr} \cdot t) \tag{8}$$

Thus it is clarified from Eq. 8 that the change in the amount of bone depends on remodeling activity, R_{act}, and the characteristic of the trabecular structure, K_{tr}. Generally, such a cancellous bone is an anisotropic material, and the trabecular bone volume fraction V_{frac} is not equal to the sectional area fraction A_{frac}. Thus, to clarify the relation between V_{frac} and A_{frac}, the authors have utilized a simple stereology method and found that V_{frac} is equal to the mean value of A_{frac} in each section. Therefore, the change of bone volume fraction, V_{frac}, is given as follows:

$$V_{frac} = 1/n \cdot \sum_{i=1}^{n} A_{frac_i}(t) \tag{9}$$

The value of V_{frac} obtained by Eq. 9 may show in which portion of the proximal end of the femur osteoporosis is progressing.

2.2 Experimental

2.2.1 Materials

Twenty-five rabbits (2–150 weeks old) were used to measure the progression rate of osteoporosis at the proximal end of the femur. The bone volume fraction in cancellous bone was measured, in five human femurs without osteoporosis and five osteoporotic femoral heads. Normal bones were collected from cadavers, and osteoporotic heads were obtained from patients who underwent total hip arthroplasty as the result of femoral neck fracture.

2.2.2 Bone Labeling and Preparation of Hard Tissue Specimens

The double-labeling technique developed by Frost [9] was employed tetracycline (Nippon Lederle, Tokyo, Japan), 30 mg/kg, was intravenously injected into each rabbit for 3 days. After an interval of 7 days, the injections were resumed in the same manner. Several days after the injections, the rabbits were killed, the femur was resected, and its proximal and distal ends were subjected to preparation of hard tissue specimens. Resected bones were soaked for 1 week each in a series of 70%, 80%, and 95% ethanol. The bones were then embedded in polymethylmetacrylate (PMMA) resin and sectioned at 30–50 μm. After polishing, the sections were examined.

2.2.3 Bone Histomorphometry

For each specimen, histomorphometric measurements of the remodeling activity, R_{act}, and the trabecular structure parameter, K_{tr}, were taken under a fluorescence microscope (model BHS-RFC-N1, Olympus, Tokyo, Japan) at ×50 magnification.

2.2.4 Measurement of Bone Volume Fraction of Human Cancellous Bone

Bones were defatted and deproteinized with 5% NaOH for 30 min at 50°C. The tissue was then embedded in resin and sectioned serially in 200-μm slices. Each slice was subjected to image analysis to estimate the trabecular area per unit area, A_{frac}, and the circumferential length, l_{frac}.

2.3 Results

2.3.1 Metabolic Activity, R_{act}, and Structure Parameter, K_{tr}

Double-labeled images of tetracycline incorporated into the proximal end of the femur in immature (4-week-old), mature (16-week-old), and old (96-week-old) rabbits are shown in Fig. 3. These images were obtained from the central regions of the femoral head. Individual parameters for metabolic activities in rabbits at different growth stages are shown in Fig. 4.

The trabecular structure of cancellous bone can be classified into three types (Fig. 5) [10]. These structures are characterized by the intensity of orientation and the bone volume fraction. We measured the values of K_{tr} in cross sections showing typical trabecular structures in five normal and five osteoporotic human femurs and 20 rabbits (16 and 110 weeks old) (Table 1).

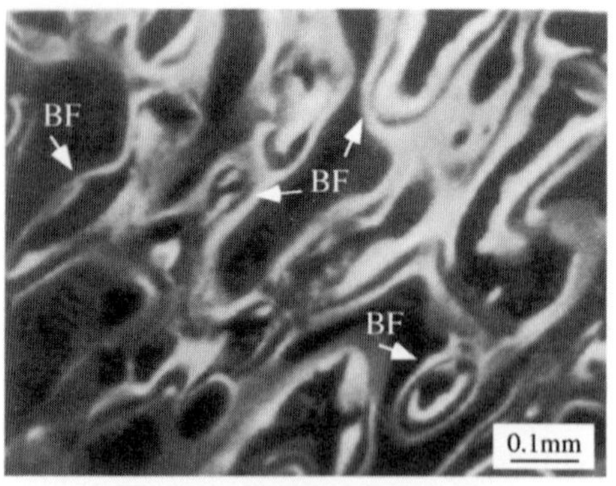

a Immature (4 week old)

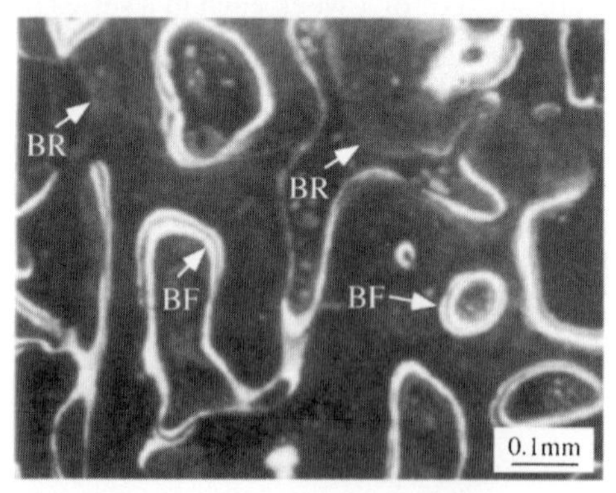

b Mature (16 week old)

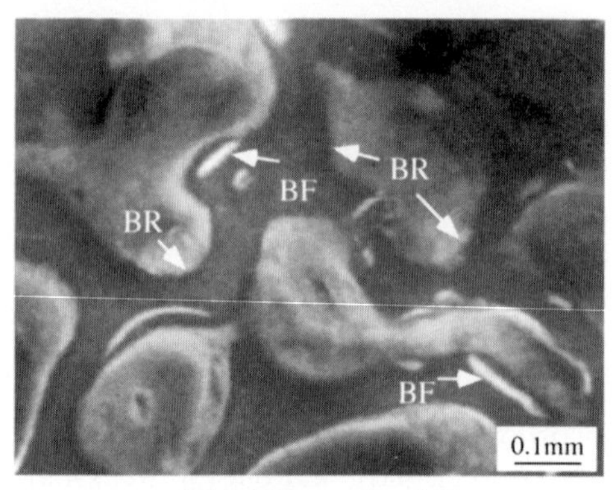

c Old (96 week old)

FIG. 3a–c. Tetracycline double-labeled images at different growth stages of rabbits in experiment. Lines observed in the peripheral area of the trabecular bone show the incorporation of tetracycline. **a** Immature stage (4 week old); **b** mature stage (16 week old); **c** old stage (96 week old). Fringe lines in the trabecular bone are tetracycline-labeled images and show bone formation area during the test period. *BF*, Bone formation; *BR*, bone resorption

2.3.2 Bone Volume Fraction

To confirm that osteoporotic changes in bone volume of cancellous bone depended on the trabecular structure, we measured the bone volume fractions of cancellous bone in rabbit and human femurs in the same portions as previously.

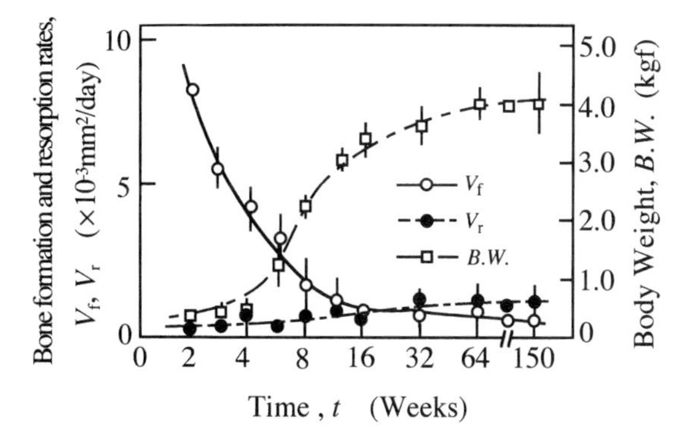

FIG. 4. Bone formation and resorption in rabbits in different growth stages. These two parameters were inverted at 16–32 weeks old, and osteoporosis may be thought to start at this stage. V_f, Bone formation activity; V_r, bone resorption activity; *B.W.*, body weight

TABLE 1. Structural parameter, K_{tr} (mm^{-1}), for each type of cancellous bone.

Type	Rabbit		Human	
	64 Weeks ($N = 4$)	110 Weeks ($N = 2$)	Normal ($N = 5$)	Osteoporotic ($N = 5$)
1	0.397 ± 0.028 ($n = 12$)	0.349 ± 0.020 ($n = 6$)	0.189 ± 0.036 ($n = 15$)	0.203 ± 0.018 ($n = 13$)
2	0.422 ± 0.021 ($n = 7$)	0.375 ± 0.016 ($n = 8$)	0.277 ± 0.010 ($n = 25$)	0.285 ± 0.031 ($n = 14$)
3	0.545 ± 0.022 ($n = 11$)	0.539 ± 0.023 ($n = 7$)	0.543 ± 0.035 ($n = 35$)	0.539 ± 0.027 ($n = 12$)

N, Total number of subjects; *n*, total number of sites for measurement. Data are given as mean ± SD.

Table 2 shows the measurement of volume fractions in normal subjects without osteoporosis and those with osteoporosis.

2.4 Discussion

2.4.1 Metabolic Changes in Bone with Aging

Cancellous bone is a highly functional material designed to tolerate stress most effectively in the usual structure, but only a slight decrease in bone volume induced by a minor metabolic disorder may result in bone fracture. As shown in Fig. 3, the image at the immature stage appears to indicate active bone metabolism in all regions, and bone remodeling occurs within a short period. In old rabbits, only slight incorporation of tetracycline was observed; thus, double-labeling was not effective in most of these cases. Figure 4 shows change in metabolic activity with aging in rabbits. These results concurred with observations in osteonal remodeling of human ribs reported by Takahashi and Frost [11]. In Fig. 3, V_f and V_r show the time-course changes in bone remodeling rate

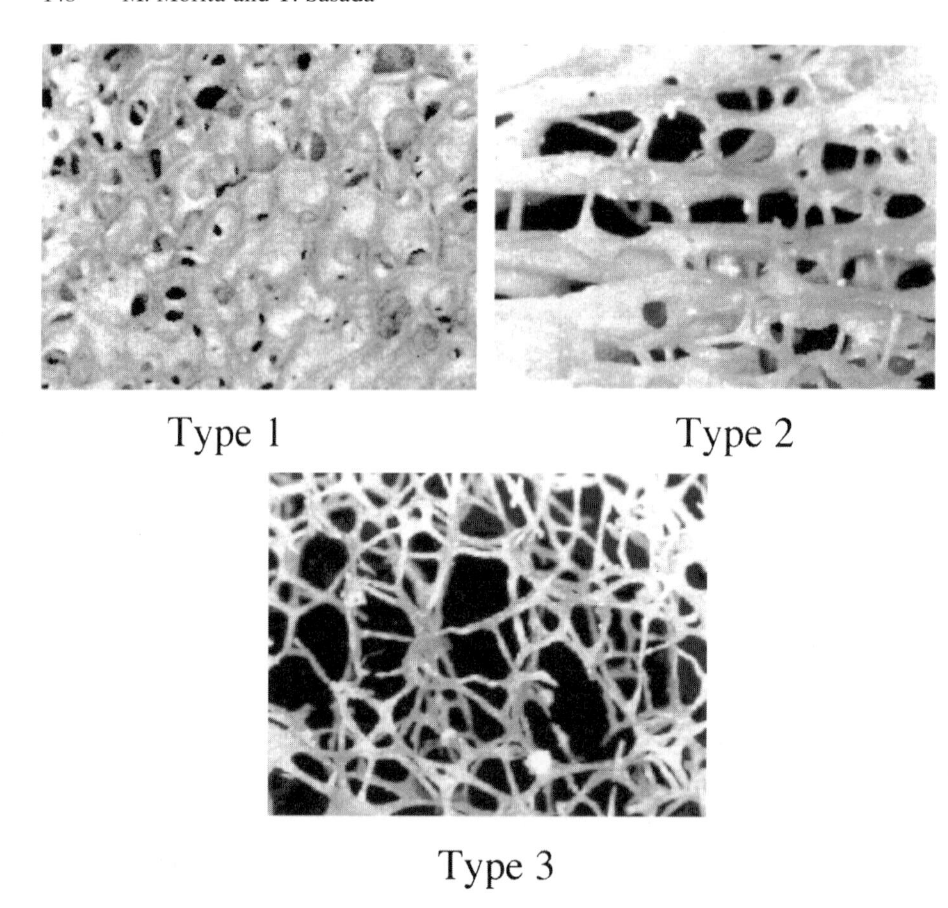

Fig. 5. Typical trabecular structures in the proximal end of femur. Type 1 bone (*top left*), composed of platelike trabecular bone elements, is generally distributed at the center of the femoral head. This type of cancellous bone is characterized by a high bone volume fraction and a low intensity of orientation. Type 2 bone (*top right*), composed of lamella plate and barlike strut trabecular bone elements, is distributed in the principal areas of tensile and compressive stress generating positions in the femoral neck. This type of cancellous bone shows a high intensity of orientation. Type 3 bone (*bottom*), composed of barlike trabecular bone elements, is frequently distributed at the greater trochanter groups of trabecular bone. This type of cancellous bone has a low bone volume fraction and a low intensity of orientation

estimated from observed parameters. The resorption rate, m_r, which cannot be measured directly, was estimated from the difference between observed changes in bone volume and the bone volume increase estimated from the rate of bone formation. In rabbits between 16 and 32 weeks old, which is near maturation, an inversion between resorption and formation activities occurs. Thus, osteoporosis may be thought to start at this inversion stage. Frost [9] measured remodeling activity in human ribs by the tetracycline labeling method. According to these

TABLE 2. Comparison of bone volume fraction, V_{frac}, between normal subjects and those with osteoporosis.

Case	Age	Gender	Type 1 V_{frac} (%)	Type 2 V_{frac} (%)	Type 3 V_{frac} (%)
Normal:					
1	38	Female	34.5 ± 2.3 (n = 6)	25.3 ± 2.1 (n = 4)	17.9 ± 1.7 (n = 4)
2	53	Female	30.9 ± 1.9 (n = 7)	23.0 ± 3.3 (n = 6)	13.6 ± 0.7 (n = 6)
3	70	Male	33.0 ± 2.1 (n = 3)	19.2 (n = 2)	14.4 ± 1.4 (n = 4)
4	74	Male	33.0 ± 1.8 (n = 4)	25.9 ± 2.6 (n = 4)	18.1 ± 0.1 (n = 3)
5	79	Male	35.2 ± 1.6 (n = 3)	27.4 ± 1.0 (n = 3)	—
Mean			33.3 ± 1.9	24.2 ± 2.3	16.0 ± 1.0
Femoral Neck Fracture (Osteoporotic):					
1	53	Female	23.8 ± 3.9 (n = 12)	13.1 ± 2.9 (n = 11)	5.9 ± 1.0 (n = 10)
2	79	Female	25.4 ± 2.7 (n = 7)	14.5 ± 4.0 (n = 12)	6.1 ± 2.2 (n = 8)
3	64	Male	27.0 ± 2.8 (n = 8)	17.7 ± 2.7 (n = 14)	9.7 ± 2.3 (n = 10)
4	65	Female	26.1 ± 3.3 (n = 10)	16.6 ± 3.6 (n = 10)	7.2 ± 2.3 (n = 9)
5	76	Female	21.3 ± 2.1 (n = 4)	13.6 ± 3.2 (n = 10)	10.5 (n = 1)
Mean			24.7 ± 3.0	15.1 ± 3.3	7.9 ± 2.0
$\Delta V_{frac}/V_{frac} \times 100$ (%)			25.8	37.6	50.6

n, Total number of samples. Data are mean ± SD.

results, the number of osteoid seams was less than the number of resorption spaces after 30 years of age. Furthermore, the percentage of the cortex replaced by new bone each year and the rate of osteon formation decreased remarkably in old subjects. In humans, osteoporotic change was considered to begin at age 30.

2.4.2 Parameter of Trabecular Structure, K_{tr}

According to Eq. 9, changes in the rate of progression of osteoporosis depend not only on remodeling activity, R_{act}, but also on the trabecular structure parameter K_{tr}. We observed three kinds of trabecular bone structures in the proximal end of the human femur (see Fig. 5). These can be classified by the morphological shapes of the trabecular bone elements, that is, the combination of platelike and barlike trabecular elements. Type 1 cancellous bone is composed of platelike trabecular bone only [12]. This kind of cancellous bone can be seen in the central area of the femoral head, an area requiring the principal tensile and compression strength of trabecular bone. These elements cross and form a honeycomb structure. The type 2 structure can be distinguished from the others by a combination of platelike and barlike trabecular bone elements showing a lamellar structure, which has a high intensity of trabecular orientation. This type of cancellous bone is observed mainly in areas requiring principal or secondary compression and tensile strength in trabecular bone, e.g., at the femoral neck and greater trochanter of the femur. Type 3 cancellous bone has a reticulate structure, which is composed entirely of barlike trabecular elements. It can be seen in the middle region of the greater and lesser trochanters, which is known as Ward's triangle. Table 1 shows the measurements of K_{tr} for these three kinds of trabecular structure.

The osteoporotic change in long bones usually appears as a spreading of the Haversian canal and a thinning of the bone. According to the measurements of bone volume in Table 2, the bone volume fraction clearly decreased more significantly in cancellous bone than in cortical bone. This can be explained by the structure parameter K_{tr}; that is, K_{tr} of cortical bone is less than that of cancellous bone. By our estimation, the value of K_{tr} in old rabbits is less than $0.5\,mm^{-1}$. Therefore, the rate of osteoporosis in cortical bone may be markedly delayed compared with that in cancellous bone if R_{act} is not affected by the region.

2.4.3 Changes in Bone Volume with Aging

Figure 6 shows the changes in bone volume fraction with aging obtained for the three types of trabecular structures using Eq. 9. To calculate the results, the values of R_{act} and K_{tr} must be determined. From the tetracycline labeling test in rabbits, $R_{act} = -1.0 \times 10^{-3}\,mm/day$ in the senior stage. This figure was drawn with the parameter of K_{tr} shown in Table 1. Here, $R_{act} = -1.0 \times 10^{-3}\,mm/day$ was tentatively substituted in Eq. 9 for human bone. To determine the actual osteoporotic rate clinically, R_{act} must be evaluated in each patient. The assumption that R_{act} is independent of the trabecular structure and equal in all three types of structure is supported by the following two facts. First, the tetracycline fluorescence image in old rabbits had uniformly disappeared in the cancellous bone at the upper end of femur, and second, the epiphysial growth line showing vigorous bone proliferation in an immature stage had disappeared in the old stage and was indistinguishable from other cancellous bone. According to this hypothesis, the progression of osteoporosis was most accelerated in type 3 cancellous bone, followed by type 2, and then type 1. Because these trabecular structures, type 3 and type 2, are concentrated in the neck of the femur, it is expected

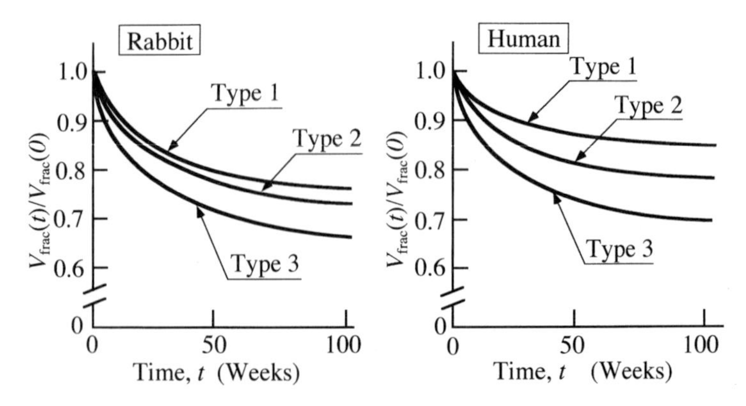

FIG. 6. Changes in the progression rate of osteoporosis caused by differences in trabecular structure in rabbits (*left*) and humans (*right*) for the three bone structure types. All bone volume fractions were calculated by Eq. 9. using the remodeling activity, $R_{act} = -1.0 \times 10^{-3}\,mm/day$

that bone strength at this portion is reduced most markedly by the progression of osteoporosis.

2.4.4 Actual Bone Volume Change in Human Bone

To examine whether these calculations are comparable to actual findings in osteoporosis, the volume fraction of cancellous bone was measured for each type of trabecular structure in human bone samples. Table 2 shows the volume fractions in normal subjects without osteoporosis and those with osteoporosis. The ratio of decrease in bone volume fraction, $\Delta V_{frac}/V_{frac}(0)$, was represented by the difference between the mean value in normal cases and the mean value in patients with osteoporosis for the same types of trabecular structures. A remarkable decrease in bone volume was noted in type 3 cancellous bone, and it was confirmed that the other two types also showed a similar tendency to decrease (Fig. 6).

Singh et al. [13] tried to standardize the progression of osteoporosis using X-rays of the proximal ends of femurs, and the degree of osteoporosis was classified in six stages. According to their study, during the first stage, areas of secondary tensile and compression groups of cancellous bone, which are composed mainly of type 3 trabecular structure, are reduced in grade 5. The trabeculae providing principal tensile strength are markedly reduced in grade 4. In grade 3, there is a break in the continuity of the trabeculae opposite the greater trochanter, which provide principal tensile strength. This progressive behavior of osteoporosis in cancellous bone concurs with the results of the theoretical analysis in our study. As shown in Singh's study, it is important to determine the progression rate of osteoporosis and to predict the occurrence of femoral neck fracture. Using our analysis, the progression of osteoporosis can be calculated for the individual patient.

2.4.5 Mechanical Stimulation and Remodeling Activity Change

Many investigators have reported that the remodeling activity in bone depends on mechanical stimulation. Almost all of them calculated bone mass change based on Wolff's law. The principle of remodeling behavior period is the functional adaptation of mature bone in response to its strain history and distribution [14,15]. Turner [16] investigated that relation, comparing the mechanical properties and the trabecular structures, and demonstrated remodeling (bone adaptation) using a mathematical model. Carter et al. [17] described bone volume change with loading and showed that trabecular structure was determined by the loading history.

Bone adaptation or resorption occurs as the result of self-optimization of the mechanical condition to which the bone is exposed, that is, the shape and the density distribution are designed by external loading conditions and history. For example, Huiskes [18] explained the rate of bone mass change (dM/dt) by the equation $dM/dt = C(S - k)$, where C is a physical parameter, and S and k are a mechanical signal and a "prearranged value". This equation shows that bone mass change is determined by the difference between parameters S and k. If $S - k > 0$, the bone mass increases; if $S - k < 0$, the bone is resorbed.

This bone adaptation mechanism concurs with findings in our study. Such an effect of mechanical stimulation on remodeling activity is included in the parameter R_{act} in our model of bone metabolism. If mechanical stimulation is applied uniformly to the bone, the bone volume fraction change can be calculated by Eq. 9, the same as Huiskes' equation. The equation suggests that the adaptation of bone will occur more rapidly in type 3 strutures than in other types of trabecular bone or cortical bone. Furthermore, the rate of adaptation in the cortical bone will be delayed compared to that in cancellous bone, because of the low K_{tr}. It is considered that the value of R_{act} in normal subjects will be increased depending on the magnitude of the mechanical stimulation, as described by many investigators. Such bone remodeling occurs in osteoporotic patients because of the low R_{act}. Furthermore, R_{act}, may not become positive even if mechanical stimulation is applied, and osteoporosis will continue to progress. Thus, both bone adaptation and resorption behavior can also be explained by our model of trabecular bone metabolism.

2.4.6 Mechanical Characteristics of Osteoporotic Bone

Cancellous bone is a porous material, and its stiffness and strength are dependent on the bone volume fraction, V_{frac}, and the intensity of trabecular orientation, I_{tr}. We [19] have defined the elasticity of cancellous bone perpendicular to the trabecular orientation, $E_p(V_{frac}, I_{tr})$, as follows:

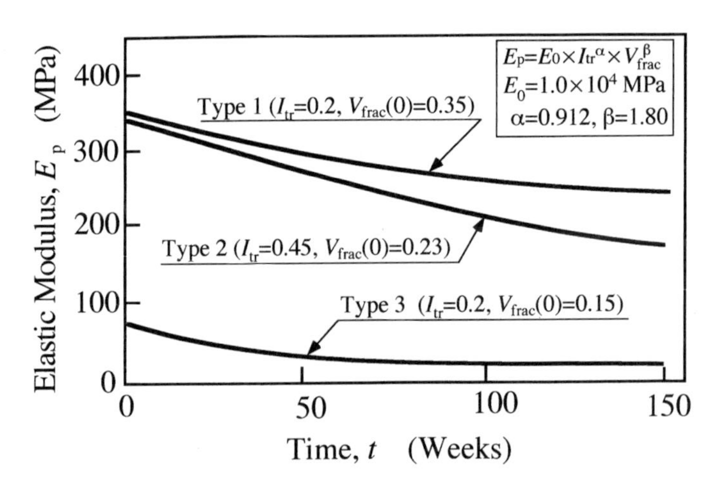

FIG. 7. Elastic modulus degradation of the three different types of cancellous bone caused by the progression of osteoporosis. These moduli can be calculated by Eq. 10. Both α and β are determined by trabecular structure analysis and mechanical loading tests for each bone type $V_{frac}(0)$ represents the initial bone volume fraction of each cancellous bone at $t = 0$, and it decreases according to Eq. 9

$$E_p\left(V_{frac}, I_{tr}\right) = E_0 \cdot I_{tr}^{\alpha} \cdot V_{frac}^{\beta} \tag{10}$$

where α and β are the constants that show coefficients of I_{tr} and V_{frac} on the elastic modulus, E_p, respectively. Both coefficients are determined for the three typical trabecular structures mentioned previously by image analysis. E_0 shows the elastic modulus of bone matrix. Figure 7 shows that the elastic modulus E_p changes depending on the structure with aging calculated from Eqs. 9 and 10.

3 Fatigue Fracture of Living Bone

3.1 Fatigue Strength of Cadaveric Bone

The fatigue process of living bone is presumed to be distinct from that of common industrial materials at any point at which the damage sustained by mechanical cyclic loading can be repaired though a bone metabolic turnover process. Lafferty and Raju [20] have reported that the fatigue strength of a cadaveric bovine cortical bone is not greater than that of common industrial materials. That is, in their experimental results, the endurance limit did not appear in the bone even if the stress ratio (applied stress amplitude/ultimate strength) was less than 0.2 and the final number of loading cycles to fatigue fracture was $N = 10^4$. However, it is well known empirically that this type of bone shows a high durability in living bodies. This inconsistency between the fatigue strength of cadaveric and living bone was caused by the fact that the mechanical damage which is accumulated by cyclic loads is removed momentarily in living bone. In other words, real fatigue bone fractures are observed only when the metabolic activity does not exceed the accumulation rate of the mechanical damage. The excellent fatigue strength of living bone will be maintained only when sufficient metabolic activity continues. Here it must be pointed out that metabolic activity in bone will differ depending on age or the presence of metabolic diseases.

3.2 Self-Healing Effect on Damage Accumulation

Mechanical damages are generally accumulated in a solid body when it suffers an affective cyclic load. A fatigue fracture will occur when the damages are saturated in the body in these artificial materials; damage can be stored in living bone in such loading conditions. Miner [21] defined the amount of damage accumulation caused by external forces in the following equation:

$$D = \sum_{i=1}^{m} n_i / N_i \tag{11}$$

where n is the number of loading cycles and N is the total number of cycles when the fatigue fracture occurs. In this equation, the damage reaches saturation and the solid is fractured by fatigue at $D = 1$. On the other hand, damage is healed by the metabolic turnover process and a high fatigue dunability can be obtained [22,23]. Here, we define the bone volume that is remodeled by the turnover by the following equation:

$$R = V_f(t)/V_0 \tag{12}$$

where V_f is the newly remodeled bone volume and V_0 is the bone volume as a whole. According to this self-healing process, the amount of actual damage accumulated in the bone is expressed by the following equation:

$$D_t = D - R \tag{13}$$

Hence, the accumulation rate of damage is written as follows:

$$dD_t/dt = d(D - R)/dt \tag{14}$$

Here, if $dD_t/dt < 0$, fatigue will not occur. If $dD_t/dt > 0$, the damage is accumulated so that the fatigue is propagated.

3.3 Experimental Findings

3.3.1 In Vivo Fatigue Test

To understand the effect of such self-healing activity on the fatigue life of bone, a cyclic loading test was carried out in vivo, using a servo-controlled hydraulic

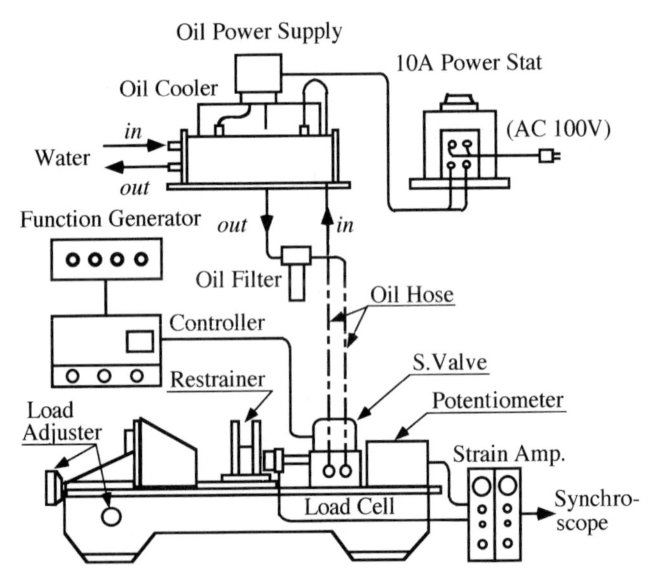

FIG. 8. Schematic diagram of the cyclic loading system driven with a servo-controlled hydraulic power generator. A rat was restrained on the test device, and the distal end of the left tibia was fixed in a fetter under in vivo anesthesia condition. A reciprocating load was applied horizontally to the tibia with a load cell attached on the top of the driver rod; the reaction force generated was monitored with a synchroscope

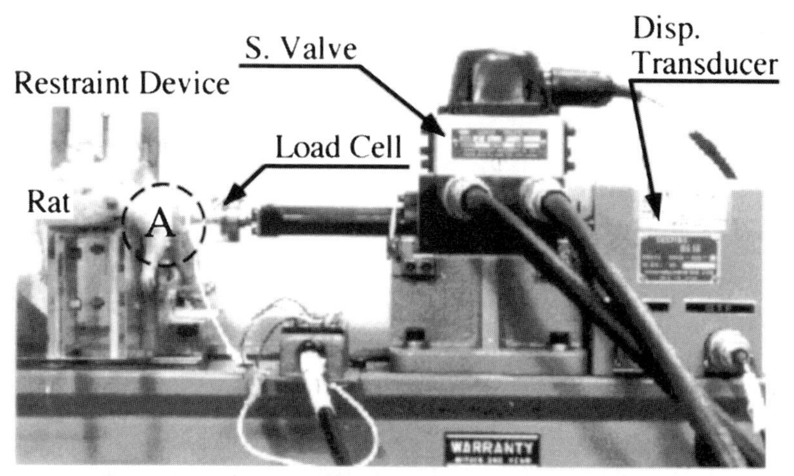

In vivo bone fatigue test apparatus

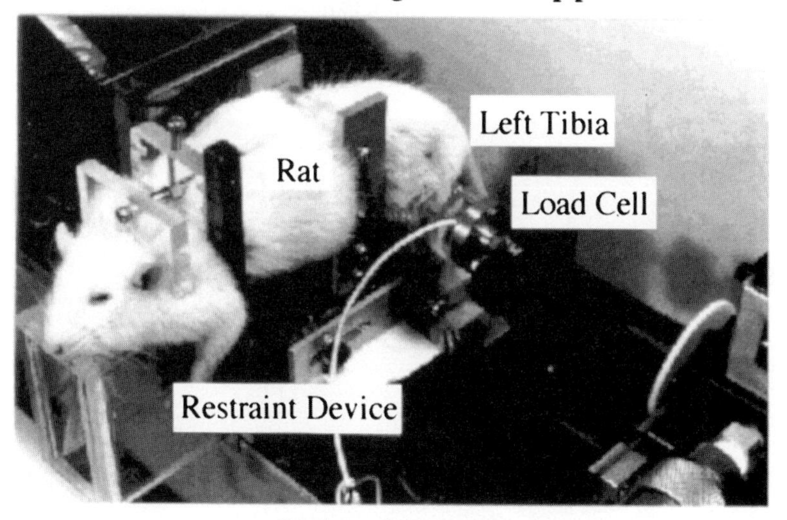

Side view of "A"

FIG. 9. Cyclical loading system and the actual loading experiment using a living rat's tibia

testing machine (Figs. 8 and 9). Ten female Wistar rats (body weight approximately 200g) were used. Each of the rats was restrained with the test device and the distal end of the left tibia was fixed in a fetter (Fig. 10). The screw-end of a Kirschner-wire (K-wire) pin with a diameter of 1mm was inserted in the tibia at a position 20mm from the distal end, and the pin was pushed cyclically in the lateromedial direction through a load cell that was attached on the top of the driver rod. Because the tibiofemoral joint is free to move, the test tibia makes a cantilever (lever arm length = 20mm) fixed with the fetter and loaded with the K-wire pin. The load wave *P* drawn in Fig. 11 is expressed by the following formula:

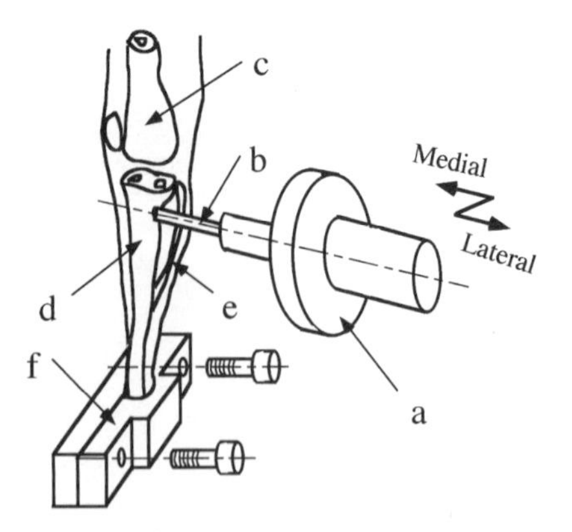

FIG. 10. Schematic of in vivo bone fatigue test. A Kirschner wire (K-wire) pin inserted in the tibia was pushed cyclically in the lateromedial direction with a load cell attached on the top of the rod. The test tibia forms a cantilever fixed by the fetter and loaded with the K-wire pin. Bending moment distributed on the tibial bone was calculated as the product of the loading amplitude, P_0, and the lever arm length from the top surface of the fetter. *a*, Load cell; *b*, K-wire; *c*, femur; *d*, tibia; *e*, fibula; *f*, fetter

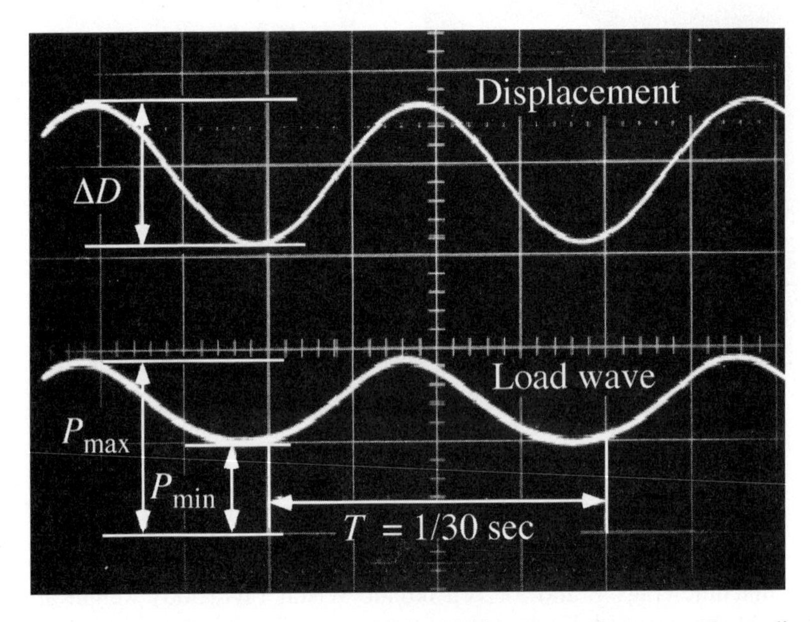

FIG. 11. An applied waveform monitored by a synchroscope. *Upper* waveform, displacement of the rod driven with the function generator and the controller; *lower* waveform, reaction force waveform generated on the tibia. The amplitude of the load was adjusted to make the rat fixed on the restrainer move with adjustments of the load

$$P = P_0 \cdot \{a + \sin(\omega t)\} \tag{15}$$

where P_0 is the amplitude of the load and ω is the angular velocity. The frequency of the load was kept constant at $f = 30$ Hz in this test series. The fatigue test was carried out under this loading conditions until fracture occurred in the tibia.

3.3.2 Measurement of Remodeling Activity of Cortical Bone

Three Wistar rats were used to measure bone remodeling activity. The cyclic load $P_{max} = 1.5$ kgf was applied to the tibia of the rats for 10h in the same manner as discussed previously. After loading exercise, 0.6mg of tetracycline powder dissolved in 0.1 ml of normal saline solution was injected into the abdominal cavity of the rats. One week later, the same volume of tetracycline was injected again. Several days after the last administration the rats were killed, and the right and left tibiae were removed from the body. These bones were defatted for 1 week in 95% ethanol, embedded in methyl methacrylate resin, and sliced at a thickness of 50 μm. After being polished, hard tissue specimens of bone were prepared from these slices and the bone volume that appeared as new or reformed tissue was measured by a fluorescent microscope. Thus, the remodeling activity of bone was measured quantitatively. The remodeled bone volume is determined from the product of the doubly labeled surface and the mean distance between the fringes. Thus, the remodeling bone volume ($V_f/V_0/t_0$) was obtained from each of the tetracycline-labeled bone specimens.

3.3.3 Measurement of Remodeling Activity of Cancellous Bone

To measure the remodeling activity change of cancellous bone by mechanical stimulation, a cyclic load ($P_{max} = 1.5$ kgf, $f = 30$ Hz) was applied to the proximal end of the tibia of living rats by a indenter attached to the loading device shown in Fig. 12. A total of nine rats were used in this test. Loading stimulation was applied to the rats for 4h. The same amount of tetraycline as used previously was injected for the first 7 days after the loading stimulation and the last 7 days before the rat was killed. After the loading test, each rat was fed with normal food for a total of 24 days. The slice specimens were prepared in the same manner as mentioned previously. The new bone appositional rate, m_f, of endosteal bone formation was determined as the mean distance between the tetracycline-labeled fringes. The surface-calcified surface area, S_f/S_0, and the amount of bone metabolic changes per day, $(V_f/V_0)/t_0$, in the trabecular bone were measured for each specimen.

3.4 Results

3.4.1 Fatigue Strength of Tibial Bone

All fractures occurred at the same position, 1mm from the upper surface of the fetter. Figure 13 shows the results of the fatigue test, that is, the relation between the applied load and the number of cycles in which the tibia was fractured. All

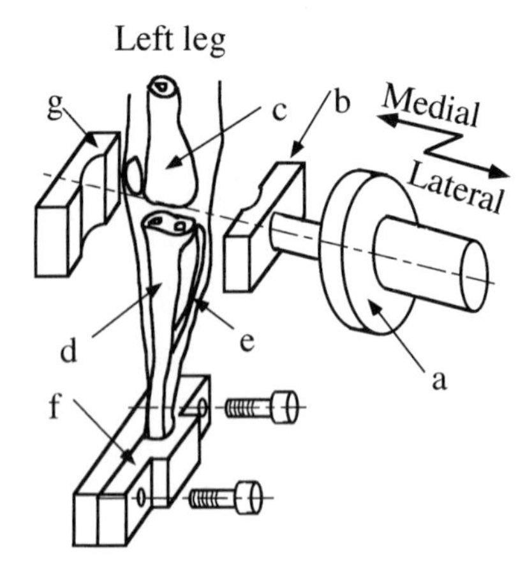

FIG. 12. Schematic of device to apply mechanical stimulation on in vivo cancellous bone. A cyclic load was applied to the proximal end of the rat tibia by an indenter attached on the top of the driver rod. *a*, Load cell; *b*, indenter; *c*, femur; *d*, tibia; *e*, fibula; *f*, fetter; *g*, guide plate

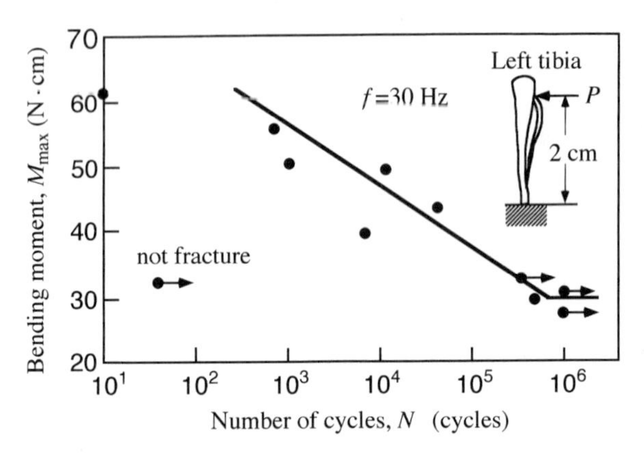

FIG. 13. Fatigue strength properties of rat tibial bone at $f = 30$ Hz. This loading frequency can be regarded as being high in comparison with the physiological condition, and the effect of the self-healing activity was negligible on the fatigue strength throughout the test period

results were obtained at $f = 30$ Hz. The vertical axis is the maximum bending moment, M_{max}, generated by the applied load, and the horizontal axis is the number of cycles.

3.4.2 Tetracycline-Labeled Bone Images

Figure 14 shows microscopic photographs of bone specimens labeled with tetracycline. The image on the left is a cut section of a tibial bone to which a cyclic load

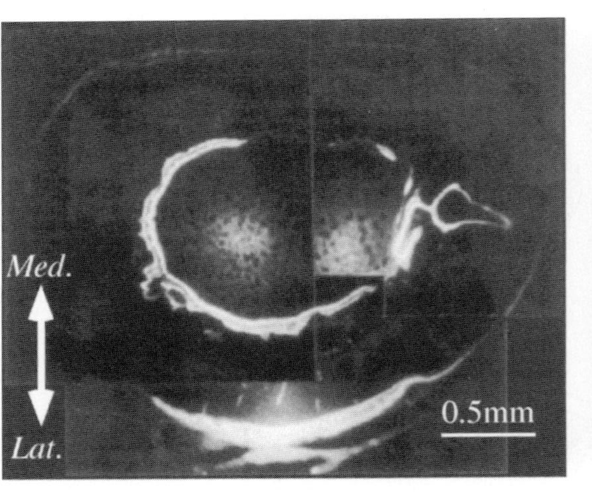

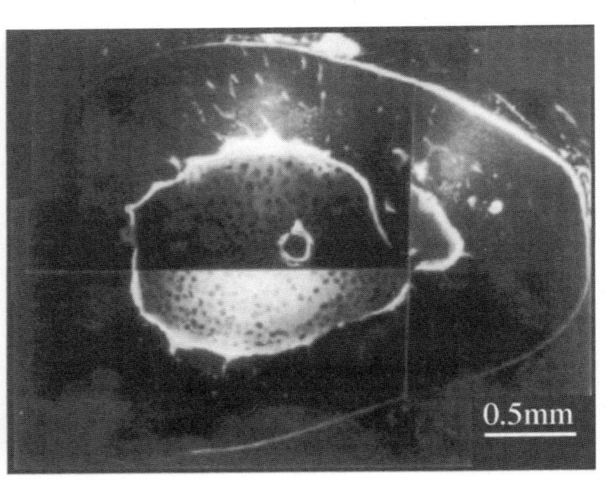

Fig. 14. Tetracycline-labeled images of tibial bones. *Left* and *right* images show the experimental and the control sides, respectively. A doubly labeled image was clearly observed on the intermediary side of the bone and a callus formed on the periosteal side, showing an active self-healing process caused by the mechanical damage generated in the bone tissue. *Arrow*, direction of loading

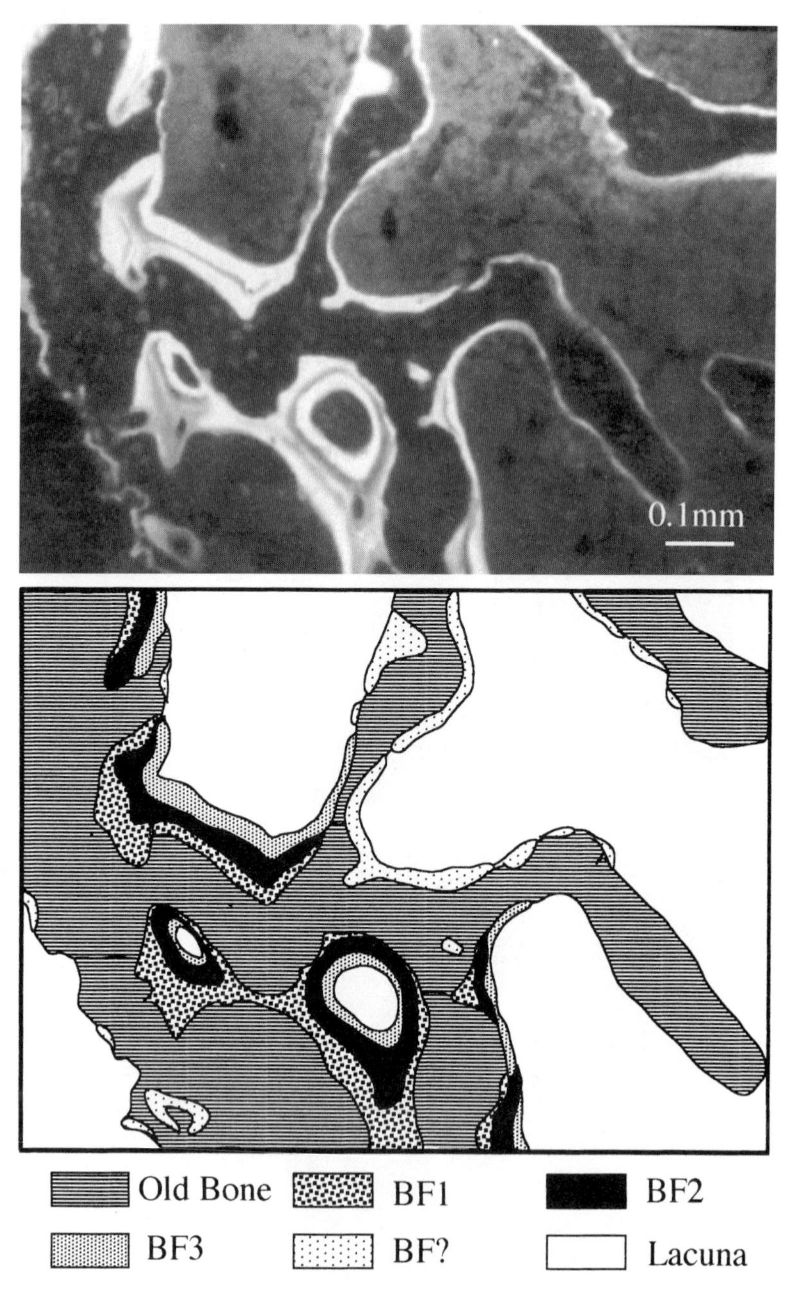

Fig. 15a,b. Tetracycline-labeled images of cancellous bone. *Top*, photomicrograph; *bottom*, explanatory diagram show the new bones formed within the first, the second, and the third 7 days, respectively, after the loading stimulation. *BF?* shows the new bone for which the formation time was unknown. **a** Control site (right tibia); no loading stimulation. **b** Experimental site (left tibia)

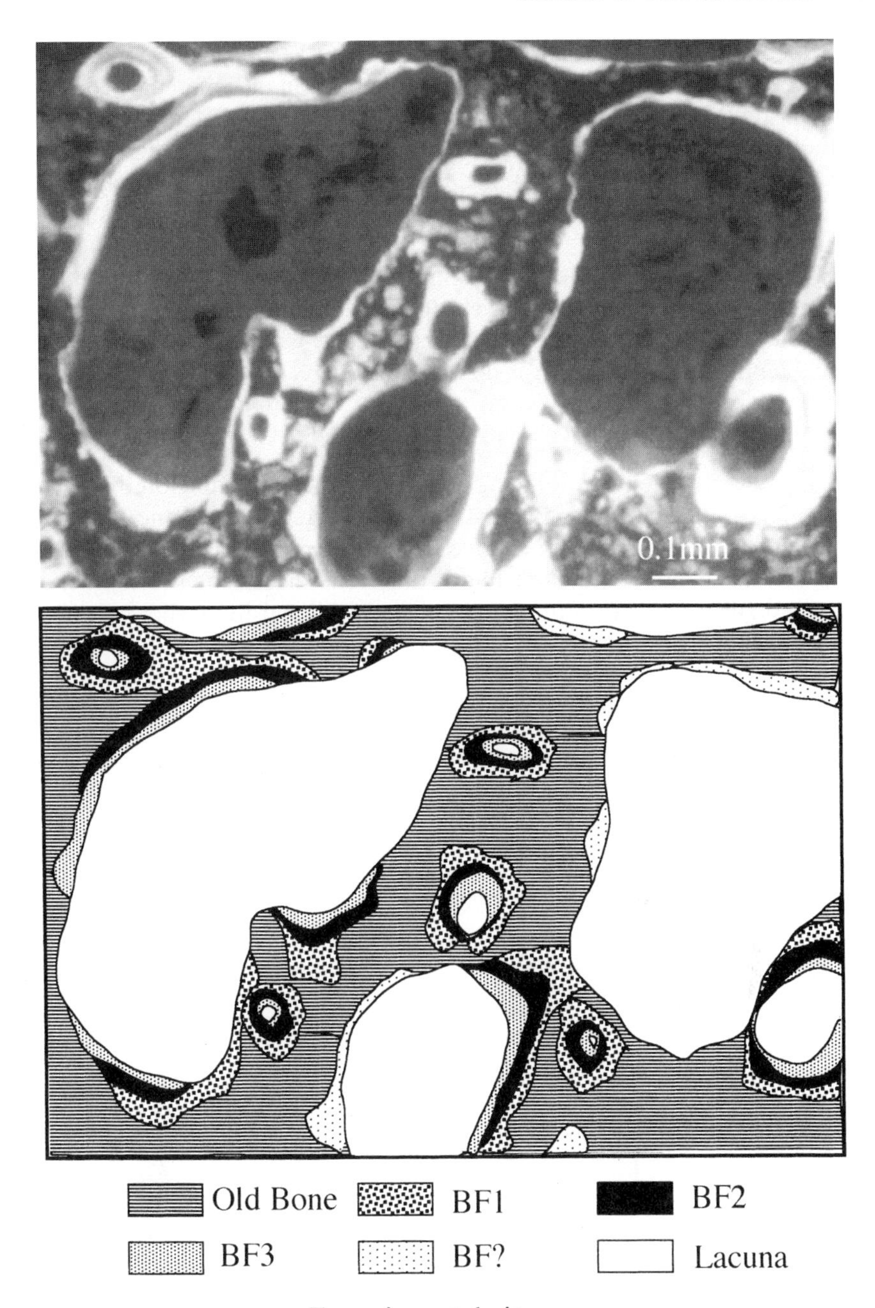

FIG. 15. *Continued*

was applied, and the image on the right is a bone section of the control side. These two specimens are taken from the same portion of bone, marked as A in Fig. 16, in the left and right tibia. No tetracycline doubly labeled portion can be found on the control side. In contrast, a vigorous tetracycline-labeled image is observed in the left experimental bone. Furthermore, a callus was shown to have formed on the outside of the cortex. The arrow marked in this figure shows the direction of loading.

Figure 15 shows tetracycline-labeled images of cancellous bone collected from the proximal ends of tibia after the cyclic loading stimulation. In this photograph, the left and right sides are the cut sections of bone with loading (experimental side) and without loading (control side), respectively.

3.4.3 Bone Remodeling Activity Change

Figure 16 shows the bone remodeling activity change depending on the bending moment. Here, the self-healing activities are indicated as the bone remodeling rate, which is defined as the percentage of new bone area labeled with tetracycline in the cut sectional area per unit of time. The measuring positions (A–D) are shown on the right side of this figure; point *P* is the position where the K-wire pin was inserted.

3.4.4 Effect of Mechanical Stimulation on Bone Remodeling Activity

Figure 17 shows the remodeling activity changes as measured with the tetracycline bone labeling method. Because the stress distribution in each trabecular

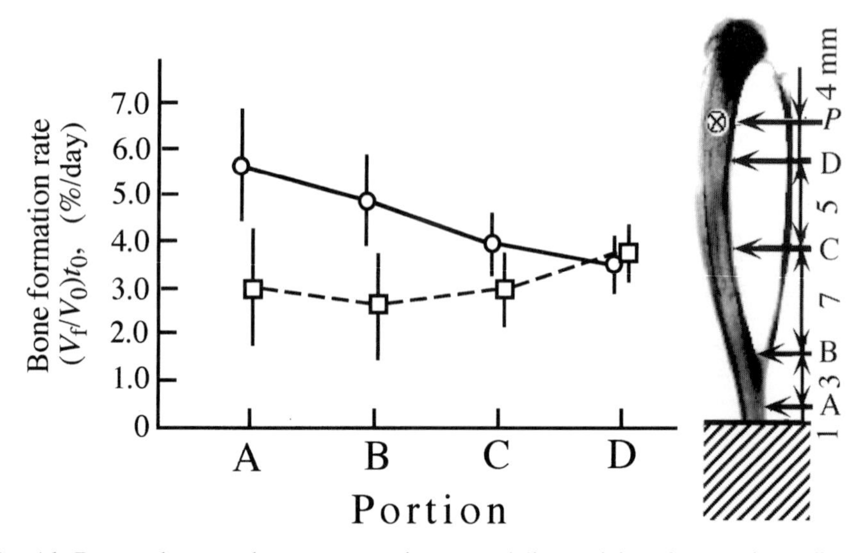

Fig. 16. Bone volume replacement rate, i.e., remodeling activity, changes depending on the amount of the mechanical stimulation, which was measured from the tetracycline-labeled images of tibial bone. *Circles*, experimental (stimulation); *squares*, control

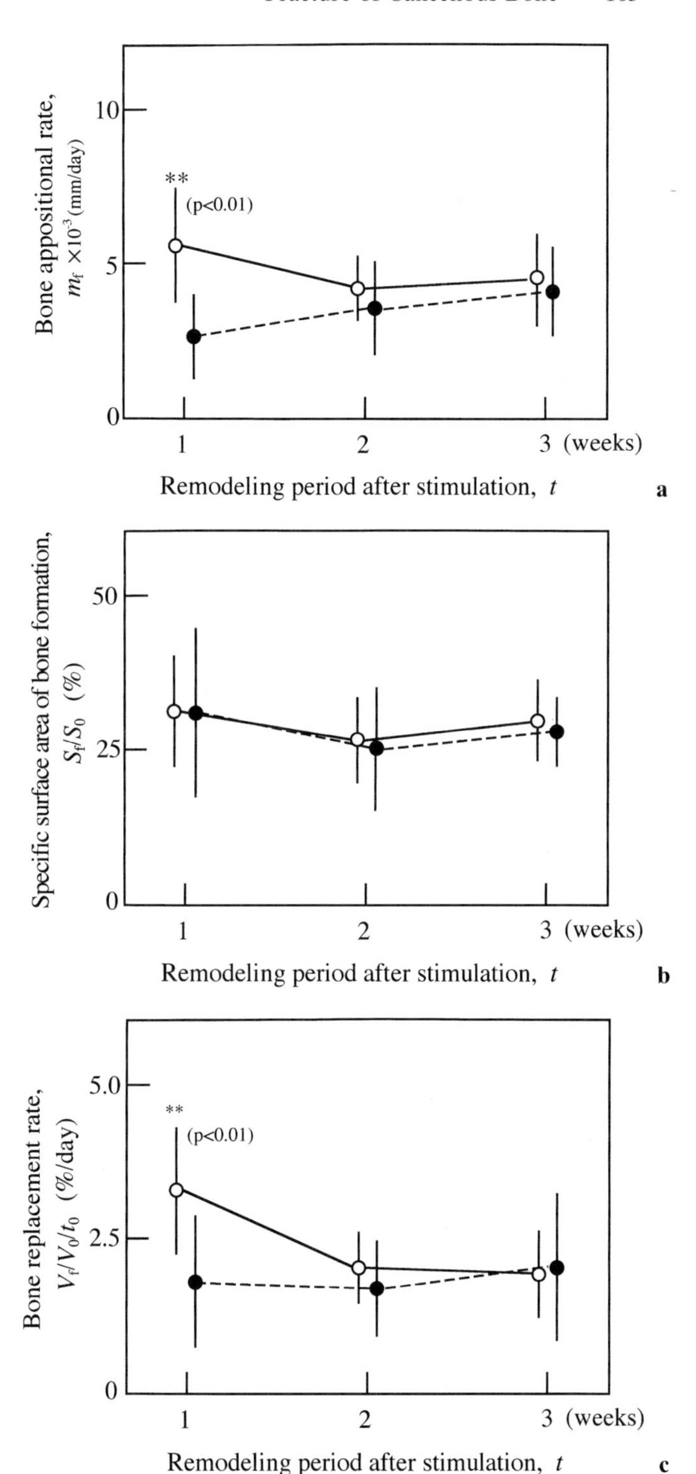

FIG. 17. **a** New bone apposition rate, m_f. **b** Specific surface area of bone apposition, S_f/S_0. **c** Amount of bone metabolic activity, $\{(V_f/V_0)/t_0\}$. These parameters change after loading stimulation for 4h. *Open circles*, left tibia (stimulation); *solid circles*, right tibia (control)

TABLE 3. Remodeling activity parameters in the first 7 days of experiment.

Bone	m_f	S_f/S_o	$(V_f/V_o)/t_o$
	10^{-3} mm/day	%	%/day
Control	3.02 ± 1.81	32.10 ± 13.65	1.70 ± 0.91
Experimental	5.72 ± 1.82	28.3 ± 8.61	3.38 ± 0.88

Total number of samples, $n = 9$. Data are mean ± SD. m_f, rate of bone formation obtained from the tetracycline double-labeled images; S_f/S_o, specific calcified surface area; $(V_f/V_o)/t_o$, remodeling rate of bone volume. The change per day, from tetracyline double-labeled images.

bone element was unknown in our loading stimulation system, the relation between the amount of mechanical stimulation and the change in bone remodeling activity was not clarified quantitatively. In this figure, it is found that the parameter of the bone appositional rate, m_f, was activated significantly for the first 7 days in all groups, and the specific calcified surface area, S_f/S_0, did not change markedly throughout the test period.

From these experimental findings, it is clear that mechanical stimulation caused by such cyclic loading improves the appositional rate of endosteal bone formation, and also that the state shown as high remodeling activity disappeared with 1 week after the mechanical stimulation. The parameters of remodeling activity in the first 7 days are listed in Table 3.

3.5 Discussion

3.5.1 Fatigue Properties of Bone

Several studies on bone fatigue fracture have been conducted [24–26]. Evans and Riolo [27] carried out a fatigue test of the cortical bone obtained from human cadavers. According to their study, the final number of cycles, $N = (2.35 \pm 2.03) \times 10^6$, was obtained at the load amplitude $P_{max} = 34.5$ MPa. Lafferty and Raju [20] carried out a fatigue test using bone specimens collected from a bovine femoral shaft. The number of cycles when the fatigue fracture occurred was 10^4 at a stress ratio of less than 0.2. Unfortunately, almost all these studies have been carried out using bone obtained from cadavers, not living bone.

The fatigue strength of living bone was measured formerly only by Seireg and Kempki [28] using the tibia of living rats. They carried out two kinds of fatigue tests, that is, high- and low-frequency tests. In the former short period test, they obtained a relation between the stress amplitude and the final number of loading cycles [29]. On the other hand, in the latter long period test, they studied the change of bone mineral content with applied load [30,31]. However, their interest was limited only to studying the quantitative changes in bone morphology, not those of bone histology depending on a mechanical stimulation, so that the relation between aging or other histological changes and the fatigue strength of bone was not clarified.

We conducted an experimental and an analytical study to clarify the influence of mechanical damage on the self-healing behavior of living bone. As shown in Fig. 13, the endurance limit might appear in $M_{max} = 3.0\,\mathrm{kgf \cdot cm}$ and the stress ratio might be evaluated as approximately 0.5. However, the appearance of such an endurance limit would have obliged us to stop the test, as it is impossible to restrain the rat for more than 10 h ($N = 10^6$ cycles). The actual endurance limit of fatigue may be decreased even more.

3.5.2 Influence of Mechanical Stimulation on Remodeling Activity

As shown in Fig. 16, a different magnitude of bending moment is generated in each position. In the control bone, no essential differences in the bone remodeling rate were noted with changes in the bone portion. In the experimental bone, however, it is clear that this remodeling rate changed depending on the bone portion, that is, the magnitude of the bending moment. Thus, it is obvious that the self-healing activities changed as a function of mechanical stimulation.

3.5.3 Fatigue Life of Living Bone

From Eq. 14, it can be easily recognized that the fatigue life of bone depends on the frequency of the cyclic load. Fatigue fracture ought not to occur if the frequency is less than a critical value. This critical value would be obtained as for the experimental conditions mentioned in Section 3.2. Similar to Miner's concept, let us assume that only a load greater than a specific magnitude can be accumulated in the bone to cause mechanical fatigue damage. The number of cycles of such loading is written as

$$n(t) = f \cdot t \tag{16}$$

We assume that the bone remodeling activity is unchanged during the loading term. The rate of bone remodeling is defined as the volume fraction of bone that is reformed during a time period unit.

Using Eqs. 14 and 15, the actual accumulation rate of fatigue damage is given as follows:

$$dD_t/dt = f/N - (V_f/V_0)/t_0 \tag{17}$$

Integrating Eq. 17 with time t, we get the following equation:

$$D_t = (f/N) \cdot T - (V_f/V_0/t_0) \cdot T \tag{18}$$

In conclusion, the fatigue life, T, is expressed by the following equation:

$$T = 1/\{(f/N) - (V_f/V_0/t_0)\} \tag{19}$$

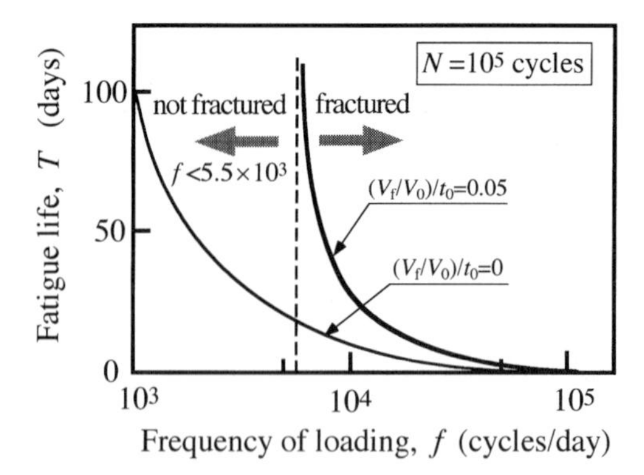

FIG. 18. Relations between the fatigue life, T, and the frequency of cyclic load, f, as calculated by Eq. 19 at the total number of loading ($N = 10^5$) cycles. Under this loading condition, fatigue life is determined at 100 days, if $(V_f/V_0)/t_0$ is negligible. If $(V_f/V_0)/t_0 > 0$, bone fracture occurs only when the loading frequency f is greater than $(V_f/V_0)\cdot t_0/N$. *Dotted line*, limit of fatigue fracture occurrence at $(V_f/V_0)/t_0 = 0.05$

From Eq. 19, it is found that bone fracture occurs only when the frequency of load, f, is greater than $(V_f/V_0)\cdot t_0/N$. The relation between the frequency of load and the fatigue life, T is shown in Fig. 18 for an example in which the total number of cycles is $N = 10^5$ until the fatigue fracture. The fatigue life in cases in which bone reformation does not occur is determined by the following equation:

$$T = N/f \tag{20}$$

On the other hand, in cases in which the loading frequency f is less than 5.5×10^3 cycles/day and the bone remodeling rate $(V_f/V_0)/t_0$ is more than 5.0×10^{-2} (day^{-1}), fatigue fracture does not occur. To avoid fatigue fracture, in practical daily life situation, the bone must not be exposed to a cyclic loading that has a frequency higher than such a safety value.

References

1. Avioli LV (1977) Osteoporosis: metabolic bone disease. Academic Press, New York
2. Rice JC, Cowin SC, Bowman JA (1988) On the dependency of the elasticity and strength of cancellous bone on apparent density. J Biomech 21:155–168
3. Richardson SP, Winck A, Black J (1986) Mechanical behavior of osteoporotic bone structural approach. Trans Orthop Res Soc 11:462
4. Frankel VH (1960) The femoral neck: function, fracture mechanism, internal fixation. Thomas, Springfield, pp 105–125
5. Todd RC, Freeman MAR, Pirie CJ (1972) Isolated trabecular fatigue fracture in the femoral head. J Bone Joint Surg 54B:723–728
6. Recker RR (1993) Architecture and vertebral fracture. Calcif Tissue Int 53(suppl 1):S139–S142
7. Gibson LJ (1985) The mechanical behavior of cancellous bone. J Biomech 18:317–328

8. Milch RA, Rall DP, Toble JE (1957) Bone localization of the tetracycline. J Natl Cancer Inst 19:87–93
9. Frost HM (1966) Bone dynamics in metabolic bone disease. J Bone Joint Surg 48A:1192–1203
10. Morita M, Ebihara A, Itoman M, Sasada T (1994) Progression of osteoporosis in cancellous bone depending on trabecular structure. Ann Biomed Eng 22:532–539
11. Takahashi H, Frost HM (1966) A tetracycline-based comparision of the number of cortical bone forming sites in normal and diabetic persons. J Jpn Orthop Assoc 39:1115–1121
12. Kock JC (1917) The laws of bone architecture. Am J Anat 21:177
13. Singh M, Nagrath AR, Maini PS (1970) Changes in trabecular pattern of upper end of the femur as an index of osteoporosis. J Bone Joint Surg 52A:457–467
14. Cowin SC, Hegedus DH (1976) Bone remodeling. I. Theory of adaptive elasticity. J Elasticity 6:313–326
15. Hart RT, Davy DT, Heiple KG (1984) Mathematical modeling and numerical solutions for functionally dependent bone remodeling. Calcif Tissue Int 36:S104–S109
16. Turner CH (1992) On Wolff's law of trabecular architecture. J Biomech 25:1–9
17. Carter DR, Fyhrie DP, Whalen RT (1987) Trabecular bone density and loading history: regulation of connective tissue biology by mechanical energy. J Biomech 20:785–794
18. Huiskes R (1992) Adaptive bone remodeling analysis. Chir Org Mov 77(2):121–133
19. Morita M, Sasamoto N, Yamamoto M, Sasada T (1983) Mechanical properties of cancellous bone and its trabecular structure. Proc 26th Jpn Cong Mater Res 233–240
20. Lafferty JF, Raju PVV (1979) The influence of stress frequency on the fatigue strength of cortical bone. Trans ASME 101:112–113
21. Miner MA (1945) Cumulative damage in fatigue. J Appl Mech 12:159–164
22. Nash CD Jr (1966) Fatigue of self-healing structure: a generalized theory of fatigue failure. Am Soc Mech Eng 66WA/BHF-3:1–4
23. Carter DR, Hayes WC (1977) Compact bone fatigue damage. I. Residual strength and stiffness. J Biomech 10:325–337
24. Snyder BD, Piazza S, Edwards WT, Hayes WC (1993) Role of trabecular morphology in the etiology of age-related vertebral fractures. Calcif Tissue Int 53(suppl 1):S14–S22
25. McBroom RJ, Hayes WC, Edwards WT, Goldberg RP, White AA (1985) Prediction of vertebral body compressive fracture using quantitative computed tomography. J Bone Joint Surg 67A:1206–1214
26. Griffiths WEG, Swanson SAV, Freeman MAR (1971) Experimental fatigue fracture of the human cadaveric femoral neck. J Bone Joint Surg 53B:136–143
27. Evans FG, Riolo ML (1970) Relations between the fatigue life and histology of adult human cortical bone. J Bone Joint Surg 52A:1579–1886
28. Seireg A, Kempki W (1969) Behavior of in vivo bone under cyclic loading. J Biomech 2(4):455–461
29. Lanyon LE, Rubin CT (1984) Static vs dynamic loads as influence on bone remodeling. J Biomech 7:897–905
30. Swanson SAV, Freeman MAR, Day WH (1971) The fatigue properties of human cortical bone. Med Biol Eng 9:23–32
31. Prendergast JP, Tayler D (1994) Prediction of bone adaptation using damage accumulation. J Biomech 27(8):1067–1076

Residual Stress in Bone Structure: Experimental Observation and Model Study with Uniform Stress Hypothesis

M. Tanaka[1] and T. Adachi[2]

Summary. In soft tissues, residual stress/strain have been observed experimentally and their role has been investigated from the point of view of stress/strain regulation or mechanical remodeling. In this chapter, the residual stress of bone structure is examined for the leporine tibiofibula bone and the bovine coccygeal vertebra. The fibula-cutting experiment for the leporine tibiofibula demonstrated that the axial strain induced in the tibia is distributed in the circumferential direction, suggesting that the residual stress is caused by statical indeterminacy. Numerical fibula cutting was studied for the remodeling equilibrium, which was simulated by the remodeling model considering residual stress and in which stress distribution is uniform under the working load. It was found that the strain induced by the real fibula cutting experiment coincides with that by the numerical experiments. For the bovine coccygeal vertebra, the end-plates and cancellous bone were removed sequentially, and positive strains were observed in the cephalocaudal and circumferential directions. The strain induced in the cephalocaudal direction was examined for the uniform stress state of the model of two-bar elements of the cortical and cancellous bone under working load, and the strain induced in the circumferential direction was for that of the model of two-cylinder elements as well. From these considerations, it was confirmed that the residual stress of bone structure suggested experimentally can be understood in the context of model studies based on the uniform stress hypothesis at remodeling equilibrium.

Key words: Bone mechanics—Residual stress—Stress release experiment—Uniform stress hypothesis—Remodeling equilibrium

[1] Department of Mechanical Engineering, Faculty of Engineering Science, Osaka University, 1-3 Machikaneyama-cho, Toyonaka, Osaka, 560 Japan
[2] Department of Mechanical Engineering, Faculty of Engineering, Kobe University, Nada-ku, Kobe, 657 Japan

1 Introduction

Living tissue has the capability to alter its structure and its mechanical properties, which is accomplished by remodeling as the result of tissue growth and atrophy. These processes are modulated in part by mechanical quantities in the tissue such as stress and strain. The normal steady state of the tissue maintained by remodeling is understood to be adapted to its mechanical environment, and some optimality is expected in the tissue structure and properties. In fact, many experimental observations support the influence of mechanical quantities on the tissue structure at remodeling equilibrium. Typical mathematical models for remodeling assume the existence of the optimal stress or strain in the tissue that characterizes the normal remodeling equilibrium with uniform distribution of stress or strain.

The fundamental importance of residual stress was pointed out from the standpoint of stress regulation in the tissue by Fung [1], and residual stress has been investigated experimentally in, for example, the arterial wall [2–5], left ventricle [2,6], and the trachea [7]. As a result of the mechanical consideration, it is suggested that the residual stress plays a substantial role in relaxing the stress concentration and maintains the living tissue at some optimal condition in the body during normal function [1,8,9]. Here again, uniformity of the stress or strain is considered to be an expression of mechanical optimality in the living tissue.

Structural statical indeterminacy of the tissue and nonuniformity in the natural state are inevitable in living tissue because remodeling is, at least in part, a local phenomenon and the local natural state is dependent on local growth under a certain mechanically loaded state. Therefore, residual stress, which remains in the tissue at the unloaded state, is a natural condition in living tissue. Mathematical models of the tissue remodeling process allowing for residual stress have been proposed by Rodriguez et al. [10] for soft elastic tissue. These investigations on residual stress have focused on mainly soft tissue, and little attention has been paid to residual stress in hard tissue, with the exception of wood material [11].

The authors are interested in residual stress in bone structure and the mechanical remodeling model considering residual stress [12,13]. This chapter describes residual stress in normal bone structure. Residual stresses in the leporine tibiofibula bone and in the bovine coccygeal vertebra were observed by the tissue-cutting method combined with strain gauge application. Experimental observations were considered in the context of uniform stress distribution in bone structure. Simple mechanical models were provided for the tibiofibula bone and the vertebral body, and the strains observed in the cutting experiments were studied by paying attention to the statical indeterminacy of the bone structure and the nonuniform natural state.

2 Residual Stress in the Leporine Tibiofibula

Residual stress in leporine tibiofibula bone, that is, a statically indeterminate structure of two bones, was examined by cutting the fibula.

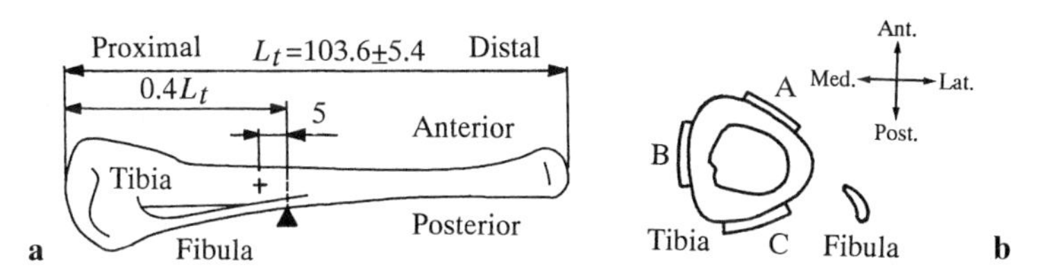

FIG. 1a,b. Leporine tibiofibula bone. **a** Gauge site (*cross*) and fibula cutting position (*triangle*). **b** Cross section and gauge sites (*A*, *B*, *C*)

2.1 Specimens and Measurement Procedure

Fresh tibiofibula bones were excised from 12 healthy Japanese white rabbits weighing 2.5 ± 0.1 kg (average ± SD). The leporine tibia and fibula are interconnected in a statically indeterminate manner (Fig. 1a); that is, the fibula branches from the tibia at the middiaphysis and meets the tibia at the tibial proximal end of the knee joint. The tibia and fibula were cleaned of muscle attachment and periosteum and kept in physiological saline bath at room temperature (21°C) for at least 5 h before the experiment. The diaphyseal surface of the tibia was scraped clean and swabbed dry, and then three uniaxial waterproof strain gauges (KFWS-2N, Kyowa Electronic Instruments, Tokyo, Japan) were bonded on the tibial surface along the longitudinal axis using cyanoacrylate adhesive. The crossed-hair mark in Fig. 1a indicates the position of the center of the gauge grid in the proximodistal direction. The positions of the three gauges are distinguished in the tibial cross section as the gauge A, anterolateral site; B, medial site; and C, posterolateral site, respectively (Fig. 1b).

The specimen was immediately returned to the physiological saline bath for at least 2 h to obtain the resting state, which was used as the reference state of the gauge system. All the measurements were performed in this bath. A dummy gauge was also placed in the same saline bath to compensate for the drift from change in temperature. The resolution of the measuring instrument (UCAM70A, Kyowa Electronic Instruments) was 1 μ-strain, and the value of the zero drift was satisfactorily small compared with the measured strains. The statical indeterminacy of the tibiofibula bone is partially released by carefully cutting the fibula with a handsaw at the position marked by the solid triangle in Fig. 1a. After cutting, the specimen is immediately returned to the saline bath and kept at rest until the strain level becomes steady, and then strains induced at the gauge sites are measured [12].

2.2 Results

Averaged values of the strain change of 12 specimens along the longitudinal direction measured using gauges A, B, and C are shown in Table 1. Positive strains were recorded by gauges A at the anterolateral site and C at the posterolateral site, and a negative strain was recorded by gauge B at the medial site. Observed strain scattered because of individual variation; however, strain at the right tibia correlated significantly with that at the left tibia of the same individual.

TABLE 1. Strain observed by cutting fibula (μ-strain).

Gauge site	Right ($n = 12$)	Left ($n = 12$)	All ($n = 24$)
A	15.9 ± 8.2	11.6 ± 6.4	13.9 ± 7.7
B	−21.8 ± 7.0	−18.9 ± 9.4	−20.3 ± 8.4
C	19.3 ± 8.5	16.1 ± 9.0	17.8 ± 8.9

Because statical indeterminacy between the tibia and the fibula was removed by cutting the fibula, the strain induced corresponds to the release of the residual stresses in the statically indeterminate tibiofibula bone. The result implies that the lateral side of the tibia is compressed and the medial side is stretched, and the fibula is stretched, bending the tibia toward the lateral direction. The magnitude of the residual stress released by cutting the fibula is approximately in the range from 0.3 MPa to 0.5 MPa, as Young's modulus in the longitudinal direction is 24.5 ± 7.3 GPa, which was obtained by the compression test of tibial bone pieces with a single strain gauge. This value of released residual stress is comparable to the representative stress, ≃0.8 MPa, calculated by the resultant cross-sectional areas of left and right tibias and body weight.

3 Cylinder Model of the Leporine Tibiofibula Bone

Residual stress observed in the tibiofibula bone was investigated in terms of the remodeling equilibrium simulated by using the model of bone remodeling considering residual stress based on the uniform stress hypothesis. The numerical fibula-cutting experiment was compared with the real observation reported in the previous section.

3.1 Tibiofibula Model and Remodeling Rate Equation

The diaphyses of the tibia and fibula were simplified as hollow and solid cylinders, rigidly connected at both ends making a statically indeterminate structure. The remodeling equilibrium was examined by simulation based on the model of bone remodeling considering residual stress [13,14]. At the initial of the simulation, the tibia has a uniform wall thickness of $W = 1.0$ mm and a diameter of $D_t = 5.0$ mm at the wall center, and the fibula has the diameter of $D_f = 1.0$ mm with the distance of $d = 6.0$ mm from the center of the tibia (Fig. 2a). This initial of the state has a uniform stress distribution under the compression of load $P = -15$ N acting at the resultant geometric center of the cross sections of tibia and fibula, which is 1 mm from the center of the tibia toward the fibula. Neither tibia or fibula has initial strain, which represents the nonuniformity of the natural state in the model of remodeling, and Young's modulus is 15 GPa. When attention is focused on the tibia and the fibula is assumed to be of conventional elasticity, the remodeling rate equation provided for the tibial wall thickness $W(\theta,t)$ is

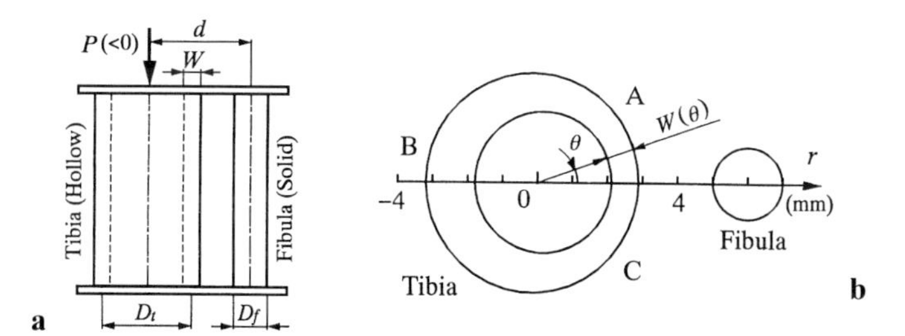

FIG. 2a,b. Cylinder model of tibiofibula bone. **a** Tibiofibula model. W, Wall thickness; D_t, representative tibia diameter; D_f, fibula diameter. **b** Cross section at remodeling equilibrium

$$\frac{1}{W}\frac{\partial W}{\partial t} = C_t \frac{|\sigma| - |\sigma^m|}{|\sigma^m|} - C_s \frac{\partial^2 |\sigma|}{\partial \theta^2} \tag{1}$$

and the rate equation of the initial strain $\varepsilon^0(\theta,t)$ accompanies as follows:

$$\frac{\partial \varepsilon^0}{\partial t} = \frac{1}{W}\frac{\partial W}{\partial t}\left(\varepsilon - \varepsilon^0\right) \tag{2}$$

where θ represents the circumferential direction of the tibia, σ and σ^m denote the instantaneous stress and the representative of the past stress history, and C_t and C_s are the remodeling rate constants in time and circumferential axes. The first term of the right-hand side of Eq. 1 expresses the stress nonuniformity in time and the second term that in space. The detail of the remodeling model is found in the literature [13] [15].

3.2 Remodeling Equilibrium and the Fibula-Cutting Experiment

When the position of the external load of P is shifted to the center of the tibia at initiation of simulation, the tibial stress distributes along the circumferential direction of θ. As the result of remodeling induced by the nonuniformity of stress distribution, the tibial wall thickness is altered (Fig. 2b), and the stress distribution becomes uniform again in the tibia. As the counterpart, the initial strain is not uniform along the circumferential direction of θ in the tibia, and the residual stress σ_r remains even when the external load P is removed and distributed along the circumferential direction (Table 2). At the same time, the tensile residual stress of 0.15 MPa remains in the fibula.

Cutting the fibula was carried out numerically for the remodeling equilibrium obtained by the remodeling simulation, and the result was compared with the strain change observed in the tibia by cutting of the real fibula reported in the previous section. The external load was numerically removed from the remodel-

TABLE 2. Residual stress at remodeling equilibrium and strain induced by cutting fibula numerically.

Position (θ/π)	Residual stress (MPa)	Induced strain (μ-strain)	Gauge site
0	−0.198	16.32	
1/3	−0.124	10.47	A
2/3	0.025	−0.11	
1	0.099	−7.00	B
4/3	0.025	−0.11	
5/3	−0.124	10.47	C
2	−0.198	16.32	

ing equilibrium and used as the reference for the strain change in the fibula-cutting study. When the fibula of Fig. 2b was cut numerically, the residual stress of the tibia was released, and a positive strain was induced at the lateral site of $\theta = 0$ and a negative strain at the medial site of $\theta = \pi$ (see Table 2); the gauge sites A, B, and C correspond to the real gauge sites in Table 1. Comparing the results in Tables 1 and 2, we confirmed that the strains induced at the tibia by the real experiment show reasonable agreement with those achieved by the numerical one for remodeling equilibrium based on the uniform stress hypothesis. That is, it is expected that the residual stress in the tibiofibula bone is effective to determine the uniform stress distribution in the bone structure.

4 Residual Stress in the Bovine Coccygeal Vertebra

The vertebral body consists of cortical bone outside and cancellous bone inside, and its bone structure has statical indeterminacy. In this section, the residual stress was again examined by using bovine coccygeal vertebra in a cutting experiment.

4.1 Specimen and Measurement Procedure

The most cranial vertebrae were excised from the fresh tails of 2-year-old Holsteins, purchased from the meat market immediately after slaughter. Each coccygeal vertebra specimen was cleaned of all soft tissues such as muscle, ligament, disk, and periosteum, and kept in the physiological saline bath at a constant temperature of 20°C for at least 4h before performing the following experimental procedure. The specimen was swabbed dry, and two biaxial gauges were bonded on the cortical surface at symmetrical positions with respect to the sagittal plane using cyanoacrylate adhesive. The gauge sites were between the spinous process and the transverse processes on the midtransverse plane in the cephalocaudal direction, as is indicated with cross-hair marks in Fig. 3a. The principal axes of the rectangular biaxial strain gauge (SKF-20250, Kyowa Electronic Instruments) were arranged in the cephalocaudal and circumferential directions. The measuring instruments were the same as those used for the leporine tibiofibula experiment (Section 2.1).

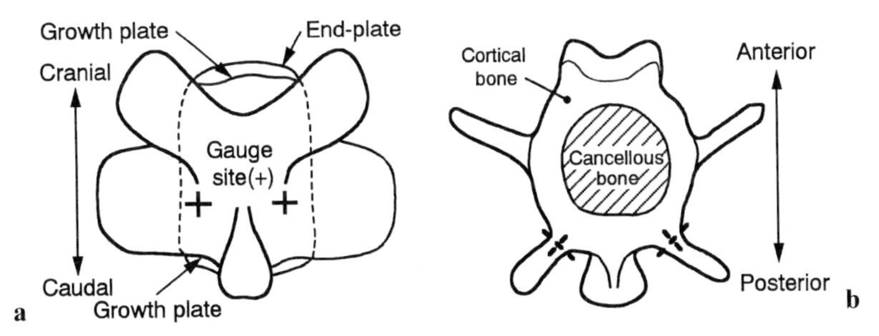

FIG. 3. Bovine coccygeal vertebra body. **a** Gauge sites and growth plates (posterior view). **b** Cranial view after cutting end-plates

TABLE 3. Strain induced at cortical surface of bovine vertebra ($n = 7$).

Cutting operation	Cephalocaudal direction $\Delta\varepsilon_z^{exp}$ (μ-strain)	Circumferential direction $\Delta\varepsilon_\theta^{exp}$ (μ-strain)
End-plates: $(\ )_1$	33.3 ± 10.9	22.7 ± 10.0
Cancellous bone: $(\ )_2$	53.0 ± 21.5	96.9 ± 24.3
Resultant	86.3 ± 25.8	119.6 ± 25.3

The specimen with gauges was kept at rest for at least 2h in the saline bath to confirm that the output of the gauges had become steady. This was used as the reference state for the following measurements. As the first stage, both end-plates were carefully cut off using a handsaw at the cranial and caudal growth plates (Fig. 3a), and the strain induced at the gauges was measured at steady state in the saline bath 1h after the cutting operation. As the second stage, using a light-duty rotary cutter (AC-D7, Sunhayato, Tokyo, Japan), the cancellous bone inside was removed (Fig. 3b). The strain induced by removing the cancellous bone was again observed at the steady state in the saline bath almost 1h after the cutting operation. The output of the gauge was continuously recorded even during the operations, and this confirmed that the maximum strain instantaneously induced by the operation was not significant compared with the yielding strain reported in the literature. It is noted that the gauge sites used here are appropriate to observe the effect of the cutting operations, because the bovine coccygeal vertebra has no vertebral foramen [13].

4.2 Results

The strains induced at the gauges are summarized in Table 3 for the individual operation and the resultant. In the cephalocaudal direction, the positive (tensile) strain of $\Delta\varepsilon_{z1}^{exp} = 33.3$ μ-strain on average was induced by cutting the end-plates and the positive strain of $\Delta\varepsilon_{z2}^{exp} = 53.0$ μ-strain on average by removing the cancellous bone inside. The strains induced by these two operations are comparable in magnitude, and the resultant is the tensile strain of $\Delta\varepsilon_z^{exp} = 86.3$ μ-strain on average. The deviation of the observed strains is not small among specimens, but the two gauges at the right and left sites showed acceptable correlation with each other for an individual specimen.

In the circumferential direction, a positive (tensile) strain of $\Delta\varepsilon_{\theta 1}^{exp} = 22.7$ μ-strain was induced on average by cutting the end-plates, and a positive strain of $\Delta\varepsilon_{\theta 2}^{exp} = 96.9$ μ-strain was observed on average by removing cancellous bone. In this case, the strain induced by cutting the end-plates is small, and that induced by removing cancellous bone is dominant. The resultant, $\Delta\varepsilon_{\theta}^{exp} = 119.6$ μ-strain on average, was comparable to or slightly larger than that in the cephalocaudal direction.

The positive strains observed in the cephalocaudal direction by these two cutting operations imply that compressive stress is present in the cortical bone on the outside, and tensile stress is present in the cancellous bone inside, as the self-equilibrium stresses caused by statical indeterminacy in the cephalocaudal direction. The positive strains in the circumferential direction mean that the cortical bone outside is compressed in the circumferential direction because of tensile radial stress at the boundary surface with the cancellous bone inside, which is tensioned in the opposite sense. Both observations suggest to us that self-equilibrium stress in the corticocancellous bone structure of the vertebral body remains at the unloaded state. The magnitude of the observed strain is not negligible in comparison with several hundred microstrains reported as the threshold strain values of bone remodeling.

5 Two-Bar Model of the Bovine Coccygeal Vertebra

The behavior of the bovine coccygeal vertebra was modeled as a two-bar structure in the cephalocaudal direction, and the strain observed by the experiment in the previous section and the suggested residual stress were considered in the context of the uniformity of the effective stress in the cortical and cancellous elements.

5.1 Two-Bar Model

The cortical bone of the coccygeal vertebra was simplified as a hollow circular cylinder and the cancellous bone inside as a solid cylinder filling the hollow part of the cortical bone. They were connected to the rigid plate at both ends, making a statically indeterminate structure (Fig. 4a). When discussion is focused on the centric force, this cylindrical model of the vertebra reduces to a lumped parameter system of indeterminate two bar elements. The cancellous and cortical members were distinguished by indices $i = 1$ and 2, respectively, and the apparent Young's modulus and the cross-sectional area are denoted by E_i and A_i ($i = 1, 2$).

By assuming linear elastic behavior for both members, the equilibrium condition is written as

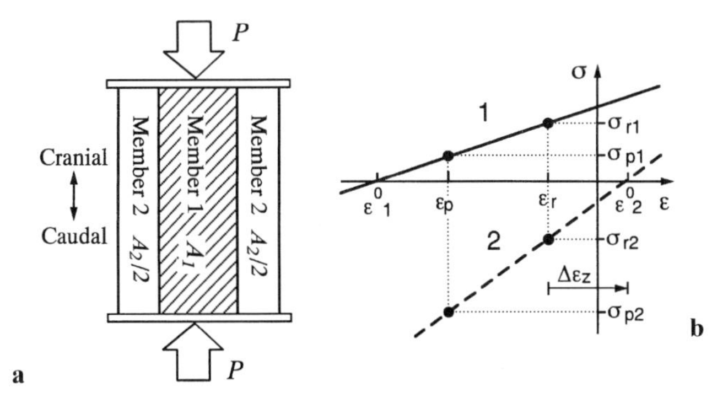

FIG. 4. Two-bar model of bovine coccygeal vertebra in cephalocaudal direction. **a** Two-bar model. A, Cross-sectional area; P, external load (*large arrows*). **b** Stress–strain relation expected at remodeling equilibrium. σ, Stress; ε, strain

$$\sigma_{p1}A_1 + \sigma_{p2}A_2 = P \tag{3}$$

under a certain external load of P, where σ_{pi} denotes the apparent stress of the individual element i. If the members have no initial stress nor strain in the unloaded state as the whole structure, the member strains ε_{p1} and ε_{p2} become identical to ε_p because of the statically indeterminate constraint. That is, the stress σ_{p1} is different from σ_{p2} in general, and the effective stresses σ_{pi}/A_{ri} (i = 1, 2) are also not the same, where A_{ri} denotes the area fraction of $0 \le A_{ri} \le 1$ and the product A_iA_{ri} gives the effective cross-sectional area of the member i.

The uniform stress state is considered as an expression of the structural optimality maintained by mechanical adaptation by remodeling, and the remodeling equilibrium state in this two-bar model is characterized as

$$\frac{|\sigma_{p1}|}{A_{r1}} = \frac{|\sigma_{p2}|}{A_{r2}} \tag{4}$$

At the same time, the constitutive relationship is written as

$$\sigma_i = E_i\left(\varepsilon - \varepsilon_i^0\right) \tag{5}$$

by using the initial strain ε_i^0, that is, the results of the volumetric change of the elements by remodeling [15,16], and this represents the nonuniformity of the natural state between the elements. Figure 4b illustrates the expected stress–strain state of the cancellous and cortical members at remodeling equilibrium of the uniform effective stresses. Thus, the residual stress σ_{ri}, which satisfies the equilibrium condition

$$\sigma_{r1}A_1 + \sigma_{r2}A_2 = 0 \tag{6}$$

and the residual strain ε_r remain at the unloaded state (see Fig. 4b).

5.2 Residual Stress and Working Load

As the result of cutting off the end-plates and the cancellous bone, the cortical bone was isolated in the experiment of the previous section. When the cortical element was isolated from the two-bar model of the vertebra, the cortical element returned to the zero stress state. That is, the deformation of the cortical element $i = 2$ is the difference between the residual strain of ε_r at the unloaded state as the structure and the initial strain of ε_2^0 at the zero stress state:

$$\Delta\varepsilon_z = \varepsilon_2^0 - \varepsilon_r \tag{7}$$

as shown in Fig. 4b. This is the stain induced by the cutting operation in the experiment in Section 4, and enables us to calculate the stresses σ_{pi} at which the remodeling equilibrium is expected with the uniform stress condition as

$$\sigma_{p1} = \frac{A_{r1}\left(E_1 A_1 + E_2 A_2\right)}{A_1\left(E_1 A_{r2} + E_2 A_{r1}\right)} E_2 \Delta\varepsilon_z \tag{8}$$

and

$$\sigma_{p2} = -\frac{A_{r2}\left(E_1 A_1 + E_2 A_2\right)}{A_1\left(E_1 A_{r2} + E_2 A_{r1}\right)} E_2 \Delta\varepsilon_z \tag{9}$$

Of course, it also enables us to calculate the residual stresses σ_{ri} as

$$\sigma_{r1} = \frac{A_2}{A_1} E_2 \Delta\varepsilon_z, \qquad \sigma_{r2} = -E_2 \Delta\varepsilon_z \tag{10}$$

Based on the observation on the transverse section of the vertebrae at the gauge site, the area fraction for the cancellous member is determined as $A_{r1} = 0.28$ as the average of ratios of the area occupied by trabecular bone in the inspection area. The cortical bone is assumed as the fully dense material compared with the cancellous bone, and the area fraction is set as $A_{r2} = 1.0$. By referring to a general approximation that Young's modulus is proportional to a power function of the apparent density [17], Young's modulus E_1 of the cancellous member is related to that of the cortical member as

$$E_1 = E_2 A_{r1}^3 \tag{11}$$

For the numerical study, Young's modulus of the cortical member is assumed as $E_2 = 20\,\text{GPa}$, the cross-sectional areas of the cancellous as $A_1 = 168.0\,\text{mm}^2$ and of the cortical as $A_2 = 139.5\,\text{mm}^2$. The experimental observation of the induced strain of $\Delta\varepsilon_z^{exp} = 86.3$ μ-strain and the area fraction $A_{r1} = 0.28$ are substituted into Eqs. 8–11; the working stresses at the remodeling equilibrium are obtained as $\sigma_{p1} = 1.36\,\text{MPa}$ for cancellous bone and $\sigma_{p2} = -4.87\,\text{MPa}$ for cortical. The magnitude of the effective stress is 4.87 MPa and the working load is $P = -450.6\,\text{N}$. With these working stresses, the residual stresses are calculated as $\sigma_{r1} = 1.43\,\mathrm{MPa}$ and $\sigma_{r2} = -1.73\,\mathrm{MPa}$. The working load calculated is moderate. This is acceptable considering that the tail is not responsible for supporting its own weight and that the working force is the resultant of the surrounding soft tissues such as muscles and ligaments.

6 Cylinder Model of the Bovine Coccygeal Vertebra

The cylinder model of the bovine coccygeal vertebra was used to investigate behavior in the circumferential direction, and the strains induced by the cutting operations and the suggested residual stress were considered in the context of the fitting phenomenon of axisymmetrical elements and the uniform distribution of the effective circumferential stress.

6.1 Cylinder Model

As is similar to the two-bar model in the previous section, the cortical bone of the coccygeal vertebra was simplified as a hollow circular cylinder with inner and outer diameters of $d_2 = 2r_2$ and $d_3 = 2r_3$, and the cancellous bone inside as a solid circular cylinder with the diameter of $d_1 = 2r_1$ (Fig. 5). In this case, however, the diameter of the cancellous cylinder inside is not identical to the inner diameter of the cortical cylinder outside. The statical indeterminacy is represented as the compatibility at the interface between the solid and hollow cylinders, and this problem falls in a class of the shrink-fit problem [18]. The origin of the cylindrical coordinates system (r, θ, z) is placed on the axis of vertebral model. Again, the cancellous and cortical elements are distinguished by indices $i = 1$ and 2, respectively, and Young's modulus and Poisson's ratio are denoted by E_i and ν_i for the cylinder element $i = 1,2$. The cross-sectional areas are $A_1 = \pi r_1^2$ and $A_2 = \pi(r_3^2 - r_2^2)$.

Consider that two members are fitted at the interface with the fitting pressure of p, under longitudinal external load P with longitudinal strain ε_z. The circumferential stresses $\sigma_{\theta 1}(r)$ and $\sigma_{\theta 2}(r)$ of the cancellous and cortical members are written as

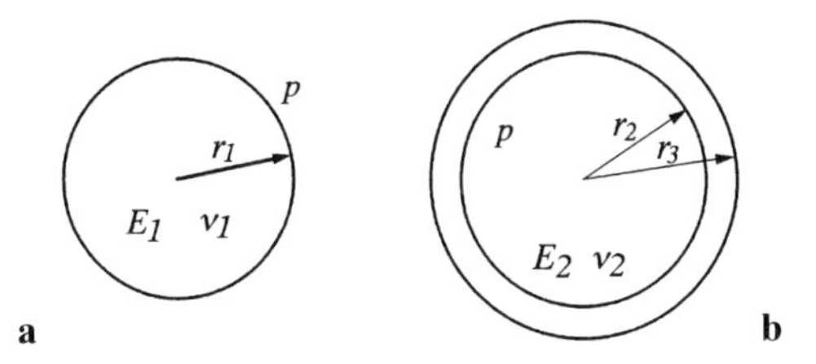

FIG. 5. Cylinder model of vertebra body. **a** Solid cylinder of inner cancellous bone. E, Young's modulus; r, radius; p, interface pressure; ν, Poisson's ratio. **b** Hollow cylinder of outer cortical bone ($r_1 \neq r_2$)

$$\sigma_{\theta 1} = p, \quad \sigma_{\theta 2}(r) = -\frac{r_2^2}{r_3^2 - r_2^2}\left(\frac{r_3^2}{r^2} + 1\right)p \tag{12}$$

and the longitudinal stresses σ_{z1} and σ_{z2} as

$$\sigma_{z1} = 2\nu_1 p + E_1\varepsilon_z, \quad \sigma_{z2} = -2\nu_2 \frac{r_2^2}{r_3^2 - r_2^2} p + E_2\varepsilon_z \tag{13}$$

The radial displacement $u_1(r)$ in the cancellous and $u_2(r)$ in the cortical bone are written as

$$u_1(r) = \left\{\frac{(1+\nu_1)(1-2\nu_1)}{E_1} p - \nu_1\varepsilon_z\right\} r \tag{14}$$

and

$$u_2(r) = -\frac{1+\nu_2}{E_2}\frac{r_2^2}{r_3^2 - r_2^2}\left\{(1-2\nu_2)r + \frac{r_3^2}{r}\right\}p - \nu_2\varepsilon_z r \tag{15}$$

respectively. The continuity of the radial displacement at the interface between the elements

$$u_1(r_1) + r_1 = u_2(r_2) + r_2 \tag{16}$$

gives us the interface pressure p as

$$p = \frac{E_1 E_2 r_1^2 (r_3^2 - r_2^2)\{(r_2 - r_1) - (\nu_2 r_2 - \nu_1 r_1)\varepsilon_z\}}{E_1(1+\nu_2)\{r_3^2 + (1-2\nu_2)r_2^2\}r_1^2 r_2 + E_2(1+\nu_1)(1-2\nu_1)r_1^3(r_3^2 - r_2^2)} \tag{17}$$

and the longitudinal external load P as

$$P = \sigma_{z1}A_1 + \sigma_{z2}A_2 = \pi\left\{E_1 r_1^2 + E_2(r_3^2 - r_2^2)\right\}\varepsilon_z$$
$$-\frac{2\pi E_1 E_2 r_1^2 (r_3^2 - r_2^2)(\nu_2 r_2^2 - \nu_1 r_1^2)\{(r_2 - r_1) - (\nu_2 r_2 - \nu_1 r_1)\varepsilon_z\}}{E_1(1+\nu_2)\{(1-2\nu_2)r_2^2 + r_3^2\}r_1^2 r_2 + E_2(1+\nu_1)(1-2\nu_1)(r_3^2 - r_2^2)r_1^3} \tag{18}$$

That is, the fitting pressure p and the longitudinal load P are linear functions of the longitudinal strain ε_z, and are abbreviated as $p = K_1 + K_2\varepsilon_z$ and $P = K_3\varepsilon_z + K_4$, where K_i ($i = 1, \ldots, 4$) are coefficients determined by E_i and ν_i ($i = 1, 2$), and r_i ($i = 1, 2, 3$).

6.2 *Uniform Stress State and Residual Stress*

Uniform stress distribution was again considered as the characteristic of mechanical optimality maintained through adaptation by remodeling. Because the circumferential stress is distributed in the radial direction, and the macroscopic uniformity of the effective stress $\sigma^e_{\theta i} = \sigma_{\theta i}/A_{ri}$ is written as

$$\frac{|\bar{\sigma}_{\theta 1}|}{A_{r1}} = \frac{|\bar{\sigma}_{\theta 2}|}{A_{r2}} \tag{19}$$

in terms of the representative stresses $\bar{\sigma}_{\theta 1}$ and $\bar{\sigma}_{\theta 2}$ of each member defined as

$$\bar{\sigma}_{\theta 1} = \frac{1}{r_1}\int_0^{r_1} \sigma_{\theta 1} dr = p, \quad \text{then} \quad \bar{\sigma}_{\theta 2} = \frac{1}{r_3 - r_2}\int_{r_2}^{r_3} \sigma_{\theta 2} dr = -\frac{r_2}{r_3 - r_2} p \tag{20}$$

Equation 20 reduces Eq. 19 to

$$A_{r2}\left(r_3 - r_2\right) = A_{r1} r_2 \tag{21}$$

which is independent of the interface pressure p and the longitudinal load P.

The circumferential strain induced at the outer surface of the isolated cortical cylinder is $\Delta\varepsilon_\theta = -u_2(r_3)/r_3$ in this model. Because the cutting experiment is initiated from the unloaded state as the structure of $P = 0$, Eq. 18 gives the longitudinal strain $\varepsilon_z = -K_4/K_3$. Therefore, the fitting pressure $p = K_1 - K_2K_4/K_3$ is calculated from Eq. 17, and the experimentally observed strain becomes

$$\Delta\varepsilon_\theta = -\frac{u_2(r_3)}{r_3} = \frac{2(K_2K_4 - K_1K_3)(1 - v_2^2)r_2^2 + E_2K_4v_2(r_3^2 - r_2^2)}{E_2K_3(r_3^2 - r_2^2)} \tag{22}$$

by using Eq. 15. The uniform stress condition of Eq. 21 and the experimental observation of Eq. 22 determine the outer radius r_1 of the cancellous element and the inner radius r_2 of the cortical element so as to satisfy the uniform stress condition of Eq. 21.

For the numerical consideration, Young's modulus E_i and area fractions A_{ri} are assumed to be the same as those used in Section 5.2, and Poisson's ratios are $v_1 = v_2 = 0.3$. The real cutting experiment gives us the resultant induced strain at the outer surface as $\Delta\varepsilon_\theta^{exp} = 119.6$ μ-strain, and the observation of cross section of the isolated cortical layer gives the outer radius of the cortical cylinder as $r_3 = 9.89$ mm. Then, the radius of the cancellous member is calculated as $r_1 = 7.72$ mm and the inner radius of the cortical member as $r_2 = 7.73$ mm. These calculated radii approximately coincide with $r_2 = 7.31$ mm, obtained by the cross-sectional observation of the isolated cortical layer. The representative working stresses calculated are $\bar{\sigma}_{\theta 1} = p = 0.84$ MPa and $\bar{\sigma}_{\theta 2} = -3.00$ MPa, corresponding to the effective stress of $|\sigma^e_{\theta 1}| = |\sigma^e_{\theta 2}| = 3.00$ MPa. This magnitude of the effective working stress in circumferential direction is of the same order as those in the cephalocaudal direction obtained in Section 5.

7 Conclusions

This chapter has discussed residual stress in bone structure by means of two experimental observations and corresponding model studies from the point of view of the uniform stress hypothesis. The fibula-cutting experiment was conducted for the leporine tibiofibula bone, and the strain induced in the tibia was observed by using a strain gauge showing the self-equilibrium stress remaining in the unloaded state. The observed strain in the axial direction was examined by comparison with that of the numerical fibula-cutting experiment for the simulated remodeling equilibrium by a mathematical model in which the mechanical remodeling was driven by the nonuniformity of the stress distribution, allowing the existence of the residual stress. The reasonable correlation between the real and numerical experiments suggests the role of residual stress in uniform stress distribution under working loads.

The cutting experiment was also applied to the bovine coccygeal vertebral body, which consists of cortical and cancellous bone tissues. The cutting operations removed the cranial and caudal end-plates first, and then the cancellous bone inside. The strains induced in the cephalocaudal and circumferential directions were observed by the strain gauges bonded on the cortical surface. The observed strains were all positive, which suggests a compressive stress remained in the cortical bone and a tensile stress in the cancellous bone in the unloaded state. The role of the residual stresses in these directions was considered by using the statically indeterminate models of two-bar elements and of two-cylinder elements for cortical and cancellous bone, respectively. In both models, the experimentally observed strains are understood in the context of the hypothesis of uniform stress distribution maintained through the adaptation by remodeling.

As the result of these consideration, it is suggested that bone structure has residual stress because of its statical indeterminacy, and that the existence of residual stress contributes to maintain the uniformity of stress distribution under the working load. A general mathematical model allowing for the residual stress is needed in bone mechanics for further understanding of the mechanical adaptation resulting from remodeling. The authors are starting a study on cancellous bone remodeling that considers residual stress in which bone tissue is modeled as a lattice continuum [19], with several preliminary results for the bovine experiment [15,20]. However, much further effort is expected in this field.

Acknowledgment. This work was financially supported in part by the Grant-in-Aid for Scientific Research on Priority Area [Biomechanics] (Nos. 04237101 and

06213225) and for Encouragement of Young Scientists (No. 06855015) from the Ministry of Education, Science, Sports and Culture, Japan.

References

1. Fung YC (1984) Biodynamics: circulation. Springer, Berlin Heidelberg New York, p 64
2. Fung YC (1983) What principle governs the stress distribution in living organs? In: Fung YC, Fukada E, Junjian W (eds) Biomechanis in China, Japan, and U.S.A. Science Press, Beijing, pp 1–13
3. Chuong CJ, Fung YC (1986) On residual stresses in arteries. Trans ASME J Biomech Eng 108:189–192
4. Vaishnav RN, Vossoughi J (1987) Residual stress and strain in aortic segments. J Biomech 20(3):235–239
5. Liu SQ, Fung YC (1989) Relationship between hypertension, hypertrophy, and opening angle of zero-stress state of arteries following aortic construction. Trans ASME J Biomech Eng 111:325–335
6. Nevo E, Lanir Y (1994) The effect of residual strain on the diastolic function of the left ventricle as predicted by a structural model. J Biomech 27(12):1433–1446
7. Han HC, Fung YC (1991) Residual strains in porcine and canine trachea. J Biomech 24(5):307–315
8. Takamizawa K, Hayashi K (1987) Strain energy density function and uniform strain hypothesis for arterial mechanics. J Biomech 20(1):7–17
9. Hayashi K (1992) Residual stress in living materials. In: Fujiwara H, Abe T, Tanaka K (eds) Residual stress. III: Science and technology, vol 1. Elsevier, New York, pp 121–127
10. Rodriguez EK, Hoger A, McCulloch AD (1994) Stress-dependent finite growth in soft elastic tissues. J Biomech 27(4):455–467
11. Okuyama T, Yamamoto H (1992) Residual stresses in living tree. In: Fujiwara H, Abe T, Tanaka K (eds) Residual stress. III: Science and technology, vol 1. Elsevier, pp 128–133
12. Tanaka M, Adachi T, Seguchi Y, Morimoto Y (1992) Residual stress of biological tissues and model of adaptation by remodeling. In: Fujiwara H, Abe T, Tanaka K (eds) Residual stress. III: Science and technology, vol 1. Elsevier, pp 134–139
13. Tanaka M, Adachi T, Tomita Y (1993) Bone remodeling considering residual stress: preliminary experimental observation and theoretical model development. In: Held DK, Brebbia CA, Ciskowski RD, Power H (eds) Computational biomedicine. Computational Mechanics, Southampton, pp 239–246
14. Tanaka M, Adachi T (1994) Preliminary study on mechanical bone remodeling permitting residual stress. Jpn Soc Mech Eng Int J A37(1):87–95
15. Tanaka M, Adachi T (1996) Model and simulation of bone remodeling considering residual stress. In: Hayashi K, Ishikawa H (eds) Computational Biomechanics. Springer, Tokyo, pp 3–21
16. Seguchi Y (1989) Preliminary study on adaptation by remodeling. In: Woo SL-Y, Seguchi Y (eds) Tissue engineering 1989. BED-14, American Society of Mechanical Engineers, New York, pp 75–78

17. Carter DR, Hayes WC (1977) The compressive behavior of bone as a two-phase porous structure. J Bone Jt Surg 59A:954–962
18. Timoshenko S (1956) Strength of material. Part II. Advanced theory and problems. Norstrand, New York, pp 205–213
19. Adachi T, Tomita Y, Tanaka M (1994) Cosserat continuum model of bone structure and simulation. In: Proceedings of 37th Japan congress on material research, Society of Material Science, Kyoto, pp 215–221
20. Adachi T, Tomita Y, Tanaka M (1994) Mechanical bone remodeling considering residual stress: a lattice continuum model. In: Askew MJ (ed) Advances in bioengineering 1994. BED-28, American Society of Mechanical Engineers, New York, pp 255–256

Response of Knee Joint Tendons and Ligaments to Mechanical Stress

KOZABURO HAYASHI[1], NORITAKA YAMAMOTO[1], and KAZUNORI YASUDA[2]

Summary. Effects of stress deprivation, resumption of stress after stress deprivation, and stress enhancement on the biomechanical properties and dimensions of the patellar tendon and anterior cruciate ligament were studied in the rabbit and dog. For this purpose, we developed several unique techniques and methods by which in vivo stresses applied to the tendons and ligaments could be changed directly and quantitatively. Complete stress shielding (100% stress deprivation) decreased the structural strength of the rabbit patellar tendon rapidly and markedly, while partial stress shielding (70% stress deprivation) induced no significant decrease in its strength. The canine anterior cruciate ligament showed biomechanical response to complete stress shielding essentially similar to that of the rabbit patellar tendon; however, the response appeared more slowly and moderately in the ligament than in the tendon. Stress resumption caused gradual recovery of tendon strength that was reduced by stress shielding, although strength did not return to the control level. Overstressing to 133% of the normal stress (33% overstressing), which was induced by decreasing the cross-sectional area, increased the structural strength of the patellar tendon to that of the whole intact tendon. In the case of 100% overstressing, strength increased in some tendons but decreased in others. These results indicate that tendons and ligaments respond well to change in mechanical stresses and have the ability to adapt to stress change within a certain range, although they cannot adapt if stress exceeds this limit.

Key words: Stress shielding—Restressing—Overstressing—Patellar tendon—Anterior cruciate ligament

---

[1] Biomechanics Laboratory, Department of Mechanical Engineering, Faculty of Engineering Science, Osaka University, 1-3 Machikaneyama-cho, Toyonaka, Osaka, 560 Japan
[2] Department of Orthopaedic Surgery, Hokkaido University School of Medicine, N15 W7, Kita-ku, Sapporo, 060 Japan

1 Introduction

Living tissues grow or resorb in response to change in mechanical stresses exerted on them and adapt mechanically and structurally to mechanical demands. This phenomenon, which is unique to living tissues and organs, is called functional adaptation of tissue or tissue remodeling [1,2]. This adaptation of living tissues to mechanical stress, or tissue remodeling, has been studied rather extensively in the bone [3–5]. In addition, several recent studies on the effects of physical activities on tendons and ligaments have shown that these fibrous connective tissues also have the ability of remodeling and respond to mechanical stress [6–8]. There is however a paucity of quantitative information related to the effects of stress on the characteristics of biological soft tissues such as tendons and ligaments. In physical activities like exercise and immobilization, it is hard to know the exact amount of stress that is applied to the tissues.

We have done a series of studies on the biomechanical response of knee joint tendons and ligaments to mechanical stress [8–14]. In these experimental studies, we have paid special attention to the quantification of the stress exerted on tissues. For this purpose, we have developed several unique experimental techniques and methods for stress deprivation and stress enhancement. Some of these techniques have been applied to autograft models as well [15–18] and the results obtained were summarized elsewhere [19] (also see chapter by Yasuda and Hayashi, this volume).

This chapter is primarily concerned with our experimental findings on the changes of the biomechanical characteristics of rabbit and canine knee joint tendons and ligaments induced by a wide variety of treatment modalities, including stress shielding, restressing after stress shielding, and overstressing.

2 Effects of Stress Shielding on Rabbit Patellar Tendons

Effects of stress deprivation on the biomechanical properties of knee joint tendons and ligaments have so far been studied by experiments of immobilizing animal knees [6–8]. Although immobilization may decrease stress applied to the tendons and ligaments, we cannot state the exact amount of stress reduction because of difficulty in this measurement.

As a direct and quantitative method for stress deprivation, we developed an experimental technique of stress shielding by which tension applied to the patellar tendon is chronically released or decreased during regular knee motion. This method has been applied to study the effects of stress deprivation on the biomechanical properties of the rabbit patellar tendon.

2.1 Experimental Technique and Method

After surgically exposing the patellar tendon in one side of a rabbit knee, with the knee being flexed 90°, two stainless steel pin markers (0.7 mm in diameter and

3 mm in length) were embedded into the patellar and the tibial tubercle (Fig. 1); the distance between the two markers was measured with calipers. In addition, a stainless steel pin (1 mm in diameter) and a stainless steel screw (3 mm in diameter) were transversely inserted into the patella and in the tibial tubercle, respectively. Then, a 10-mm-wide, commercially available Leeds-Keio (LK) artificial ligament (Neoligaments Systems, UK) was rolled, and hooked between the pin and the screw [14,16].

For complete stress shielding (CSS), the patellar tendon was slackened by drawing the patella toward the tibial tubercle by use of the LK ligament, and this position was maintained by firmly fixing both ends of the LK ligament with a stainless steel clamp. In the first series of experiments, a flexible stainless steel wire (0.97 mm in diameter) was used for stress shielding instead of the LK ligament [10]. In both cases, the distance between the two pin markers was shortened by about 6 mm (approximately 30% of the tendon length), and it was confirmed that the patellar tendon was relaxed at all knee angles.

To know the effects of the extent of stress reduction, tension applied to the patellar tendon was partially reduced to about 30% of the peak tension developed by normal activities (partial stress shielding, PSS). For this purpose, the LK ligament was pulled in the same manner as in the CSS group, but fixed when the load in the patellar tendon was reduced to about 30% of the normal force [14,17]. Reduction of the load was monitored with a buckle transducer [20] temporarily attached to the patellar tendon. At the time of sacrifice, the percentage ratio of the load shared by the partially stress-shielded tendon was measured again.

Postoperatively, no immobilization treatment was applied, and all animals were allowed unrestricted daily activities in cages. With the technique described here, stress in the patellar tendon was completely or partially shielded during normal motion of the knee joint. The nontreated patellar tendon in the contralateral knee supplied control data.

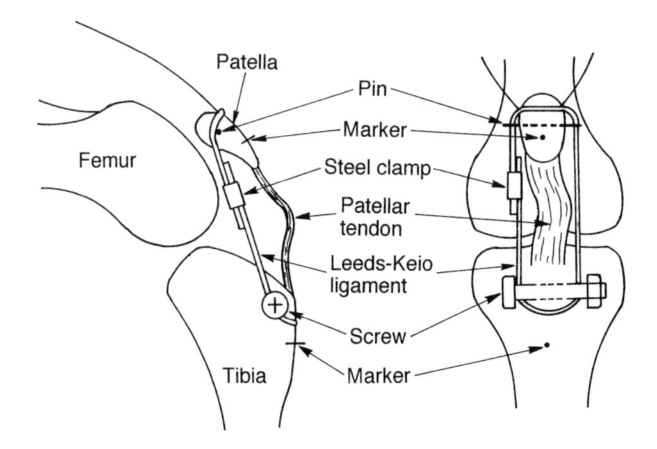

Fig. 1. Experimental arrangement for stress shielding in the rabbit patellar tendon. (From [17], with permission)

One to 6 weeks (CSS group) or 12 weeks (PSS group) after surgery, a patella–patellar tendon–tibia complex was excised from each knee for tensile testing and histological observation. Data for tendons completely stress-shielded for 12 weeks could not be obtained, mainly because the patella was fractured inside the body by this time. After measuring the cross-sectional area of the patellar tendon with an area micrometer, tensile testing was performed at a crosshead speed of 20 mm/min, with the tibia inclined at 45° in the sagittal plane against the patella and the patellar tendon [21]. The specimen and the grips for mounting the specimen to a tensile tester were immersed in 37°C saline solution during testing. Strain in the midsubstance of the tendon was determined using a video dimension analyzer [22].

2.2 Results and Discussion

In the CSS group, it was found at the time of sacrifice that the distance between the two pin markers embedded into the patella and the tibial tubercle was still shortened by more than 4 mm from that measured before the stress-shielding treatment [14]. This shortening was more than 20% of the original tendon length, which indicates that no load was applied to the tendon until sacrifice. The percentage of the load shared by the partially stress-shielded tendons in the PSS group were 28.7% ± 2.6%, 32.2% ± 3.9%, 32.8% ± 4.9%, 41.2% ± 5.6%, and 48.0% ± 5.5% at 1, 2, 3, 6, and 12 weeks, respectively (means ± SE; $n = 8$ for each period) [14]. The loads measured at 6 and 12 weeks were significantly higher than those at the time of the operation. These load changes were attributable to the creep of the LK ligament or shortening of tendon length induced by tissue remodeling.

The cross-sectional area of the stress-shielded patellar tendons was significantly larger than that of the nontreated (control) tendons and the sham-operated tendons throughout the experimental period (Fig. 2). In the CSS group, the area increased rapidly and markedly for the first 3 weeks, reaching a maximum value (approximately 230% of the control value) at 3 weeks. Area in the PSS group was also increased by stress shielding, although the increase was somewhat less at 2 and 3 weeks as compared to the CSS group. The maximum cross-sectional area in the PSS group, which was observed at 2 weeks, was approximately 170% of the control value. In both groups, the cross-sectional area decreased after reaching each maximum value, although it had not returned to the control value at 6–12 weeks. The completely stress-shielded patellar tendons were slightly but significantly shorter than the partially stress-shielded tendons and the sham-operated ones at 3 and 6 weeks (Fig. 3). There was no significant difference in length between the PSS and sham-operated groups.

The tensile strength of the patellar tendon in the CSS group was decreased very rapidly and markedly by stress shielding. For example, it decreased to about 50% at 1 week and to less than 10% at 3 weeks compared to that of the control tendon (Fig. 4) [10,14]. However, it became almost constant thereafter. As in the CSS group, tendon strength in the PSS group was also significantly decreased,

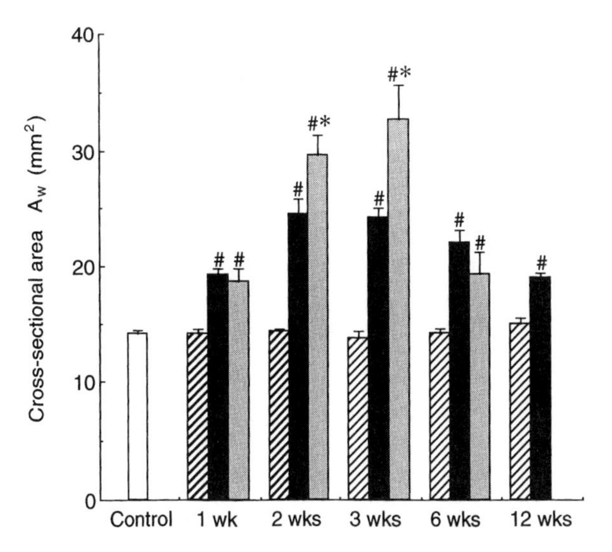

FIG. 2. Cross-sectional area (A_W, mm^2) of control (*open bar*), sham-operated (*hatched*), partially stress-shielded (PSS) (*solid*), and completely stress-shielded (CSS) (*shaded*) patellar tendons in the rabbit [14]. For each experiment, $n = 6$; control, $n = 45$. (Data are means ± SE at time after surgery; $P < 0.05$ ($^{\#}$sham; * PSS)

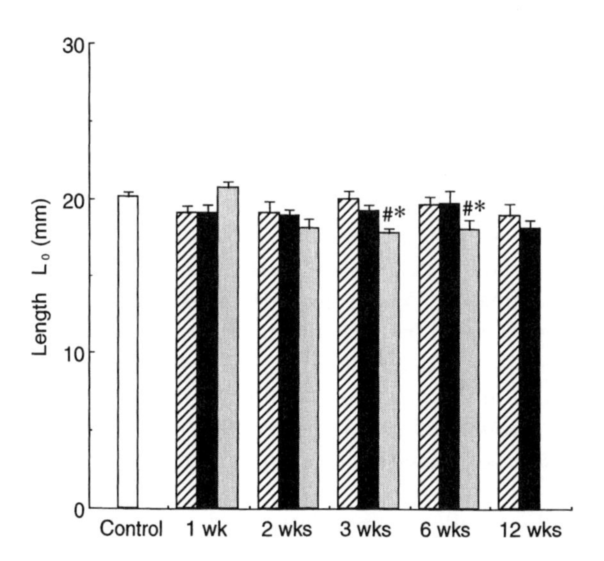

FIG. 3. Length (L_O, mm) of control (*open bar*), sham-operated (*hatched*), partially stress-shielded (PSS) (*solid*), and completely stress-shielded (CSS) (*shaded*) patellar tendons in the rabbit [14]. For each experiment, $n = 6$; control, $n = 45$. Data are means ± SE at time after surgery; $P < 0.05$ ($^{\#}$sham; * PSS)

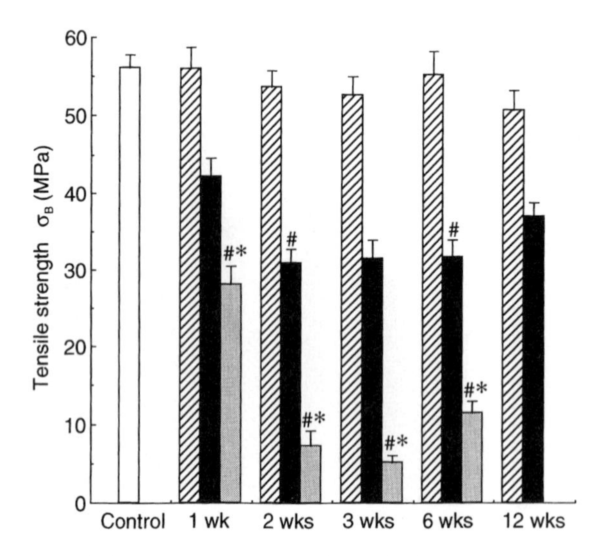

FIG. 4. Tensile strength (σ_B, MPa) of control (*open bar*), sham-operated (*hatched*), partially stress-shielded (PSS) (*solid*), and completely stress-shielded (CSS) (*shaded*) patellar tendons in the rabbit [14]. For each experiment, $n = 6$; control, $n = 45$. Data are means ± SE at time after surgery; $P < 0.05$ ($^{\#}$sham; *PSS)

although the decrease was much less than in the case of complete stress shielding. For example, tensile strength of the partially stress-shielded tendons at 2 and 6 weeks was approximately 55% and 56%, respectively, of that of the control tendon [14]. There was no significant effect of either complete or partial stress shielding on elongation to failure.

The maximum load, which is the failure load of a whole patellar tendon, was calculated from tensile strength and cross-sectional area (Fig. 5). The maximum load of the patellar tendon decreased greatly in the CSS group; for example, it decreased to approximately 25% of the control value at 2 weeks [10]. However, because of the considerable increase in the cross-sectional area, the maximum load decreased less than did tensile strength. In the PSS group, there was no significant difference in the maximum load when comparing control and sham-operated tendons throughout the experimental period [14]. The patellar tendon seems to compensate for decreased tensile strength by increasing cross-sectional area [10,14].

Histologically, the number of fibroblasts increased markedly between 1 and 2 weeks in the CSS group, and it reached four- to five-fold that in the control group at 2–3 weeks [10,14]. However, there was almost no change in the number between 2 and 6 weeks after the complete stress-shielding treatment. In the PSS group, the fibroblast number progressively increased until 3 weeks and reached about threefold that in the control tendons [14]. Then, it gradually decreased until 12 weeks had elapsed.

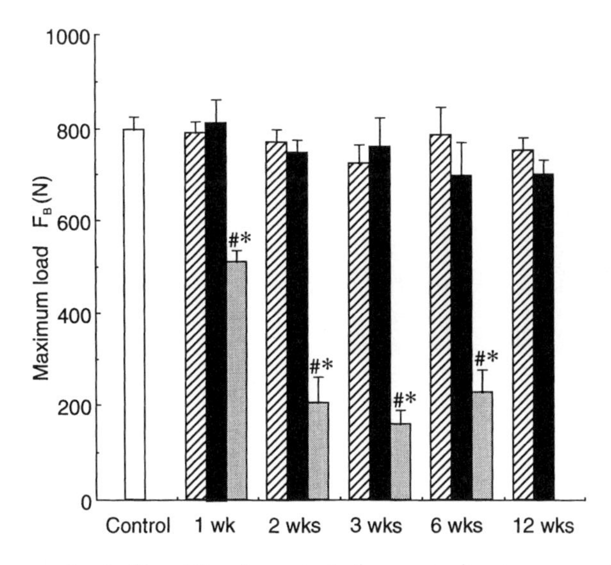

FIG. 5. Maximum load (F_B, N) of control (*open bar*), sham-operated (*hatched*), partially stress-shielded (PSS) (*solid*), and completely stress-shielded (CSS) (*shaded*) patellar tendons in the rabbit. For each experiment, $n = 6$; control, $n = 45$. Data are means ± SE at time after surgery; $P < 0.05$ ($^{\#}$sham; *PSS)

To study the effects of fibroblasts on the mechanical properties and cross-sectional area, we wrapped patellar tendons with polytetrafluoroethylene (PTFE) membrane filters (0.1-μm pore size) for 2 weeks of complete stress shielding with the intention of inhibiting the invasion of fibroblasts from surrounding tissue into the tendon [11]. In fact, the number of fibroblasts in tendons thus treated was almost unchanged from that in the control tendons. The increase in cross-sectional area of the tendon was only about 40% of the control area, which was much less than that in the nonfiltered, completely stress-shielded tendons (110%) (Fig. 6). The reduction of tensile strength was significantly less in the filtered tendons than in the nonfiltered ones (Fig. 7). These results may indicate that the increased cross-sectional area and decreased tensile strength in completely stress-shielded patellar tendons are attributable to the immature collagen newly synthesized by the fibroblasts that invaded from the surrounding tissue.

Effects of complete stress shielding on the ultramicro-structure of the patellar tendon were studied by means of transmission electron microscopy [23]. In control specimens, the diameter of collagen fibrils ranged from 30 to 360 nm and showed a bimodal distribution. The stress shielding for 6 weeks markedly increased the number of thin collagen fibrils having a diameter of 30–60 nm. In addition, stress shielding decreased the average number of collagen fibrils in a unit area at 3 weeks (to 17.6 ± 2.1/μm^2 [Mean ± SD] from control, 24.4 ± 1.4/μm^2) and then increased it at 6 weeks (26.5 ± 3.5/μm^2), possibly because of the in-

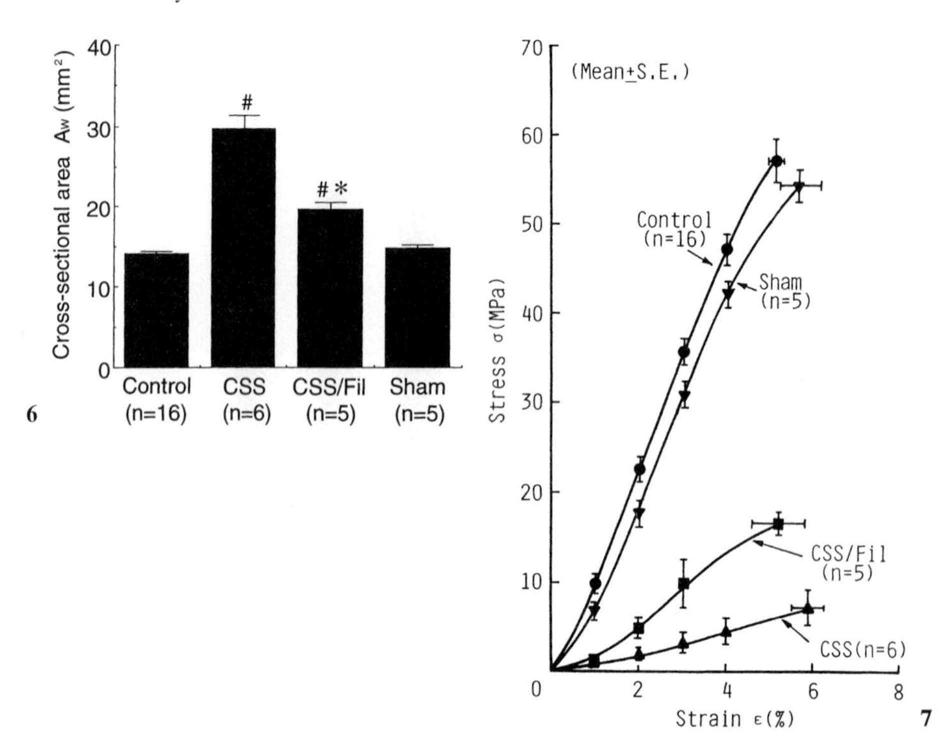

Fig. 6. Cross-sectional area (A_w, mm²) of control (n = 16), completely stress-shielded (*CSS*) (n = 6), completely stress-shielded/filtered (*CSS/Fil*) (n = 5), and sham-operated (n = 5) patellar tendons in the rabbit. Data are means ± SE; $P < 0.05$ (#control; * CSS). (From [11], with permission)

Fig. 7. Stress–strain curves of control (n = 16), completely stress-shielded (*CSS*) (n = 6), completely stress-shielded/filtered (*CSS/Fil*) (n = 5), and sham-operated (n = 5) patellar tendons in the rabbit. Data are means ± SE. (From [11], with permission)

creased number of thin fibrils. The ratio of the total area of collagen fibrils to the whole visualized area changed with time during stress shielding in a fashion similar to the fibril number (49.0 ± 3.2% and 55.1 ± 0.9% at 3 and 6 weeks, respectively), although the area ratio at 6 weeks was still significantly less than the control value (68.0 ± 1.1%). These results suggest that decrease in tensile strength of the stress-shielded patellar tendon was induced by reduction of the total area of fibrils in tendon cross section and the increase of thin and immature fibrils.

Many experimental studies have been done on the effects of stress deprivation on the biomechanical properties of knee joint ligaments and tendons by immobilizing or disusing animal knees [6–8]. These biomechanical data also indicate that the strength of tendons and ligaments is significantly reduced by immobilization. However, strength reduction induced by immobilization was between that observed in the CSS and PSS groups of our study. These results may indicate that stress applied to tendons and ligaments during immobilization or disuse is not always zero.

3 Effects of Resumption of Loading on Stress-Shielded Rabbit Patellar Tendons

As mentioned, stress shielding markedly decreased the mechanical strength of the rabbit patellar tendon while increasing the cross-sectional area. When normal tension was resumed on the stress-shielded patellar tendon, how do the mechanical characteristics and dimensions change? The study of recovery from stress deprivation is very important for obtaining basic knowledge on remodeling of tendons and ligaments as well as for determining optimal rehabilitation procedures after surgery.

3.1 Method

We used a method of reapplying normal stress to previously stress-shielded patellar tendons. In the first series of this study, the Leeds-Keio (LK) artificial ligament installed between the patella and the tibial tubercle (see Fig. 1) was cut and removed after partially (70%) or completely (100%) shielding the patellar tendon from stress for 2 weeks; then, normal stress was again applied to the tendon for the subsequent 3–12 weeks. To study the effect of resumption of loading in more detail, we performed a second series of experiments on completely stress-shielded patellar tendons. In these experiments, tension in the rabbit patellar tendon was first completely released for 1, 2, or 3 weeks (CSS1, CSS2, or CSS3) by tightening a stainless steel wire hooked between the patella and the tibial tubercle using the technique described earlier. For stress resumption, this stainless steel wire was cut and removed [12]. All rabbits were then allowed unrestricted activities in cages for the subsequent 3, 6, or 12 weeks (RS3, RS6, or RS12).

After sacrifice, the patella–patellar tendon–tibia complex was excised from each knee for tensile testing and histological observation. Patellar tendons removed from nontreated, contralateral knees were used to obtain control data.

3.2 Results and Discussion

Tensile strength of the stress-shielded patellar tendons was progressively increased by restressing; the recovery of tensile strength was larger and faster in completely stress-shielded tendons than in partially stress-shielded tendons (Fig. 8). However, tensile strength in the partially and completely stress-shielded tendons was about 76% and 57% of that of the control tendons, respectively, even if normal stress was resumed for 12 weeks. In contrast to the increase of tensile strength, the cross-sectional area of the tendons progressively decreased

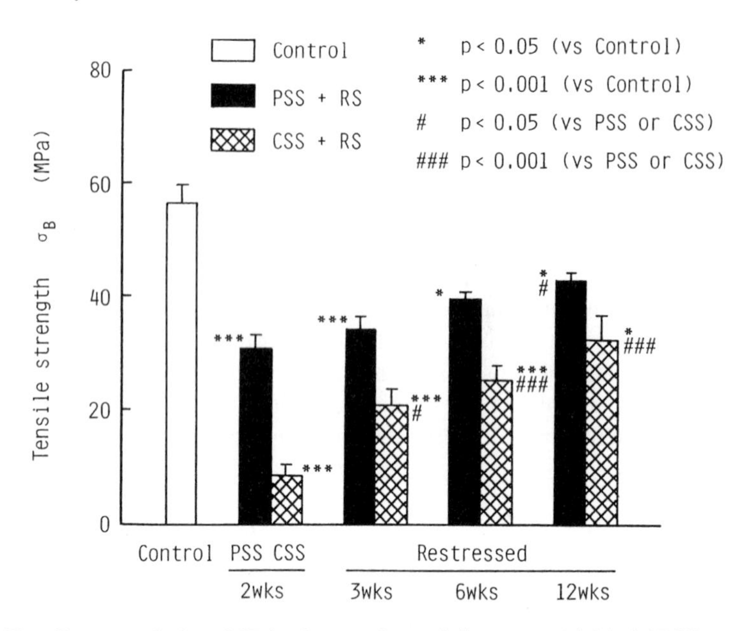

Fig. 8. Tensile strength (σ_B, *MPa*) of control, partially stress-shielded (*PSS*), completely stress-shielded (*CSS*), and restressed (*RS*) patellar tendons in the rabbit. PSS and CSS, $n = 6$; RS, $n = 5$; control, $n = 10$. Data are means ± SE

with time (Fig. 9). Concomitantly, the maximum failure load of the partially stress-shielded tendons remained at control level throughout the period of stress resumption (Fig. 10). In the completely stress-shielded tendons, the recovery of tensile strength was very fast, and the maximum failure load finally returned to about 80% of the control level after stressing for 12 weeks.

The results obtained from the detailed study of the effects of stress resumption on the biomechanical properties and histology of completely stress-shielded rabbit patellar tendons are shown in Figs. 11–13 [12]. The cross-sectional area of the patellar tendon completely stress-shielded for 1 week, which was already significantly increased by the stress shielding, was not significantly changed by restressing (Fig. 11). As shown previously, complete stress shielding for 2 and 3 weeks markedly increased the cross-sectional area, that is, to more than 200% of the control value. Stress resumption after complete stress shielding gradually reduced the area. With regard to tendon length, there were no significant changes after 1-week stress shielding or after the subsequent stress resumption. Although length was significantly decreased by 2- and 3-week stress shielding, it was then greatly increased by 3–12 weeks of restressing and became significantly longer than the length of the control tendons [12].

As can be seen from Fig. 12, the tangent modulus and tensile strength that were significantly reduced by 1-week stress shielding were not changed by the subsequent restressing. Complete stress shielding for 2 and 3 weeks markedly decreased both parameters to as much as about 10% of their control values, as mentioned previously. However, the subsequent restressing increased tangent

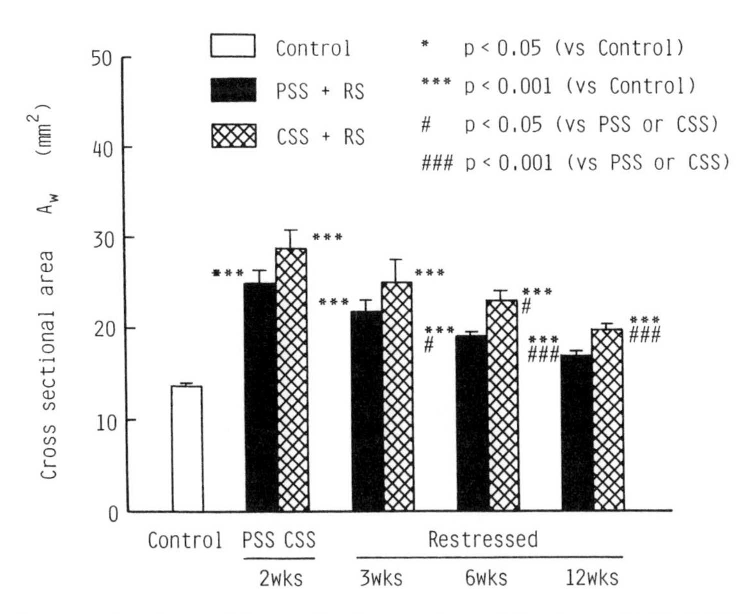

Fig. 9. Cross-sectional area (A_w, mm^2) of control, partially stress-shielded (*PSS*), completely stress-shielded (*CSS*), and restressed (*RS*) patellar tendons in the rabbit. PSS and CSS, $n = 6$; RS, $n = 5$; control, $n = 10$. Data are means ± SE

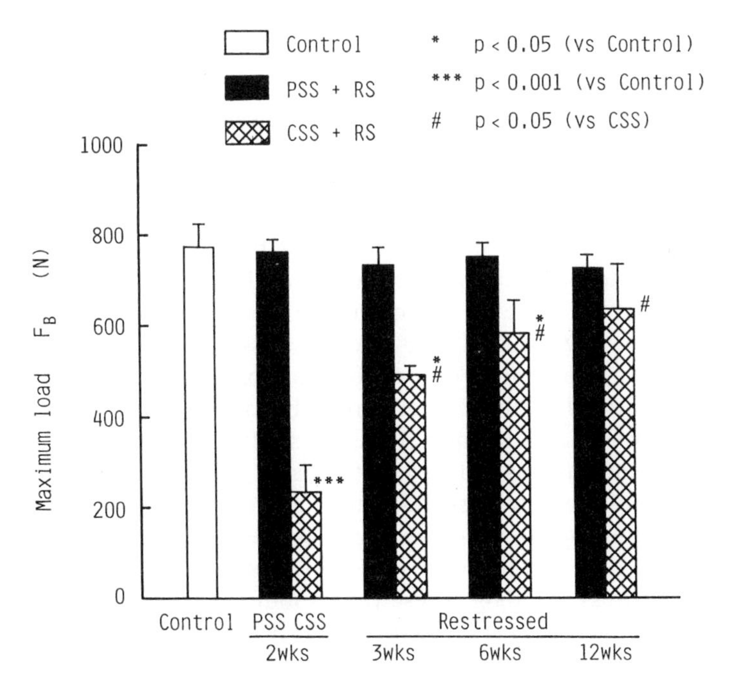

Fig. 10. Maximum load (F_B, N) of control, partially stress-shielded (*PSS*), completely stress-shielded (*CSS*), and restressed (*RS*) patellar tendons in the rabbit. PSS and CSS, $n = 6$; RS, $n = 5$; control, $n = 10$. Data are means ± SE

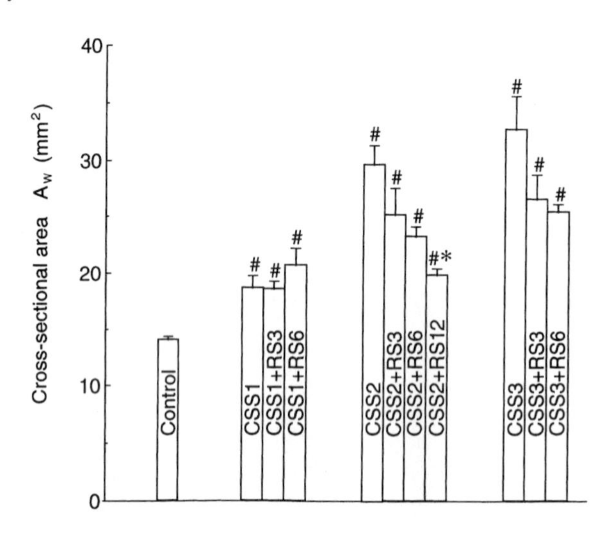

FIG. 11. Cross-sectional area (A_w, mm^2) of control, completely stress-shielded (*CSS*), and restressed (*RS*) patellar tendons in the rabbit [12]. *CSS1*, *CSS2*, and *CSS3* indicate complete stress shielding for 1, 2, and 3 weeks, respectively; *RS3*, *RS6*, and *RS12* are stress resumption for 3, 6, and 12 weeks, respectively. Means ± SE; $P < 0.05$ ($^{\#}$control; *CSS2)

modulus and tensile strength gradually; for example, tangent modulus and tensile strength recovered to 64% and 60% of the control values, respectively, after 12 weeks of stress resumption following 2-week stress shielding (CSS2 + RS12). However, these values were still significantly lower than those of the control patellar tendons. No significant difference was observed for elongation at failure except that between the control tendons and the tendons restressed for 6 weeks after 3-week stress shielding (CSS3 + RS6).

Temporal changes of the maximum load of the completely stress-shielded and restressed patellar tendons are summarized in Fig. 13. Stress resumption had no significant influence on the maximum load of the tendons completely stress-shielded for 1 week. In 2- and 3-week stress-shielded tendons, however, maximum load was markedly recovered by the subsequent restressing: recovery of the maximum load of the 2-week stress-shielded tendons was fairly rapid for the first 3 weeks and became slower thereafter. A rapid increase in the maximum load such as was induced at the early stage of restressing was also observed in the 3-week stress-shielded tendons, although it continued during 6 weeks of restressing. It is interesting that the maximum load seemed to finally converge to a certain level (70%–80% of the control), regardless of the duration of stress shielding. These results indicated that it takes much time for the tendon to recover the reduced mechanical strength if it is once exposed to a nonstress condition even for a short period of time, and that tendon strength does not completely recover even after a prolonged period of stress resumption.

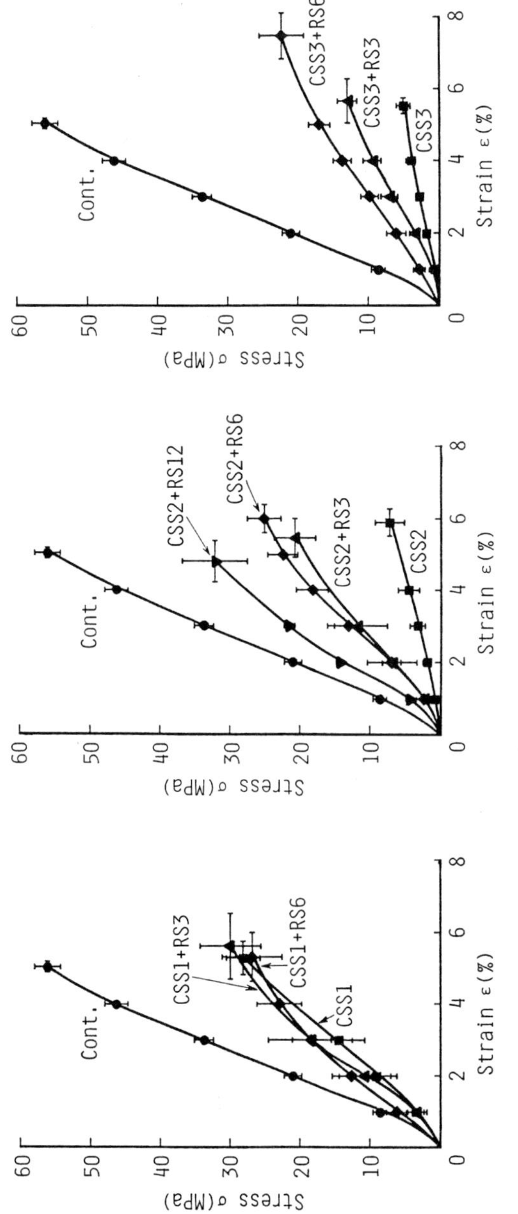

FIG. 12. Stress–strain curves of control, completely stress-shielded (*CSS*), and restressed (*RS*) patellar tendons [12]. *CSS1*, *CSS2*, and *CSS3* indicate complete stress shielding for 1, 2, and 3 weeks, respectively; *RS3*, *RS6*, and *RS12* are stress resumption for 3, 6, and 12 weeks, respectively

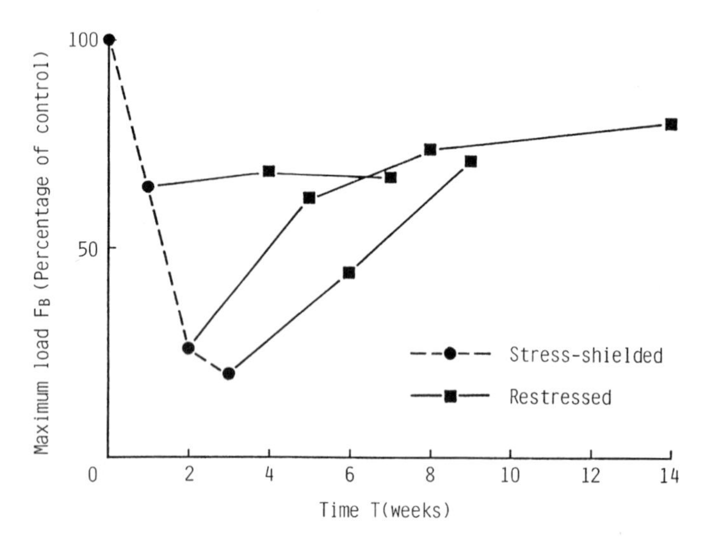

FIG. 13. Temporal changes of the maximum load (F_B) of the completely stress-shielded (*circles, dashed line*) and restressed (*squares, solid line*) patellar tendons in the rabbit. (From [12], with permission)

Histologically, complete stress shielding for 2–3 weeks increased the number of round-shaped fibroblasts and decreased the number of longitudinally aligned collagen fibers. Stress resumption restored microstructure toward that of the control, although the number of fibroblasts did not reach the control level [12]. Such a change in the microstructure corresponded well with the recovery of tensile strength induced by restressing. These results may imply that the collagen fibers newly synthesized from fibroblasts during complete stress shielding were gradually matured and therefore strengthened by subsequent stress resumption.

4 Effects of Chronic Relaxation in the Canine Anterior Cruciate Ligament

We believe there is little information related to the effects of mechanical environment on the biomechanical properties of ligament tissues. In particular, only a few biomechanical studies have been performed on the anterior cruciate ligament although it is functionally the most important fibrous tissue in the knee joint.

To study the biomechanical effects of stress deprivation on the anterior cruciate ligament, we developed a novel method of directly releasing tension in the canine ligament by translocating the tibial insertion proximally along the ligament axis [13].

4.1 Experimental Design

Fifty-eight mature mongrel dogs weighing 10.0 ± 1.8 kg (mean ± SD) were used for the experiment, divided into two groups: relaxed group (n = 30) and sham group (n = 28). The tibial insertion of the anterior cruciate ligament in the right knee of each animal was made free from the tibia as a cylindrical bone block by penetrating a specially designed trephine (9.5 mm, external diameter; 8 mm, internal diameter; 40 mm, length) from a location 15 mm distal to the articular surface and 7 mm medial to the medial edge of the patellar tendon on the anteromedial aspect of the tibia to the center of the tibial insertion of the anterior cruciate ligament. In the relaxed group, the bone block was translocated toward the femoral attachment site of the anterior cruciate ligament by 3 mm and then fixed by inserting a cortical screw (3.5 mm in diameter, 24 mm in length) between the bone block and the wall of the bone tunnel. We confirmed that the elevation of the bone block did not cause any incongruity of the joint surface or impingement during joint motion. In the sham group, the bone block was reduced to the anatomical position and then fixed with a screw. The left knee of each dog was given no treatment and was used to obtain control data.

No postoperative immobilization was applied to the knee joints. All dogs were allowed unrestricted activities in cages for 6 or 12 weeks.

After sacrifice, femur–anterior cruciate ligament–tibia complexes were excised from the knees for biomechanical testing and histological observation. After surrounding tissues were carefully removed from the anterior cruciate ligament, the cross-sectional area of each ligament was measured using an area micrometer [21]. The medial and lateral one-third of the ligament was cut off along the longitudinal axis, and the remaining central one-third was used for tensile testing. When mounting these prepared bone–ligament complexes onto a tensile tester, the tibia was flexed at 90° toward the internal direction to remove the distortion of the bundles in the ligament substance. The specimens were stretched to failure at a crosshead speed of 20 mm/min. Strain in the ligament substance was determined using a video dimension analyzer [22].

The biomechanical effects of the treatments (ligament relaxation and sham operation) were evaluated using the treat/nontreat ratio, which was defined by dividing each datum obtained from the treated (right) side by that from the non-treated, intact (left) side in the same animal. The reason we used this ratio was that individual variations of the data among the mongrel dogs were fairly large compared to those in the rabbits used for the patellar tendon studies.

4.2 Results and Discussion

The whole cross-sectional area of the anterior cruciate ligament significantly increased in both groups (Table 1) [13]. The increase of the treat/nontreat ratio in the sham group was rather small but was significant both at 6 and at 12 weeks. In the relaxed group, the ratio markedly increased at 6 weeks and then slightly decreased at 12 weeks. If the ratios between the sham and the relaxed groups were compared, a significant difference was observed only at 6 weeks. The change in the cross-sectional area induced by stress deprivation in the canine anterior cruciate ligament was essentially similar to that observed in the rabbit patellar tendon mentioned previously.

The averaged treat/nontreat ratio of the tensile strength in the sham group was significantly reduced at 6 weeks and then slightly increased at 12 weeks (Table 2) [13]. In the relaxed group, however, it was significantly decreased at 6 weeks and remained at a low level until 12 weeks. There was a significant difference in the ratio between the sham and the relaxed groups only at 12 weeks.

Histologically, no marked changes were observed in the morphology or the number of cell nuclei in the sham-operated anterior cruciate ligaments [13]. The only difference found in comparison with the nontreated ligaments was that intervals were larger between crimped collagen bundles in limited areas; the areas seemed to be more common at 12 weeks than at 6 weeks. In the relaxed ligaments, the intervals between collagen bundles were obviously larger compared to the sham group. In addition, partial fragmentation of collagen fibers was observed only in the posterior bundles of the relaxed ligaments; fragmentation was observed more frequently at 12 weeks than at 6 weeks.

As mentioned previously, the effects of stress shielding on the mechanical properties of the rabbit patellar tendon appeared within 1 week after treatment, and mechanical strength decreased progressively and markedly for the duration

TABLE 1. Treat/nontreat ratio of the cross-sectional area of the whole anterior cruciate ligament in the dog [13].

Group	6 Weeks	12 Weeks
Sham group	1.16 ± 0.12*,** (n = 11)	1.19 ± 0.10* (n = 9)
Relaxed group	1.37 ± 0.10*,** (n = 9)	1.29 ± 0.17* (n = 12)

Data are mean ± SD.
* Significantly different between treated and nontreated ligaments ($P < 0.01$).
** Significantly different between sham and relaxed groups ($P < 0.01$).

TABLE 2. Treat/nontreat ratio of the tensile strength of the anterior cruciate ligament in the dog [13].

Group	6 Weeks	12 Weeks
Sham group	0.79 ± 0.15* (n = 6)	0.87 ± 0.12** (n = 6)
Relaxed group	0.67 ± 0.17* (n = 5)	0.58 ± 0.08*,** (n = 5)

Data are means ± SD.
* Significantly different between treated and nontreated ligaments ($P < 0.05$).
** Significantly different between sham and relaxed groups ($P < 0.01$).

of the 6-week experiment. On the other hand, reduction of tensile strength seemed to appear more slowly in the canine anterior cruciate ligament, and the decrease was more moderate in comparison to the rabbit patellar tendon. Both the tendon and the ligament tissues are similarly composed of collagen fibers, fibroblasts, and matrix. However, Amiel et al. [24] showed that there are many differences in morphology and biochemistry between the two tissues. In addition, these tissues are located in a different environment, that is, intraarticularly and extraarticularly, respectively. Also, these tissues are exposed to different mechanical conditions. Therefore, the phenomena and mechanisms of remodeling induced by stress deprivation may differ between the tendon and the ligament.

5 Effects of Overstressing in the Rabbit Patellar Tendon

Several studies have been done on the effects of stress enhancement on the structural and biomechanical properties of knee joint tendons and ligaments, mostly by applying exercise and physical training to animals [6–8,25–28]. As in immobilization studies, however, we do not know the exact amount of increase in tension induced by exercise and training. To make clear the effects of stress enhancement, we should design studies so that the force directly applied to ligaments and tendons is quantitatively increased.

Clinically, autogenous reconstruction of a damaged anterior cruciate ligament using either the central or the medial one-third portion of the patellar tendon is a common procedure [29,30]. Most experimental studies have focused on the fate of such intraarticularly transplanted patellar tendons [31–34]. However, only a few studies have considered the remaining patellar tendon after removing a portion for the reconstruction of the anterior cruciate ligament [35–37]. In addition, these studies were concerned with the patellar tendon remaining after removing only the central or the medial portion. Effects of the extent of stress enhancement induced by partial removal of the patellar tendon have not been studied.

We partially cut off both edges of the rabbit patellar tendon and reduced the cross-sectional area of the remaining tendon to about 75% or 50%. By this method, stress that is 33% or 100% higher than normal is precisely applied to the remaining tendon.

5.1 Experimental Method

After exposing the anterior part of the right knee through a straight skin incision, the subcutaneous retinacula were incised longitudinally along the patellar tendon. The posterior surface of the patellar tendon was then separated from the infrapatellar fat pad. The cross-sectional area of the middle part of the whole patellar tendon was measured with an area micrometer [21]. The medial and lateral (outer) portions of the tendon were cut off along the longitudinal axis, leaving the central three-fourths or one-half of the width. The cross-sectional area of the remaining patellar tendon was again measured with the area micrometer. By this surgical treatment, the cross-sectional area of the tendon was reduced to 75% or 50% of its original area, which increased stress in the remaining tendon to 133% or 200% of the normal stress. No immobilization treatment was applied postoperatively, and all rabbits were allowed their daily activities in cages.

In several selected animals, the tension applied to the patellar tendon which remained after the partial cut/removal was measured in vivo during normal walking using a buckle transducer [20], and was compared to that measured from the nontreated, intact patellar tendon. We confirmed that there were no significant differences in the mean peak tension between the remaining patellar tendon and the intact tendon, which indicates that the stress in the remaining tendon was increased to 133% or 200% of that in the intact tendon by the partial cut removal.

After sacrifice at 3–12 weeks postoperatively, patella–patellar tendon–tibia complexes were excised from the knees for tensile testing and histological observation as mentioned previously. The patellar tendons removed from the nontreated left knees were used to obtain control data.

5.2 Results and Discussion

There were no significant differences in the stress-strain relation and tensile strength among the control tendons and the tendons stress-elevated to 133% of normal stress for 3–12 weeks. However, the cross-sectional area of the treated tendons significantly increased for the first 3 weeks, although it remained almost constant thereafter (Table 3). Concomitantly, the structural strength of the whole patellar tendon, that is, the maximum load, increased with increase in the cross-sectional area, and at 6 and 12 weeks there were no significant differences in the maximum load between the 33% overstressed tendons and the control whole tendons. The microstructure of the 33%-overstressed tendons was essentially similar to that of the control tendons.

The patellar tendons in which stress was elevated to 200% of the normal stress were divided into two groups, namely groups N and D, from their stress–strain curves (Fig. 14) [8]. In group N, there was almost no change in the stress-strain behavior for the duration of 12 weeks. In group D, on the other hand, the slope of the stress–strain curve and tensile strength progressively and markedly decreased with time. Histologically, there was essentially no difference between the

TABLE 3. Cross-sectional area of control patellar tendons and tendons stress-elevated to 133% of normal stress in the rabbit.

Control (n = 18)	3 Weeks (n = 6)		6 Weeks (n = 6)		12 Weeks (n = 6)	
	At operation	At sacrifice	At operation	At sacrifice	At operation	At sacrifice
13.3 ± 0.3	10.4 ± 0.4	12.6 ± 0.3*	11.2 ± 0.3	13.1 ± 0.3*	9.70 ± 0.38	12.2 ± 0.6*

Data are in mm^2; mean ± SE.
*Significantly different from the value at opeation ($P < 0.05$).

TABLE 4. Cross-sectional area of control patellar tendons and tendons stress-elevated to 200% of normal stress in the rabbit. See the text for groups N and D.

Group	3 Weeks		6 Weeks		12 Weeks	
	At operation	At sacrifice	At operation	At sacrifice	At operation	At sacrifice
Control (13.3 ± 0.3; n = 18)						
N	7.30 ± 0.47	8.90 ± 0.48*	7.15 ± 0.23	9.19 ± 0.75*	7.80 ± 0.27	9.16 ± 0.34*
	(n = 5)		(n = 5)		(n = 4)	
D	7.19 ± 0.24	10.8 ± 0.9*	7.21 ± 0.32	14.5 ± 3.9	7.07 ± 0.40	14.9 ± 4.3
	(n = 4)		(n = 4)		(n = 5)	

Data are in mm^2; mean ± SE.
*Significantly different from the value at operation ($P < 0.05$).

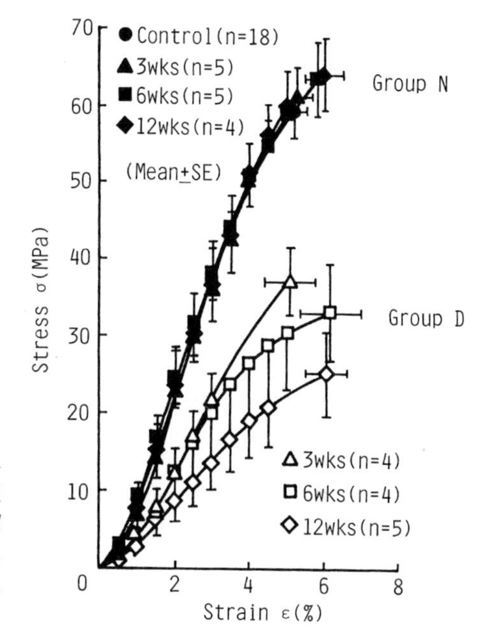

FIG. 14. Stress–strain curves of control patellar tendons and tendons stress-elevated to 200% of normal stress in the rabbit. *Solid symbols*, group N; *open symbols*, group D. See the text for groups N and D. (From [8], with permission)

tendons of group N and the control tendons. In group D, on the other hand, there were a great number of fibroblasts, and breakage of collagen bundles was observed occasionally.

In group N, the cross-sectional area of the patellar tendon significantly increased for the first 3 weeks, and was almost constant thereafter (Table 4). Because of the increase in cross-sectional area, the maximum load increased by about 40% at 12 weeks in comparison to that obtained immediately after the partial removal of tissue, and reached more than 70% of the magnitude in the control whole tendon. In group D, on the other hand, the cross-sectional area

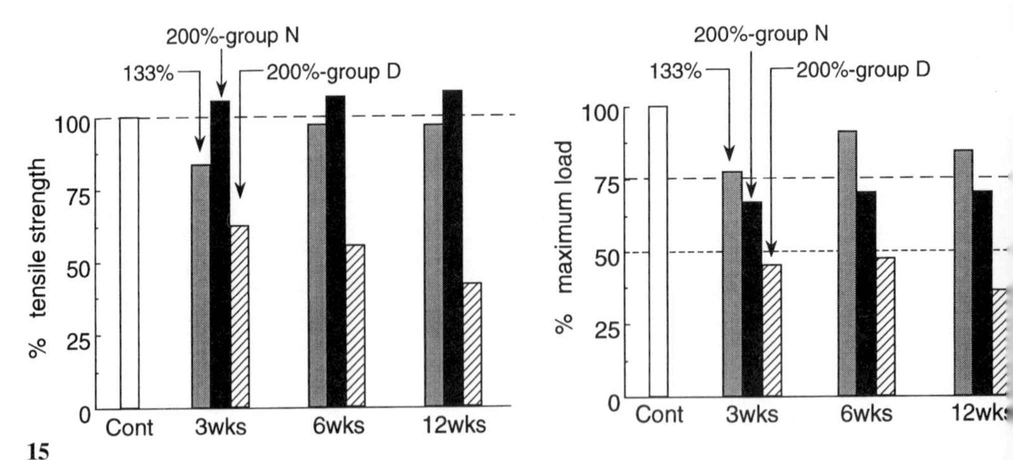

FIG. 15. Percent change of tensile strength of overstressed patellar tendons in the rabbit at time after surgery. *Open bar*, control. See the text for groups N and D

FIG. 16. Percent change of maximum load of overstressed patellar tendons in the rabbit at time after surgery. *Open bar*, control. See the text for groups N and D

increased progressively and markedly for the first 6 weeks, and reached twice the area measured at the time of operation, although there was no further change until 12 weeks. Because of the increases in cross-sectional area, the maximum load remained at the level obtained immediately after tissue removal, that is, about 50% of the maximum load in the whole normal tendon.

In the case of 100% overstressing, tensile strength and tangent modulus did not change in some tendons, but decreased in others. In tendons in which the mechanical properties did not change, the cross-sectional area increased, and therefore the maximum load representing the structural strength increased. On the other hand, in tendons in which tensile strength and tangent modulus decreased, the cross-sectional area increased markedly so as to maintain the structural strength at the level at the time of tissue removal.

The experimental results on the effects of overstressing on the biomechanical properties of the rabbit patellar tendon are summarized in Figs. 15–18. Although the tensile strength of 33% overstressed tendons seemed to slightly decrease at 3 weeks, there was essentially no significant change in tensile strength in this group (Fig. 15). In the case of 100% overstressed tendons, there was almost no change in tensile strength in group N, while that of group D decreased progressively and markedly with time.

The maximum load of 33% overstressed tendons theoretically should decrease to 75% of the control value as the result of 25% reduction of the cross-sectional area. The experimental results demonstrated that the maximum load was about 75% or somewhat larger than 75% of that of the control whole tendons (Fig. 16). In the case of 100% overstressed tendons, the maximum load should be theoretically reduced to 50% of that of the intact whole tendon. However, the maximum load of the tendons of group N increased greatly during the course of overstressing and reached almost 75% of the control level. In the tendons of group D, on the other hand, the maximum load remained at 50% or decreased to slightly less than 50%.

As shown in Fig. 17, the cross-sectional area increased rapidly when the tendons were overstressed to 133% by 25% reduction of the area, and reached almost the same area as that of the original whole tendon by 3 weeks. Along with this change, the stress applied to the tendons decreased to a normal level (Fig. 18). Similar changes of the cross-sectional area and stress were observed in the tendons of group D. In these tendons, the cross-sectional area was initially reduced to 50%; however, it increased progressively and markedly with time and reached the original area within 6 weeks. The stress applied to the tendons also returned to the control level; that is, the stress applied to these overstressed tendons changed to the normal level because the cross-sectional area increased. Such tissue remodeling as that observed in the patellar tendons of group D may be attributable to the activity of fibroblasts, which appeared abundantly in these tendons.

However, 100% overstressed patellar tendons of group N gave quite different results. Their cross-sectional area and stress changed similarly to those in the other two groups of tendons; that is, the cross-sectional area was increased and the stress was decreased by overstressing. However, these values did not reach normal control levels. At present, we do not know why these tendons showed such different results. However, it may be explained thus: the tensile strength of

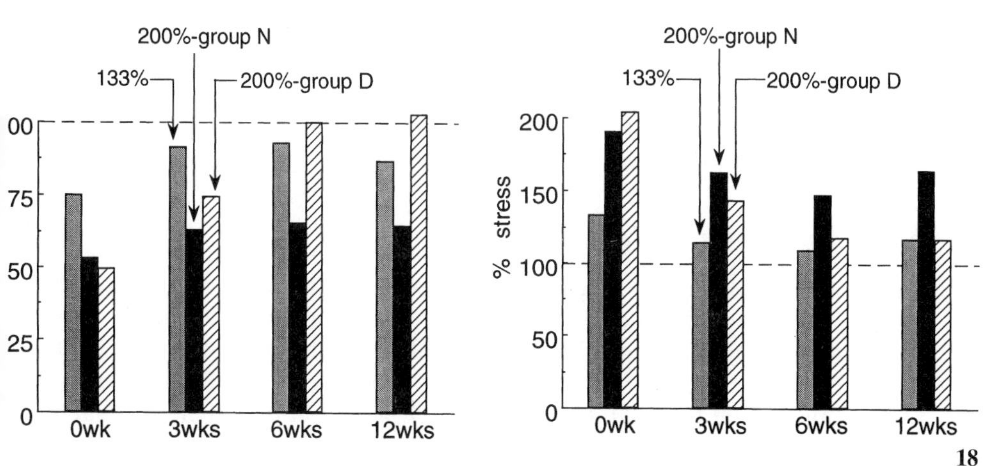

Fig. 17. Percentage of cross-sectional area of overstressed patellar tendons in the rabbit in comparison to that of the control whole tendon at time after surgery. See the text for groups N and D

Fig. 18. Percentage of stress applied to overstressed patellar tendons in the rabbit in comparison to that in the control whole tendon at time after surgery. See the text for groups N and D

these tendons showed almost no change, and the maximum load increased considerably because the cross-sectional area increased. Concomitantly, the maximum load reached about 75% of that in the original whole tendon. The level of 75% or more may be within an optimal range, and the tendons may have not needed to manifest further remodeling because the maximum load reached this range. There were not as many fibroblasts in these tendons, which also indicates a low level of remodeling.

Several animal studies have considered the biomechanical effects of stress enhancement on knee joint tendons and ligaments by exercise and physical training [25–28]. Some of these studies showed that exercise and training induced positive effects on the structural and mechanical properties of tendons and ligaments, while others indicated that there were no effects. We cannot know whether these treatment modalities increased the stress applied to tendons and ligaments or the exact amount of increased tension. In our study, the stress in the rabbit patellar tendon was directly and quantitatively increased by rigorously decreasing the cross-sectional area by removing the medial and lateral portions of the tendon. These results clearly indicated that the patellar tendon has an ability to biomechanically adapt to overstressing within a certain range, that is, if stress is less than 200% of the normal level; however, it cannot adapt if stress exceeds this limit.

6 Effects of Overstressing in the Rabbit Anterior Cruciate Ligament

As was stated previously, it is clear that the patellar tendon biomechanically adapts to overstressing if stress elevation is within a certain range. What are the effects of overstressing on the anterior cruciate ligament? To increase the stress applied to the ligament, we applied the same method to the rabbit anterior cruciate ligament as that use for the patellar tendon [9].

6.1 Experimental Method

After exposing the anterior cruciate ligament through the medial parapatellar approach in the right knee, either the medial or lateral half bundle of the ligament was cut off along the longitudinal axis. The medial bundle of the ligament in eight rabbits was left intact after the lateral bundle was removed (MB group), while the lateral bundle of the ligament in the other eight rabbits was left after the removal of the medial bundle (LB group). The left knees were used to obtain control data. No immobilization treatment was applied postoperatively, and all rabbits were allowed their daily activities in cages for 3 or 6 weeks until sacrifice.

After sacrifice, femur–ligament bundle–tibia complexes were excised from the knees for biomechanical testing. After the surrounding tissue was removed carefully, the cross-sectional area of each ligament bundle was measured using an area micrometer [21]. Each bone–ligament preparation was mounted onto a tensile tester after the ligament fascicle had been untwisted so that the ligament fascicle was uniformly loaded during tensile testing. The specimen was stretched to failure at a crosshead speed of 20 mm/min. Strain was determined with a video dimension analyzer [22].

Using the knees removed from other dead rabbits, the increase in the tension applied to the remaining medial or lateral bundle was estimated from antero-posterior drawer testing. The data obtained indicated that the tension applied to the medial bundle of ligament in the MB group was increased to 130%–150% of the normal stress, and the tension applied to the lateral bundle in the LB group was increased to 400%–800%.

6.2 Results and Discussion

Because 88% (28/32) of the specimens failed at the femur or tibia insertion sites, tensile strength of the anterior cruciate ligament could not be determined in the present study.

The cross-sectional area of the ligaments in the MB group slightly decreased at 3 weeks and then increased at 6 weeks as compared with the control value of the medial bundle (Fig. 19) [9]. The tangent modulus, which was defined as the slope of the linear portion of a stress–strain curve, was slightly increased at 3 and 6 weeks (Fig. 19) [9]. Concomitantly, the ligament stiffness, which is the slope of the linear portion of a load–strain curve, slightly increased at 6 weeks (Fig. 19). However, there were no significant differences in these parameters between the overstressed ligaments and the control ligaments.

In the lateral bundle of the ligament (LB group) in which stress was elevated to 400%–800% of normal stress, the cross-sectional area decreased to about 80% of the control value at 6 weeks, although the decrease was not significant (Fig. 20) [9]. However, the mechanical and structural properties of the bundle were changed significantly: the tangent modulus and stiffness decreased to about 60% and 45% of the control values, respectively, at 6 weeks (Fig. 20) [9].

These results indicated that the anterior cruciate ligament has an ability to biomechanically adapt to overstressing to 130%–150% of the normal stress. However, the ligament cannot adapt if the stress was increased to 400%–800% of the normal stress. As mentioned previously, the patellar tendon can biomechanically adapt to overstressing if stress is less than 200% of the normal level. Essentially similar biomechanical response to stress enhancement was observed in the anterior cruciate ligament and patellar tendon.

7 Conclusions

Biomechanical effects of stress deprivation, resumption of stress after stress deprivation, and stress enhancement have been studied in the rabbit and canine

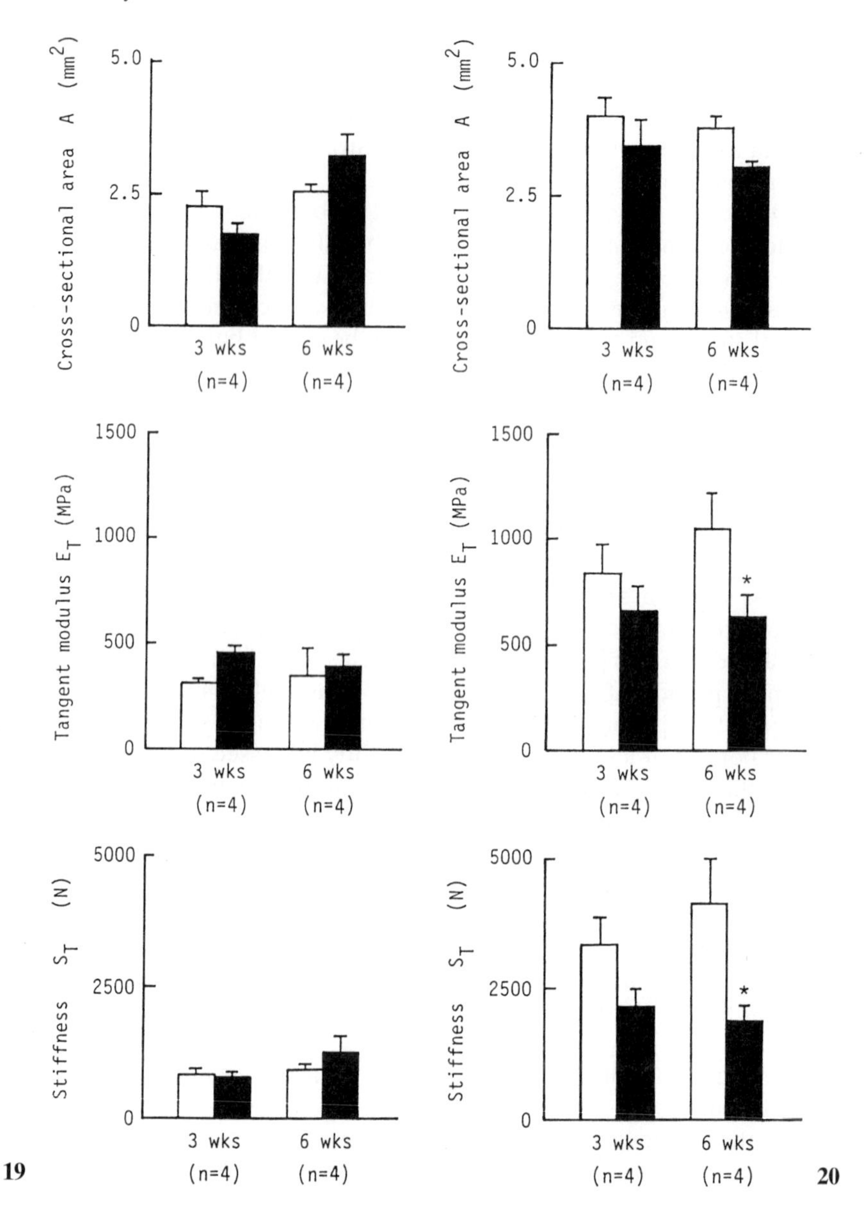

FIG. 19. Cross-sectional area, tangent modulus, and stiffness of medial bundles of control anterior cruciate ligaments (*open bars*) and bundles stress-elevated to 130%–150% of normal stress (*solid bars*) in the rabbit (MB group). Means ± SE; $P < 0.05$ (control)

FIG. 20. Cross-sectional area, tangent modulus, and stiffness of lateral bundles of control anterior cruciate ligaments (*open bars*) and bundles stress-elevated to 400–800% of normal stress (*solid bars*) in the rabbit (LB group). Means ± SE; $P < 0.05$ (control)

patellar tendons and anterior cruciate ligaments using several novel experimental techniques. Complete stress shielding markedly decreased the tensile strength of the rabbit patellar tendon. Although the cross-sectional area considerably increased, the maximum load, which is the failure load of the whole patellar tendon, decreased greatly. Similarly to the completely stress-shielded tendon, the cross-sectional area and tensile strength of the partially stress-shielded patellar tendon, in which the applied stress was reduced to 30% of the normal value, also significantly increased and decreased, respectively. However, these changes were significantly less when compared to complete stress shielding. There was no significant effect of partial stress shielding on the maximum load, which indicates that the patellar tendon can maintain the original structural strength even if stress is reduced to 30% of the normal level. Essentially similar biomechanical effects of stress deprivation were observed in the canine anterior cruciate ligament. However, the changes in mechanical strength and cross-sectional area appeared more slowly and moderately in this ligament than in the rabbit patellar tendon.

Tensile strength and cross-sectional area decreased and increased by stress shielding, respectively, were gradually recovered by stress resumption. However, much time was needed to recover the mechanical strength reduced by complete stress shielding, and strength did not return to the normal level even after a long term of stress resumption.

Overstressing the rabbit patellar tendon to 133% of normal stress did not change the mechanical properties, but did increase the cross-sectional area. As a result, structural strength was increased to the normal level of the whole intact tendon. In the case of the tendon overstressed to 200% of normal stress, structural strength increased in some tendons but decreased in others. These results indicate that the patellar tendon biomechanically adapts to overstressing unless stress does not exceed a certain level, perhaps 200% of the normal level.

In the rabbit anterior cruciate ligament, overstressing to the medial bundle to 130%–150% of normal stress induced no significant changes in the mechanical properties and dimensions. On the other hand, overstressing the lateral bundle to 400%–800% of normal stress significantly decreased tangent modulus and stiffness.

These results indicate that knee joint tendons and ligaments have the ability to adapt to stress alteration by changing the biomechanical properties or tissue dimensions if the stress level is within a certain range. The anterior cruciate ligament seems to show a more moderate and slower response than the patellar tendon.

Acknowledgment. The authors thank their collaborators, Drs. Kiyoshi Kaneda, Kazunori Ohno, Harukazu Tohyama, Motoharu Keira, Tokifumi Majima, Takamasa Tsuchida, Hirohide Ishida, Kunio Tanaka, and Messrs. Takahide Ishizaka, Kiyoshi Miyakawa, as well as their former students, Hiroyuki Kuriyama, Fumihiro Hayashi, and Takashi Fujii. These studies were financially supported in part by the Grants-in-Aid for Scientific Research on Priority Areas [Biomechanics] (K. Hayashi et al., nos. 04237102 and 04237103; K. Yasuda et al., no. 04237104; N. Yamamoto, nos. 04237201, 05221201, and 06213222), the Grant-in-Aid for Scientific Research (B) (K. Hayashi et al., no. 02452096), and the Grants-in-Aid for Encouragement of Young Scientists (N. Yamamoto, nos. 02855018, 03855013, and 04855014) from the Ministry of Education, Science and Culture, Japan.

References

1. Fung YC (1990) Biomechanics—motion, flow, stress, and growth. Springer, Berlin Heidelberg New York, pp 499–546
2. Fung YC (1993) Biomechanics—mechanical properties of living tissues. Springer, New York, pp 369–377, 514–519
3. Wolff J (1986) The law of bone remodeling. Springer, Berlin
4. Meade JB (1989) The adaptation of bone to mechanical stress: experimentation and current concepts. In: Cowin SC (ed) Bone mechanics. CRC Press, Boca Raton, FL, pp 211–251
5. Cowin SC (1993) Nature's structural engineering of bone on a daily basis. In: Bonder SR, Singer J, Dolan A, Hashin Z (eds) Theoretical and applied mechanics 1992. Elsevier, Amsterdam, pp 145–160
6. Tipton CM, Vailas AC, Matthes RD (1986) Experimental studies on the influence of physical activity on ligaments, tendons and joints: a brief review. Acta Med Scand Suppl 711:157–168
7. Woo SL-Y, Wang CW, Newton PO, Lyon RM (1990) The response of ligaments to stress deprivation and stress enhancement: biomechanical studies. In: Daniel DM, Akeson WH, O'Connor JJ (eds) Knee ligaments: structure, function, injury, and repair. Raven, New York, pp 337–350
8. Hayashi K (in press) Biomechanical studies of the remodeling of knee joint tendons and ligaments. J Biomech
9. Yamamoto N, Hayashi F, Hayashi K (1992) Mechanical response of rabbit anterior cruciate ligament to overloading. In: Goh JCH, Nather A (eds) Proceedings of the 7th international conference on biomedical engineering. Singapore, 2–4 Dec 1992, pp 110–112
10. Yamamoto N, Ohno K, Hayashi K, Kuriyama H, Yasuda K, Kaneda K (1993) Effects of stress shielding on the mechanical properties of rabbit patellar tendon. Trans ASME J Biomech Eng 115:23–28
11. Yamamoto N, Hayashi K (1995) Effects of stress shielding on the mechanical properties of rabbit patellar tendon: effects of inhibiting the invasion of fibroblasts. J Jpn Soc Clin Biomech Rel Res 16:119–122
12. Yamamoto N, Hayashi K, Kuriyama H, Ohno K, Yasuda K, Kaneda K (in press) Effects of restressing on the mechanical properties of stress-shielded patellar tendons in rabbits. Trans ASME J Biomech Eng
13. Keira M, Yasuda K, Kaneda K, Yamamoto N, Hayashi K (in press) Mechanical properties of the anterior cruciate ligament chronically relaxed by elevation of the tibial insertion. J Orthop Res
14. Majima T, Yasuda K, Fujii T, Yamamoto N, Hayashi K, Kaneda K (in press) Biomechanical effects of stress-shielding of the rabbit patellar tendon depend upon the degree of stress reduction. J Orthop Res

15. Tohyama H, Ohno K, Yamamoto N, Hayashi K, Yasuda K, Kaneda K (1992) Stress-strain characteristics of in situ frozen and stress-shielded rabbit patellar tendon. Clin Biomech 7:226–230
16. Ohno K, Yasuda K, Yamamoto N, Kaneda K, Hayashi K (1993) Effects of complete stress-shielding on the mechanical properties and histology of in situ frozen patellar tendon. J Orthop Res 11:592–602
17. Majima T, Yasuda K, Yamamoto N, Kaneda K, Hayashi K (1994) Deterioration of mechanical properties of the autograft in controlled stress-shielded augmentation procedures—an experimental study with rabbit patellar tendon. Am J Sports Med 22:821–829
18. Ishida M, Yasuda K, Kaneda K, Yamamoto N, Hayashi K (1995) Does restressing recover the mechanical strength of a patellar tendon autograft reduced by stress sheilding? An experimental study. In: Transactions of the 41st Annual Meeting of the Orthopaedic Research Society, 13–16 Feb 1995, Orlando, vol 2, Sect 2, p 611
19. Yasuda K, Hayashi K (1994) Effects of stress shielding on autografts in augmentation procedures—experimental studies using the in situ frozen patellar tendon. In: Hirasawa Y, Sledge CB, Woo SL-Y (eds) Clinical biomechanics and related research. Springer-Verlag, Tokyo, pp 363–379
20. Yamamoto N, Hayashi K, Hayashi F (1992) In vivo measurement of tension in the rabbit patellar tendon. Trans Jpn Soc Mech Eng Ser A 58:1142–1147
21. Yamamoto N, Hayashi K, Kuriyama H, Ohno K, Yasuda K, Kaneda K (1992) Mechanical properties of the rabbit patellar tendon. Trans ASME J Biomech Eng 114:332–337
22. Hayashi K, Nakamura T (1985) Material test system for the evaluation of mechanical properties of biomaterials. J Biomed Mater Res 19:133–144
23. Tsuchida T, Yasuda K, Kaneda K, Hayashi K, Yamamoto N, Miyakawa K, Tanaka K (1995) Effects of stress shielding on the ultrastructure of normal and in situ frozen rabbit patellar tendons—a role of fibroblasts. Trans Orthop Res Soc 20:613
24. Amiel D, Frank C, Harwood J, Fronek J, Akeson W (1984) Tendon and ligaments: a morphological and biochemical comparison. J Orthop Res 1:257–265
25. Cabaud HE, Chatty A, Gildengorin V, Feltman RJ (1980) Exercise effects on the strength of the rat anterior cruciate ligament. Am J Sports Med 8:79–86
26. Laros GS, Tipton CM, Cooper RR (1971) Influence of physical activity on ligament insertions in the knees of dogs. J Bone Joint Surg 53A:275–286
27. May KD, Horibe S, Woo SL-Y (1989) The effects of long-term intensive exercise and age on the biomechanical properties of canine flexor tendons. In: 1989 ASME biomechanics symposium, American Society of Mechanical Engineers, New York, pp 73–76
28. Woo SL-Y, Gomez MA, Woo YK, Akeson WH (1982) Mechanical properties of tendons and ligaments. II. The relationships of immobilization and exercise on tissue remodeling. Biorheology 19:397–408
29. Jones KG (1963) Reconstruction of the anterior cruciate ligament. A technique using the central one-third of the patellar tendon. J Bone Joint Surg 45A:925–932
30. Paterson FWN, Trickey EL (1986) Anterior cruciate ligament reconstruction using part of the patellar tendon as a free graft. J Bone Joint Surg 68B:453–457
31. Clancy WG Jr, Narechania RG, Rosenberg TD, Gmeiner JG, Wisnefske DD, Lange TA (1981) Anterior and posterior cruciate ligament reconstruction in rhesus monkeys. J Bone Joint Surg 63A:1270–1284
32. Arnoczky SP, Tarvin GB, Marshall JL (1982) Anterior cruciate ligament replacement using patellar tendon. J Bone Joint Surg 64A:217–224

33. Amiel D, Kuiper S (1990) Experimental studies on anterior cruciate ligament grafts—histology and biology. In: Daniel DM, Akeson WH, O'Connor JJ (eds) Knee ligaments: structure, function, injury, and repair. Raven, New York, pp 379–388
34. Grood ES, Walz-Hasselfeld KA, Holden JP, Noyes FR, Levy MS, Butler DL, Jackson DW, Drez DJ (1992) The correlation between anterior-posterior translation and cross-sectional area of anterior cruciate ligament reconstructions. J Orthop Res 10:878–885
35. Cabaud HE, Feagin JA, Rodkey WG (1980) Acute anterior cruciate ligament injury and augmented repair. Am J Sports Med 8:395–401
36. Burks RT, Haut RC, Lancaster RL (1990) Biomechanical and histological observations of the dog patellar tendon after removal of its central one-third. Am J Sports Med 18:146–153
37. Linder LH, Sukin DL, Burks RT, Haut RC (1994) Biomechanical and histologic properties of the canine patellar tendon after removal of its medial third. Am J Sports Med 22:136–142

Remodeling of Tendon Autograft in Ligament Reconstruction

Kazunori Yasuda[1] and Kozaburo Hayashi[2]

Summary. This review study focuses on remodeling of tendon autografts related to anterior cruciate ligament (ACL) reconstruction. When the tendon autograft is transplanted across the knee joint, immediately after surgery fibroblast necrosis occurs sequentially. Then, the graft is enveloped by a vascular synovial tissue with cellular proliferation in the graft during the first 4–6 weeks. The revascularization of the autograft can require as long as 20 weeks for completion. Biomechanically, the tensile strength of the graft decreases immediately after transplantation and gradually recovers with time as remodeling proceeds. In animal models, however, the maximum load reaches only 30%–50% of that of the control ACL after 1 year. Ultrastructural studies demonstrate the invariable predominance of small-diameter collagen fibrils in every specimen. In freeze–thaw studies using the patellar tendon, fibroblast necrosis does not change the tendon. However, the cross-sectional area increases during the period of subsequent cellular repopulation although tensile strength decreases, minimizing reduction of the maximum load. These events also occur in freeze–thaw studies using the ACL. The mechanical environment seems to significantly affect autograft remodeling. Stress deprivation reduces the strength of the in situ frozen patellar tendon, depending on the degree of stress shielding, and increases its cross-sectional area. Ultrastructurally, stress shielding reduces the total number of collagen fibrils and the ratio of the total area of the fibrils to the whole visualized area. These responses occur independently of living cells, and therefore remodeling is associated with noncellular components that include collagen. The mechanical strength of the patellar tendon autograft is essentially restored after restressing.

Key words: Remodeling—Autograft—Ligament—Tendon—Reconstruction

[1] Department of Orthopaedic Surgery, Hokkaido University School of Medicine, N15 W7, Kita-ku, Sapporo, 060 Japan
[2] Department of Mechanical Engineering, Faculty of Engineering Science, Osaka University, 1-3 Machikaneyama-cho, Toyonaka, Osaka, 560 Japan

1 Introduction

There are hundreds of ligaments in the human body, almost any of which can be injured. However, ligament injury has been studied mainly in the knee joint because this joint is both biomechanically accessible and clinically significant in terms of frequency of injury, serious disability caused by instability, and secondary impairment to the menisci and the articular cartilage [1–3]. Studies suggest that different ligaments have different healing potential [4,5]. For example, the anterior cruciate ligament (ACL) frequently shows less healing response [6–10], and the medial collateral ligament seems to have much better healing potential [2,11,12]. These differences are caused by the surrounding environment of the ligament, its course of nutrition, its functions, and other intrinsic advantages [12].

An ACL injury usually results in joint instability, which causes patients not only functional disability but also serious damage to the menisci and the articular cartilage. Therefore, the ACL has been frequently reconstructed using autogenous tissues. Clinically, intraarticular reconstruction of the ACL was first reported in 1917 by Hey Groves [13], who used the autogenous iliotibial tract as a graft material. Since that time, the patellar tendon autograft [14–21] has been most commonly used for cruciate ligament reconstruction. Therefore, remodeling of the patellar tendon autograft in ACL reconstruction has been the greatest focus in a number of recent studies.

The ligament and tendon tissues have similar components including collagen fibers, fibroblasts, and the interstitial matrix. However, these tissues are different from histological, ultrastructural, biochemical, and biomechanical aspects. Amiel et al. [22–26] documented that the grafted patellar tendon changes into the ACL ("ligamentization") in a new biological and physical environment. The time-course of the remodeling process for the patellar tendon autograft following transplantation has been morphologically, histologically, biomechanically, and ultrastructurally studied in various animal models such as the rabbit, dog, goat, and monkey [10,14,15,23,27–40]. Histological and ultrastructural data can now be obtained from human patients who have undergone arthroscopic biopsy of reconstructed ligament tissues [41,42]. Thus, it has been clarified that remodeling of the autogenous tendons grafted across the knee joint is significantly affected by many implantation variables, especially by the mechanical environment surrounding the autograft.

We review here previously published studies related to the patellar tendon autograft, and discuss remodeling of the tendon autograft in ACL reconstruction.

2 Structure and Mechanical Properties of the ACL and the Patellar Tendon

As described, the grafted patellar tendon changes to ligament tissue in ACL reconstruction (ligamentization). To understand this metamorphosis, we have to recognize the differences between the ACL and the patellar tendon.

2.1 *Structure and Mechanical Properties of the ACL*

Ligaments are short bands of tough but flexible fibrous connective tissue that bind bones together and support the organs in place. Although the gross anatomy and function of ligaments have been studied for many centuries, detailed scientific attention has been given to ligaments only within the past 20 years. To examine the histology of the ligament and tendon tissues, hematoxylin and eosin stain has been commonly used to observe both fibroblasts and collagen fibers. The degree of alignment of the collagen fibers can be visualized using polarized light microscopy. Mallory trichrome stain can delineate the interstitial matrix [43]. Ruthenium red and ruthenium hexamine trichloride also stain negatively charged matrix such as glycosaminoglycans [44]. However, the size and orientation of collagen fibrils are best observed using transmission electron microscopy [41,45–48].

Ligaments have architectural hierarchy in their structure [49]. Polarized light microscopy shows collagen crimp patterns. The human ACL is composed of collagen fibrils of 30–175 nm in diameter [26] that form fiber bundles 1–20 μm in diameter, running almost parallel to the long axis of the ligament [50]. These fiber bundles compose a subfascicular unit, which varies from 100 to 250 μm in diameter; 3–20 subfasciculi are collected together to form a collagen fasciculus that can be several millimeters in diameter [22,26,50,51]. In normal goat ACL, 80% of the collagen fibrils have a diameter between 100 and 150 nm, and no fibrils thicker than 200 nm are seen [52]. Cells (fibroblasts) are sparsely scattered in the ligament tissue. It has become increasingly obvious that ligaments are not biologically inert tissues [53]. The metabolism of the fibroblasts changes in response to environmental stimuli and functional needs in a highly regulated fashion, and the resulting metabolic changes alter the ultrastructure, chemistry, and mechanical properties of the ligament.

The mechanical and structural properties of human ligaments have been measured by many investigators (Table 1). Butler et al. [54] measured the mechanical properties of various human ligaments with bone–fascicle–bone preparations and reported that the average elastic modulus and tensile strength of the ACL were 316 and 35 MPa, respectively. With respect to the structural properties of the human ACL, Noyes et al. [55] reported that the average maximum load was 1725 N. This value has been used as a standard for many years. However, Woo et al. [56] recently documented that the normal tensile strength of the ACL has a

TABLE 1. Mechanical properties of the anterior cruciate ligament (ACL).

Authors	Species	Tensile strength (MPa)	Elastic modulus (MPa)	Strain at failure (%)
Butler et al. 1986 [54]	Human	35	316	15
Butler et al. 1989 [32]	Monkey	137	637	
Cabaud et al. 1980 [33]	Dog	75	228	36
Keira et al. 1992 [58]	Dog	150	948	
Cabaud et al. 1980 [33]	Rat	251	492	46

higher value. They performed tensile tests of the femur–ACL–tibia complex at 30° of knee flexion with the ACL aligned vertically along the direction of applied tensile load. One knee from each pair was oriented anatomically, and the contralateral knee was oriented with the tibia aligned vertically. Structural properties of the femur–ACL–tibia complex, as represented by the linear stiffness, maximum load, and energy absorbed, were found to decrease significantly with age and were also found to have higher values in specimens tested in the anatomical orientation. In the specimens obtained from younger cadavers, the average linear stiffness and maximum load were 242 N/mm and 2160 N, respectively, when the femur–ACL–tibia complex was tested in the anatomical orientation.

The mechanical properties of animal ligaments differ among species or tensile tests (see Table 1). Butler et al. [32,57] reported that the average elastic modulus and tensile strength of cynomolgus monkey ACL were 637 MPa and 137 MPa, respectively; Keira et al. [58,59] reported that the elastic modulus and tensile strength of canine ACL averaged 948 MPa and 150 MPa, respectively. Concerning structural properties, Clancy et al. [15] reported that the average maximum loads of the ACL and the posterior cruciate ligament (PCL) were 600 and 450 N, respectively, in the rhesus monkey. Butler et al. [32,57] also reported that the average maximum load of the ACL was 670 N in the cynomolgus monkey, while Ballock et al. [14] described the average maximum load of the rabbit ACL to be approximately 400 N.

2.2 Structure and Mechanical Properties of the Patellar Tendon

Microanatomical studies of tendon tissues commonly reveal the following hierarchial structures [49,60]: fascicles composed of collagen fibrils parallel to the long axis of the tendon are surrounded by visceral paratenon; one or more fascicles are surrounded by parietal paratenon; the endotenon surrounds the fascicle, blood vessels, and nerves; and the epitenon is continuous with the endotenon and bearing capillaries. The fascicle surface usually has a layer of uniform collagen fibrils and elastin. Each fascicle is capable of sliding against its neighbors. There appear to be no direct attachments or cellular communications between neighboring fascicles [49]. In the normal goat patellar tendon, 80% of the collagen fibrils have diameters in the range of 100 to 200 nm [52]. Biochemically, normal adult tendons are composed largely of genetic type I collagen (>95%) [61]. The remaining 5% consists of type III and type V collagen [62,63].

Many biomechanical studies have focused on the tensile strength of the normal patellar tendon (Table 2). It has been found that the tensile strength ranges from 58 to 93 MPa [55,64–71], while the elastic modulus ranges from 360 to 1284 MPa. These data are extremely divergent; the differences can be attributed in part to age, species, and type of specimens (obtained from different anatomical locations). In addition, experimental techniques such as the clamping of test speci-

TABLE 2. Mechanical properties of the patellar tendon.

Authors	Species	Tensile strength (MPa)	Elastic modulus (MPa)	Strain at failure (%)
Noyes et al. 1984 [55]	Human	58		
Burks et al. 1990 [64]	Dog	86	360	
Yamamoto et al. 1992 [108]	Rabbit	56	1330	5
Ohno et al. 1993 [67]	Rabbit	63	1284	5
Majima et al. 1994 [66]	Rabbit	60	1200	6

Mean values are given.

mens (for avoiding slippage), measurements of tissue deformation (for strain determination), cross-sectional area measurements (for stress calculations), length/width ratio of test specimens (for uniform stress), measurement of original length (for strain calculation), and test environment also significantly contributed to these discrepancies. Therefore, investigators should be cautious when they compare previously reported data. The maximum load of the human patellar tendon was compared with that of other tendons by Noyes et al. [72] in 1983. They reported that the 14-mm-wide patellar tendon had a maximum load of 2900N, and that of the semitendinosus was 1216N. The maximum load results indicated that the strongest graft was the bone–patellar tendon–bone specimens. The central and medial portions of this graft had 168% and 159%, respectively, of the load of the ACL. These were the only tissues that had maximum loads greater than those of the ligament being replaced.

3 Remodeling of the Patellar Tendon Grafted Across the Joint

3.1 Biological Aspect of Remodeling

Clancy et al. [15] observed the vascularity of the medial one-third of the patellar tendon after it had been grafted as a substitute for the anterior and posterior cruciate ligaments in rhesus monkeys. Both the ACL and PCL substitutes were revascularized at 8 weeks. Amiel and colleagues [22–26,38,73] described the response of the rabbit patellar tendon graft to new physical and biological (intrasynovial) milieu. Histologically, the autografts were centrally acellular with a peripheral rim of cells at 2 weeks, and had a central focal proliferation of cells at 3 weeks and cellular homogeneity by 4 weeks post operation. Necrosis followed by cellular proliferation suggested that a population of cells other than the native patellar tendon fibroblasts may be inhabiting the graft. The extrinsic contribution of cells was studied by selective destruction of native patellar tendon cells with liquid nitrogen immersion before reconstruction of the ACL [73]. Histological analysis of the tissues harvested 3 weeks postoperatively revealed fibroblastic incorporation of the graft. In contrast, no cells were observed in the sequestered autografts wrapped with semipermeable membrane. These data suggested that ACL autografts of patellar tendon origin are repopulated by cells of external origin. In vitro control studies that were carried out in parallel demonstrated that patellar tendon fibroblasts could survive in tissue culture, but not in the synovial environment of the ACL. This suggested that fibroblasts from different sources have different, tissue-specific nutritional requirements.

Autografts showed a gradual assumption of the microscopic properties of the normal ACL; by 30 weeks after transplantation the tissue characteristics were ligamentous in appearance. Histological changes paralleled the biochemical metamorphosis [22–26]. Although type III collagen was not observed in the patellar tendon, it gradually increased in the patellar tendon graft; by 30 weeks its content was the same as that in the normal ACL (10%). Similarly, glycosaminoglycan content increased from its normally low level in the patellar tendon to that found in the native ACL. Collagen-reducible cross-link analysis demonstrated that grafted tissue changed from the normal pattern of the patellar tendon having low dihydroxylysinonorleucine (DHLNL) and high histidinohydroxymerodesmosine (HHMD) to the pattern seen in the normal ACL (high DHLNL and low HHMD) by 30 weeks [22–26]. These data suggested that patellar tendon autografts undergo a process of "ligamentization" when placed in the ACL environment, and that cells responsible for this metamorphosis are of extragraft origin.

Arnoczky et al. [29] investigated the revascularization process of the patellar tendon grafts used to replace canine ACLs by histological and tissue-clearing techniques. Initially the grafts were avascular, but by 6 weeks they were completely ensheathed in a vascular synovial envelope. The soft tissues of the infrapatellar fat pad, the tibial remnant of the anterior cruciate ligament, and the posterior synovial tissues contributed to this synovial vasculature. Intrinsic revascularization of the patellar tendon graft progressed from the proximal and distal portions of the graft to the central portion and was completed by 20 weeks. At 1 year, the vascular and histological appearance of the patellar tendon graft resembled that of the normal ACL. Yasuda et al. [42] arthroscopically showed that such metamorphosis of the patellar tendon autograft occurs in human patients with ACL reconstruction.

As a conclusion obtained from a number of graft studies, when the autogenous patellar tendon is transplanted across the knee joint to reconstruct the ACL, fibroblast necrosis occurs sequentially during the early period after surgery [15,24,29,34,74]. Then, the patellar tendon graft is initially enveloped with a vascular synovial tissue that originates from the infrapatellar fat pad, synovium, and endosteal vessels emanating from the bone tunnels drilled for the graft attachment [14,15,23,29,30,32,33,35,75]. This process of synovialization occurs during the first 4–6 weeks following transplantation. The tissues that initiated the "synovialization" of the graft are also the source for intrinsic revascularization [15,23,29]. This revascularization response is accompanied by cellular proliferation in the graft. Although the exact origin of these new cells is unknown, several studies have suggested that these undifferentiated mesenchymal cells seem to originate from the synovial membrane [15,23,29,38]. The infrapatellar fat pad and remnants of the old ACL also may contribute to this process. In addition, bone marrow that is exposed during the creation of the bone tunnels may be a source for pluripotential stem cells. The cross-sectional area of the transplanted patellar tendon autograft increases during the process of synovialization and revascularization [15,23,29]. It takes at least 20 weeks before the revascularization of the patellar tendon graft can be completed; more time is required for the graft to obtain the structural and mechanical characteristics of ligament. This metamorphosis of the tendon autograft is considered to be based on the law of functional adaptation defined by Roux in 1905 [23], which states that "an organ will adapt itself structurally to an alteration, quantitative or qualitative, in function."

3.2 Mechanical Aspect of Remodeling

Clancy et al. [15] described that the average maximum load of the patellar tendon autografts expressed as percent of the average maximum load of the control medial one-third of the patellar tendon was 53% at 3 months, 52% at 6 months, and 81% at 12 months in the rhesus monkey. However, the average maximum load of these substitutes expressed as percent of the average maximum load of the control ACLs in the opposite knees was 26% at 3 months, 43% at 6 months, and only 52% at 12 months. Butler et al. [32,57] also reported a detailed study on the mechanical and structural properties of patellar tendon autografts used to replace ACLs in the cynomolgus monkey (Fig. 1). Average cross-sectional areas of the normal ACL and the grafted patellar tendon were 4.9 and 8.9 mm^2, respectively. The 12-week transplanted tissues had the largest average area (255% of the control ACL area). The areas of the 7- and 14-week groups were significantly larger than those of the 29- and 53-week groups ($P < 0.05$). The stiffness and maximum load of the grafts at 7 weeks were 71 N/mm and 105 N (24% and 16% of the control ACL values), respectively, increasing to 189 N/mm and 287 N (57% and 39% of the control ACL values), respectively, at 53 weeks. The modulus and tensile strength were 86 MPA and 10 MPa (12% and 7% of the control ACL value), respectively, increasing to only 239 MPa and 37 MPa (34% and 26% of the control ACL values), respectively, at 53 weeks.

In other animal models, Ballock et al. [14] reported assessment of the degree of success in the ACL replacement using the patellar tendon autograft in a rabbit model. The average maximum load of the patellar tendon autografts was 63 N (15% of the control ACL) at 30 weeks postoperatively. At 52 weeks, the histology of the patellar tendon autograft was apparently similar to that of the native ACL. However, this similarity did not extend to the functional properties of the autograft. Data obtained by Yoshiya et al. [21,76] from dogs revealed that the femur–patellar tendon–tibia complex had a linear stiffness and maximum load that were approximately 10% of those of the control femur–ACL–tibia complex at the time of transplantation because of weakness at the fixation site. By 3

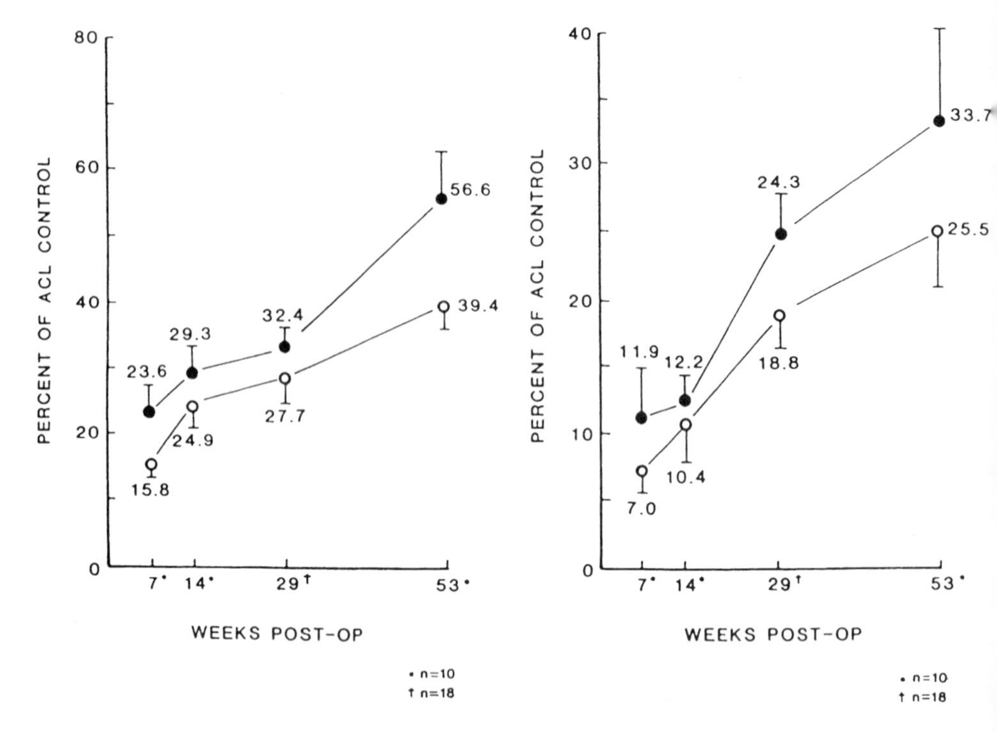

FIG. 1. Normalized graft stiffness (*left, solid circles*), maximum force (*left, open circles*), modulus (*right, solid circles*), and maximum stress (*right, open circles*) plotted against weeks post operation. *ACL*, anterior cruciale ligament. (From [32], with permission)

months postoperatively, the linear stiffness increased to 20% and maximum load increased to 30% of the control values. Slightly better biomechanical results for reconstructed ACL grafts were obtained in the goat [18]. Jackson et al. [52] also reported that the cross-sectional area of the patellar tendon autograft at 6 months was 164% of the control ACL, and the strength of the autograft was 62% of the control ACL strength.

In conclusion, patellar tendon grafts morphologically resemble the normal ACL with time. However, many biomechanical studies have demonstrated that there is a great decrease in the tensile strength of patellar tendon grafts immediately after transplantation. The reduced strength is gradually recovered with time as remodeling proceeds [14,15,32,35,37,72], although it does not completely return to normal ACL strength. Although longer term studies are necessary to determine whether the properties of patellar tendon grafts continue to improve, strength generally seems to reach a plateau at 1–2 years postoperatively. The maximum load recovers to only 30%–50% of that of the control ACL after 1 year in animal models. On the other hand, clinical outcome of ACL reconstruction has shown that a part of the patellar tendon as well as doubled semitendinosus and gracilis tendons are acceptable. The 2- to 5-year follow-up examination indicated that approximately 80%–90% of the patients who undergo ACL reconstruction regain normal knee stability in which the side-to-side anterior translation of the tibia is within 3mm [14–21,55,77–80]. However, no studies have measured the mechanical and structural properties of the reconstructed ligament tissues in the human. Moreover, Oakes [41] reported that collagen fibrils in reconstructed ligament tissues older than a few years were smaller than those in the normal ACL. Much remains unexplained about the discrepancy of results between human and animal ACL reconstruction. The reasons may involve differences in not only species but also surgical procedures, graft fixation methods, and postoperative management.

3.3 Ultrastructural Aspect of Remodeling

Jackson et al. [52] compared transmission electron micrographs of ACLs reconstructed using patellar tendon autografts with the normal patellar tendon and the ACL in the goat model. The patellar tendon autograft observed at 6 months after reconstruction demonstrated large increases in the numbers of 25- to 50-nm-diameter fibrils; these small-diameter fibrils accounted for 84% of the fibril population. On the other hand, the diameter of 80% of the collagen fibrils was in the range of 100–200nm in the normal patellar tendon, and that of 80% was 100–150nm, with no 200-nm fibrils in the normal ACL [52].

Oakes [41] observed, using transmission electron microscopy, collagen fibrils biopsied from autogenous patellar tendons grafted in human patients. In the earliest graft obtained at 6 months, a large number of small fibrils and the loss of large fibrils were observed. The most significant feature of all the specimens was the invariable predominance of small-diameter fibrils among a few larger, probably original, patellar tendon fibrils. Absence of an ordered "regular crimping" of collagen fibrils was observed by both light and electron microscopy. The collagen fibrils were not tightly packed, as in normal ligaments, and the extrafibril noncellular space was greatly increased compared to the normal ACL. The large cross section occupied by small-diameter fibrils persisted through all age groups of grafts, although the cross-sectional area of fibrils larger than 100nm seemed to increase in the old grafts, those of 5–9 years. This fact suggested that the cross-sectional area of small fibrils seemed to slowly decrease with time.

In addition, Oakes [41] also quantified collagen fibril size of the iliotibial and hamstring autografts obtained from human patients. It was noted that small fibrils less than 100nm in diameter predominated in all these autogenous grafts, which was quite different compared to the normal ACL collagen fibril profile. Some large fibrils were present in the hamstring grafts, which probably reflected remnants of the original large fibrils in normal hamstring tendons rather than the formation of large fibrils from preexisting medium-sized fibrils. This persistence of a small number of large fibrils seen in the hamstring grafts was not observed in the iliotibial band or in patellar tendon autografts.

The studies just described suggest that the predominance of small-diameter fibrils in the graft may be an essential change appearing in the remodeling of autografts. It is considered that small-diameter collagen fibrils may be synthesized at the expense of large-diameter fibrils during the repair process after ACL reconstruction. Another mechanism for the observed change to small-diameter collagen fibrils may be related to a change in the type of glycosaminoglycans synthesized by the newly infiltrated fibroblasts [81,82]. Amiel et al. [25] have shown that the rabbit cruciate ligaments have more glycosaminoglycans than the patellar tendon; this may be an important factor in determining collagen fibril populations.

3.4 Mechanism of Deterioration of the Mechanical Properties of Autografts

Deterioration of the mechanical properties of autografts was commonly observed in all the previous studies and seems to be one of the essential phenomena in the remodeling of autografts. The reasons for deterioration of tensile strength in the patellar tendon graft observed following transplantation have been a subject of speculation. One plausible explanation is that the increase in the number of vessels present within the ligament may effectively decrease the amount of ligamentous tissue, thus reducing the tensile strength of the graft [83,84]. In accordance with this theory, a novel surgical procedure that preserves the normal blood supply in a graft to eliminate the period of ischemic necrosis and revascularization following patellar tendon transplantation has been developed to minimize the reduction of graft strength following transplantation [32,72]. However, Butler et al. [32,57] demonstrated, using cynomolgus monkeys, that the mechanical properties of the vascularized patellar tendon grafts were not superior to those of free grafts. In addition, the maintenance of graft vascularity did not accelerate the improvement of graft strength in the remodeling tissue. Therefore, hypervascularity seems to play an extremely limited role in deterioration of the tensile strength of the autograft.

Another explanation of the reduction of graft strength during the remodeling process is related to an apparent change in collagen synthesis. Although type I collagen is still the predominant matrix material and collagen cross-link patterns are comparable to those of the normal ligament, the diameter of newly synthesized collagen fibrils is markedly different from that of the normal ACL; the predominance of small-diameter collagen fibrils and their poor packing and alignment in all ACL grafts are observed irrespective of the type of grafts and their age. This fact may explain the clinical and experimental evidences of decreased tensile strength of grafts. It appears that the newly infiltrated fibroblasts in the remodeled ACL grafts cannot reform the large-diameter, regularly crimped, and tightly packed collagen fibrils seen in the normal ACL, although there is a lack of scientific data to substantiate these possibilities.

In addition, we have to recognize that the degree of the reduction of mechanical strength may have been incorrectly estimated in those studies. At the time of uniaxial tensile testing, the mechanical and structural properties of the femur–autograft–tibia complex depend on the orientation of the graft fibers as well as the angle of knee flexion [56]. Extremely low values of the linear stiffness and maximum load of autografts compared with the control ACL values may be partly explained by poor direction of tensile load against ligament alignment. From this point of view, an ACL reconstruction study reported by Nikolaou et al. [85] revealed a very successful outcome using bone–real ACL–bone autografts in dogs in which the orientation of the grafted fibers was nearly normal. They reported the maximum load of the graft complex at 9 months to be 89% of the control femur–ACL–tibia complex, which is extremely different from the 30%–50% reported by others.

4 Analyses of Factors Affecting the Remodeling Process of Autografts

The data obtained from the animal studies described here have provided us with invaluable information because the mechanical data from these studies are supported by ultrastructural studies. However, we should consider that the remodeling process and the outcome of autografts are affected by many factors involving graft material selection [55,72,86], intraarticular graft position [87–90], initial graft tension at the time of fixation [91–94], the method of graft fixation to bone [95,96], and the mechanical conditions surrounding the graft [66,67,97,98]. Therefore, the data obtained from previous experimental studies using animal models should be carefully reviewed. To more precisely understand the remodeling of autografts, it is necessary to identify the effects of each factor affecting the remodeling process.

4.1 Effects of Fibroblast Necrosis on Autografts

Fibroblast necrosis has been considered to be most essential in the autografts transplanted for ACL reconstruction. Therefore, one of the most fundamental questions is: What are the effects of fibroblast necrosis on the patellar tendon autograft? To answer this question, we should make experimental models that can simulate only fibroblast necrosis without surgically disturbing the anatomical orientation, physiological tension, and anatomical attachment of the tissues. A few useful models have so far been proposed and used for this purpose.

4.1.1 Effects of Fibroblast Necrosis on the Patellar Tendon

We developed a unique rabbit model to clarify the effects of in situ freezing on the patellar tendon [67,97,99]. This model simulates an extraarticular autograft model as an alternative to intraarticular ACL reconstruction using patellar tendon autografts. After the anterior part of the rabbit's right knee was exposed through a midline longitudinal skin incision, subcutaneous retinacula were incised longitudinally along the patellar tendon edges. The posterior surface of the patellar tendon was then separated from the infrapatellar fat pad. A silicone

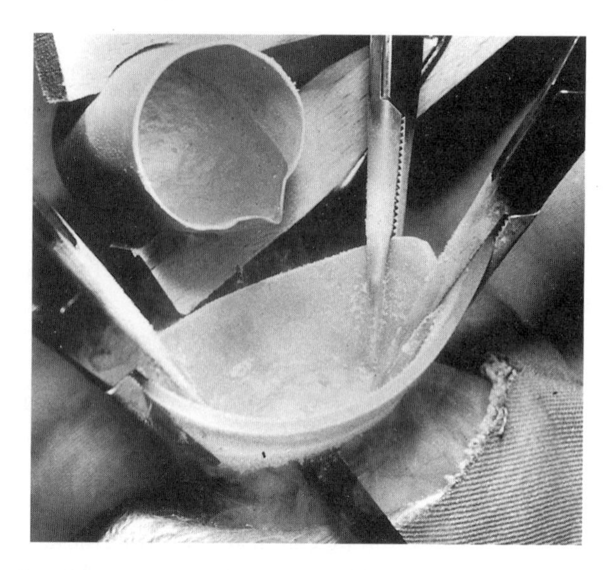

Fig. 2. In situ freezing procedure for the patellar tendon. A silicone rubber sheet is inserted between the patellar tendon and the fat pad, and the edges of the sheet are clamped to make a bath with the tendon inside it. Liquid nitrogen is then poured into this bath

rubber sheet was inserted between the patellar tendon and the fat pad, and the edges of the sheet were clamped with four graspers to make a bath with the patellar tendon inside it. Liquid nitrogen was poured into the bath; the patellar tendon was frozen for 1 min (Fig. 2). The frozen patellar tendon was then thawed by pouring a physiological saline solution into the bath. In this model, fibroblast necrosis is easily and quickly stimulated without disturbance of the tendon attachment sites and alteration of the mechanical environment of the patellar tendon.

Effects of the in situ freezing treatment (Table 3) were observed for 24 weeks [99,100]. The cross-sectional area of the patellar tendon started to increase by 3 weeks, reaching a plateau at 12 weeks. The elastic modulus and tensile strength of the tendon began decreasing by 3 weeks, and decreases then were marked at 12 and 24 weeks. In contrast to the tremendous decrease in tensile strength, the maximum load decreased only to approximately 80% and 60% of the control values at 12 and 24 weeks, respectively. Histologically, cells were absent until 2 weeks, while a large number of new fibroblasts appeared by 3 weeks. Although the previously frozen patellar tendon had apparently normal cells at 24 weeks, it did not contain such dense collagen bundles as normally seen. This study demonstrated that fibroblast necrosis itself does not change the biomechanical properties of the patellar tendon, but that the frozen patellar tendon is weakened during the subsequent period of tissue remodeling.

To clarify the effects of fibroblast necrosis on the ultrastructure of the patellar tendon, a study was conducted with transmission electron microscopy [47,48].

TABLE 3. Effects of in situ freezing on the patellar tendon. (From [100], with permission).

Parameter	Time (weeks)				
	0	2	3	12	24
Cross-sectional area (mm^2)					
Frozen	13.6	14.1	18.3*	24.6*	23.7*
Control	13.4	13.7	14.3	13.4	13.4
Elastic modulus (MPa)					
Frozen	1510	1602	932*	449*	427*
Control	1284	1490	1457	1465	1369
Tensile strength (MPa)					
Frozen	63.8	72.0	48.4*	29.8*	24.6*
Control	63.2	58.0	61.7	67.0	64.6
Maximum load (N)					
Frozen	867	1016	876	728*	540*
Control	848	803	882	894	866

Mean values ($n = 5$ for each period) are given.
*Significantly different from the control group at each corresponding period ($P < 0.05$).

Ultrathin sections were obtained from the central one-third of each patellar tendon at 3 and 6 weeks after the freezing treatment. Diameters of all collagen fibrils in a 9-μm^2 area were measured and histograms of the diameters were determined. In addition, the average of the collagen fibril numbers in a 1-μm^2 area, and the ratio of the total area of the collagen fibrils to the whole visualized area, were calculated. This study clearly demonstrated that freezing had a strong influence on the ultrastructure of the patellar tendon. In the normal patellar tendon (Fig. 3a), fibril diameters ranged from 30 to 360 nm, and the histogram was bimodal. In the frozen patellar tendon, the results were almost the same as those in the control at time zero and at 3 weeks. At 6 weeks, however, the number of thin fibrils having a diameter less than 90 nm increased (Fig. 3b), and the histogram became unimodal. These observations provide useful information for better understanding the cause of reduced tensile strength in the previously frozen patellar tendon. Reduction of the total area of fibrils in the cross section, reduction of the frictional force developed between fibrils, and decreased tensile strength of the collagen fibril itself are considered to be the reasons for the reduced tensile strength of the patellar tendon.

4.1.2 Effects of Fibroblast Necrosis on the ACL

Jackson et al. [101] reported a multiple freeze–thaw technique to kill the cells and interrupt the vascular supply of the ACL, using a special probe to freeze the goat ACL at −196°C in situ. The freeze–thaw technique represents the ideal operative placement of biological grafts in which the collagen fibers are in the proper anatomical position, with the appropriate length, as well as physiologically tensioned and biologically fixed to bones. The freeze–thaw-treated ACLs had

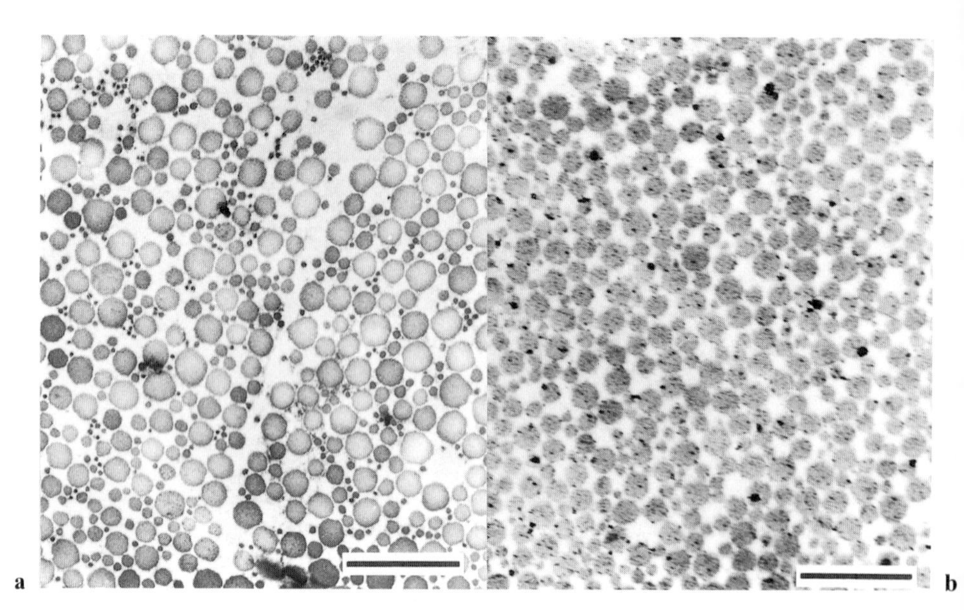

FIG. 3a,b. Ultrastructure of patellar tendons (cross section of collagen fibrils). **a** Control specimen. **b** Patellar tendon at 6 weeks after the sham operation (in situ freezing). *Bars*, 1 μm. (From [100], with permission)

less vascular response at 6 months compared with grafted tendons. At 6 months, no statistically significant differences were noted between the treated and contralateral control ligaments regarding the anteroposterior translation, maximum load, stiffness, modulus of elasticity, and strain at failure.

Significant differences observed were an increase in the cross-sectional area of the ligament and a decrease in the tensile strength (Table 4). The cross-sectional area was 42% greater than the control ligaments at 6 months. On the other hand,

TABLE 4. Structural properties of in situ frozen ACLs in the goat. (From [101], with permission).

Postoperative time (weeks)	Specimen	Area (mm²)	Maximum load (N)	Elastic modulus (MPa)	Tensile strength (MPa)	Strain at maximum stress (%)
0	Control	16.0	2201	517	133	36
	Frozen	15.5	2366	611*	157	33
26	Control	17.7	2603	578	155	32
	Frozen	25.2*	2380	403	93*	30

Mean values are given.
*$P < 0.05$ (control *vs* frozen).
Note that significant differences were observed only in the cross-sectional area and the tensile strength at 26 weeks.

the tensile strength was 40% less than that at 6 months. The increase in the size of the ligament that was treated with the freeze–thaw technique was associated with a significant change in the size of collagen fibrils; there was an increase in smaller diameter collagen fibrils. This increase in cross-sectional area was associated with a robust biological response. The authors noted that devitalized, devascularized ACLs do not lose their structural strength (maximum load) if the anatomical position and the orientation of the collagen fibers are not altered.

Nikolaou et al. [85] reported another unique experimental study on a nearly ideal canine autograft. They removed the ACL with a cylindrical bone plug at the femoral side and a triangular bone plug at the tibial side. After the ACL was placed in a 37°C saline solution for 15 min, it was reinserted into the same knee and secured with a cancellous screw in the femur and with stainless steel wire in the tibia. Histological sections showed firm union between the host and recipient bones in the femur and tibia. At 8 weeks, fibroblastic proliferation had occurred in the periligamentous portion of the graft, with a central area of diminished cellularity. An increase of fibroblasts was evident at 16 weeks, and the collagen bundles of the ligament were more oriented. At 24 and 36 weeks, the structure of the ligament insertion sites, where the four histological zones could be distinguished, appeared normal. The autograft model resembled a normal ligament with dense, longitudinally oriented collagen bundles and normal cell population.

Microangiography showed revascularization in the graft at the 8th week with a small, relatively hypovascular zone in the middle of the ligament. From the 24th week, the microangiographic appearance of the autograft was normal. The maximum load of the autograft increased significantly between the 8th and 16th weeks, from 54 N (46.2% of control) to 86 N (80.7% of control). After the 16th week, the strength of the autograft dramatically increased, and it was 91 N (87.3% of control) at 36 weeks. The stiffness of the autograft was 6.8 N/mm (84% of control) at 8 weeks, gradually increasing to 8.7 N/mm (101% of control) at 36 weeks. These results are much superior to those obtained from the other tendon autograft studies. However, it should be noted that some degree of reduction in the maximum load was observed, that these investigators did not measure the tensile strength, and that their reported value for maximum load of the control femur–ACL–tibia complexes was much lower than those reported by other investigators [58,59,75,102].

The authors of these two studies concluded that the reduction in the maximum load that occurs postoperatively in autografts used to reconstruct the ACL may not be the natural sequela of the revascularization and healing process, but rather the consequence of improper orientation and tensioning of the graft. This is a striking finding. However, we should also note that the tensile strength of the graft was decreasing even at 6 months, with an increase of smaller diameter collagen fibrils, in the study by Jackson et al., although they did not emphasize it.

4.1.3 Clinical Relevance of the Freeze-Thaw Studies

Significant reduction of the tensile strength was observed in our studies using the in situ frozen patellar tendon. Also, Jackson et al. [101] demonstrated the reduction of the tensile strength could not be avoided even in their "ideal" ACL reconstruction model, although Nikolaou et al. did not measure the strength. The reduction of strength seems to be caused by the ultrastructural changes of the collagen fibrils. We believe, therefore, that the ultrastructural changes of collagen fibrils are essential effects of the extrinsic cell proliferation following the intrinsic fibroblast necrosis. Concomitantly, reduction of the tensile strength of the autograft may be unavoidable in ACL reconstruction unless we can control the diameter and structure of the collagen fibrils.

Jackson et al. demonstrated that when an in situ frozen–thawed ACL was maintained under normal tension, a significant reduction in the maximum load did not appear during the period of revascularization and cellular repopulation. This result was supported by the study done by Nikolaou et al. on the previously removed necrotic ACLs. They pointed out that the reduction of tensile strength might be well compensated by increasing the cross-sectional area of the graft if the ACL would be "ideally" reconstructed, and that the surgical technique and the graft tension and orientation might have the most profound influence on the ultimate structural properties of the reconstructed ligament tissues in "real" transplantation surgery with patellar tendon grafts. Under the mechanical conditions surrounding the ACL graft, Roux's law seems to be effective in controlling the graft when the fiber orientation and tension are physiologically reconstructed.

On the other hand, the results of our studies implied that when the previously frozen patellar tendon is maintained under normal tension, reduction of tensile strength during the period of revascularization and cellular repopulation is not completely compensated by increase of the cross-sectional area, resulting in failure of the patellar tendon. One potential reason why the degree of the compensation is different between the two previously frozen tissues, the ACL and the patellar tendon, may be related to the difference of the surrounding biological environment; the patellar tendon in our studies was exposed to the extraarticular environment, while the ACL was used in an intraarticular environment. Another potential reason may be related to the difference of the surrounding biomechanical environment. During daily activities, for animals as well as human beings, a high load is commonly exerted on the patellar tendon, while a relatively low load is applied to the ACL. For example, during maximal isometric contraction of the quadriceps in human beings, forces of several thousand newtons are loaded on the patellar tendon, although those of only a few hundred newtons are loaded on the ACL [98]. For the previously frozen patellar tendon, reduction of tensile strength cannot be compensated by increase of the cross-sectional area because a high load is often exerted. It is considered that a range of the "appropriate" graft tension for the tendon autografts and the ligament autografts may be narrow and tissue specific.

4.2 Effects of Stress on Autografts

4.2.1 Effects of Stress Deprivation on the Previously Frozen Patellar Tendon

While biological factors (revascularization, cellular repopulation, matrix synthesis) are important in the remodeling process of transplanted tissue, results of recent studies have suggested that physical factors acting on the graft may play a significant role in determining the ability of a transplanted graft to better maintain its material properties. Several experimental studies have been conducted to clarify the effects of the physical factors on biological grafts. Stress deprivation has been one of the foci on this issue. Fowler and Amendola [103] described the normal remodeling process that occurred after augmented ACL reconstruction on the basis of their findings that the collagen content in the graft was not altered by the presence of synthetic devices. Yoshiya et al. [76] investigated the effects of the initial load of 1 and 39N on a patellar tendon graft using a canine model, and detected no significant difference in the values for the maximum load of the graft 3 months after reconstruction. However, they observed poor vascularity and focal myxoid degeneration within the graft pretensioned with a load of 39N, but not within the graft to which was applied 1N tension. Conversely, McCarthy et al. [104] found that stress shielding caused by a synthetic augmentation device prolonged graft remodeling. Andrish and Woods [105] reported that autogenous patellar tendon grafts augmented with Dacron devices used in ACL reconstruction were absorbed because of stress shielding. Yoshiya et al. [21] suggested that stress shielding caused by such augmentation had adverse effects on the strength of autografts. Thus, effects of stress deprivation on patellar tendon grafts remain controversial. We should note that those studies were performed using autogenous tendons grafted across the joint, in which many factors were considered to affect the outcome. Therefore, it is necessary to isolate the effects of stress deprivation on the autografts with normal fiber orientation and tension from those of the other factors.

We have conducted a series of precise biomechanical studies to clarify the effect of stress shielding on the patellar tendon with fibroblast necrosis [66,67,97,99,100,106–109]. First, we studied the effects of complete stress shielding on the previously frozen patellar tendons using two artificial materials, a flexible stainless steel wire 0.97mm in diameter and a 10-mm-wide, commercially available Leeds-Keio artificial ligament, as augmentation devices of different stiffness. Stress-shielding techniques [67,97,109] used in the two studies were almost identical. A stainless steel screw was inserted transversely into the tibial tubercle and a stainless steel pin was placed in the middle part of the patella (Fig. 4). The patellar tendon was slackened without immobilization of the knee joint by drawing the patella toward the tibial tubercle, and this position was maintained using one of the two artificial materials, which was installed between the screw and the pin. It was confirmed that the patellar tendon was relaxed at all knee angles not only immediately after surgery but also immediately before the animal was killed.

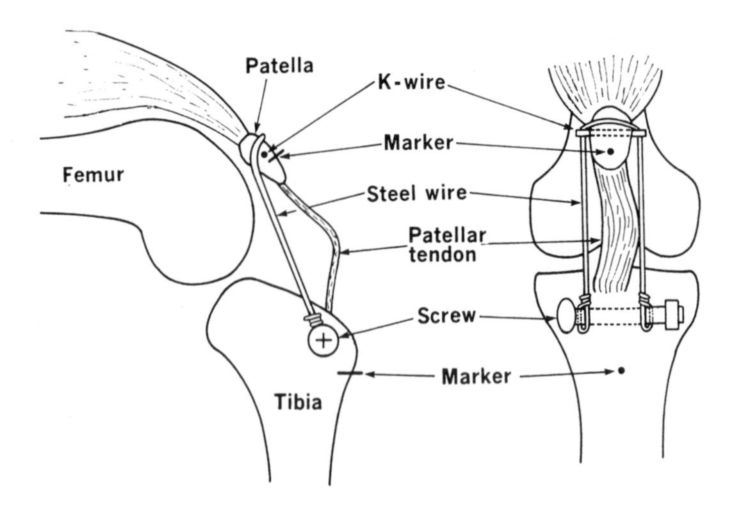

Fig. 4. Complete stress-shielding technique

We also developed a unique technique for the experiments of controlled partial stress shielding [66,106]. A buckle transducer [108,110] was attached to the patellar tendon. The transducer was calibrated by pulling the patella to the proximal direction using a load cell. Monitoring the output from the buckle transducer, the patellar tendon was slackened by drawing the patella toward the tibial tubercle with the artificial ligament until the output from the buckle transducer decreased to 30% of the reference value. Then, the artificial ligament was then fixed firmly with a clamp. After ten cycles of loading between 0 and 80 N had been applied to the patellar tendon–artificial ligament complex, a load–displacement curve was again recorded. In this study, 27.3% ± 0.9% (mean ± SE) of the total force was applied to the patellar tendon at the operation, while 72.7% of the total load was applied to the artificial ligament [66,106].

4.2.1.1 Biological Aspect of the Effects of Stress Deprivation

To clarify the effects of complete and partial stress shielding, we compared the in situ frozen patellar tendons (sham group) with the patellar tendons completely released from tension after the freezing treatment [completely shielded from stress (CSS) group] and the patellar tendons partially released from tension after the treatment [partially shielded from stress (PSS) group]. No cells were observed within the first 2 weeks in any group; new cells appeared after 3 weeks. There were essentially no differences in the light microscopic structure between the PSS (Fig. 5b) and sham groups in this period; however, the microstructure of the patellar tendon in the CSS group was obviously different from that in the other two groups (Fig. 5c); collagen fibers were fragmented in the CSS group but not in the other groups, and the nuclei of the fibroblasts were more round in the CSS group than in the other groups. These observations show that partial and complete stress shielding have different effects on patellar tendon microstructure [66,67].

4.2.1.2 Mechanical Aspect of the Effects of Stress Deprivation

Figure 6 demonstrates the averaged stress–strain curves for the control, sham (Fr in the figure), and CSS (Fr/CSS in the figure) groups. In the sham group, the slope of the curve started decreasing at 3 weeks. On the other hand, the slope for the CSS group started decreasing at the first week and reached a maximum at 3 and 6 weeks. The slope of the CSS group was significantly different from that of the sham group at each period. To quantitatively compare the initial slopes of the stress–strain curves between the CSS and sham groups, the experimental stress–strain data in strain increments of 0.5% to 3.0% were analyzed [111] using the constitutive equation derived by Gomez [112]. The material constants showed that the initial slope of the stress–strain curve of the group without stress shielding was significantly steeper than that of the group with stress shielding for 3 weeks but not for the group shielded for 6 weeks. The increase in the tangent modulus (the gradient of the stress–strain curve) of the group without stress shielding was significantly more than that of the group with complete stress shielding for both 3 and 6 weeks.

Stress shielding significantly affected the mechanical properties of the in situ frozen patellar tendons. Figure 7 demonstrates effects of complete and partial stress shielding on the tensile strength. There were no significant differences in the effects of complete stress shielding between the two experiments described earlier [66,67] using different artificial materials for augmentation devices. Partial or complete stress shielding significantly decreased the tensile strength within 1 week after treatment; the mean strength in the PSS and CSS groups at 1 week was 60% and 55% of that of the sham group, respectively. On the other hand, in the sham group to which only the freezing treatment was applied, the tensile strength started decreasing between 2 and 3 weeks. This result indicates that the stress shielding enhanced reduction of the strength in the previously frozen patellar tendon. With respect to the difference between complete and partial stress shielding, the tensile strength and tangent modulus of the partially stress-shielded patellar tendon were generally lower than those of the completely stress-shielded patellar tendon; the differences were significant at 3 and at 6 weeks. These results indicate that the strength and modulus of the patellar tendon change depending on the degree of stress shielding.

When significant reduction of the tensile strength and the tangent modulus was observed at 1 week in the PSS and CSS groups, there were no cells present. These results demonstrated that mechanical response of the patellar tendon to stress shielding occurs independent of living cells [67]. Even after cells appeared in the patellar tendon, the reduced strength and modulus did not increase, which implies that those cells were not effective in tissue remodeling.

Stress shielding significantly increased the cross-sectional area of the whole patellar tendon within 2 weeks after treatment (Fig. 8); the whole cross-sectional

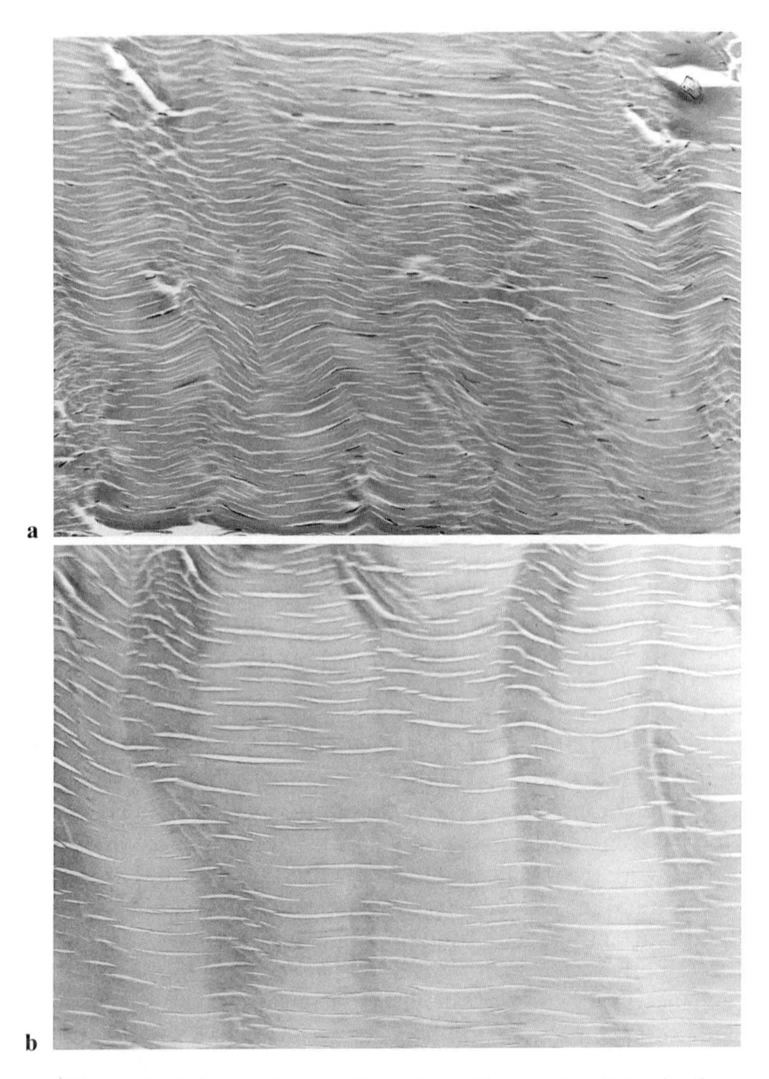

FIG. 5a–d. Histological observations. **a** Control patellar tendon. **b** In situ frozen patellar tendon partially shielded from stress for 2 weeks. **c** In situ frozen patellar tendon completely shielded from stress for 3 weeks. Note that collagen fibers are fragmented. **d** Dorsal portion of the in situ frozen patellar tendon in the 6-week, completely stress-shielded group. Most cells appeared to be ovoid. (From [66], with permission)

area in the PSS and CSS groups at 2 weeks was 133% and 156% of that in the sham group, respectively. Figure 9 demonstrates effects of complete and partial stress shielding on the estimated maximum load [66]. The maximum load in the PSS group at 1, 2, 3, 6, and 12 weeks was 68%, 78%, 63%, 85%, and 68% of the corresponding load in the sham group; however, there were no significant differences between the PSS and sham groups at 1, 2, and 6 weeks. The increase in area

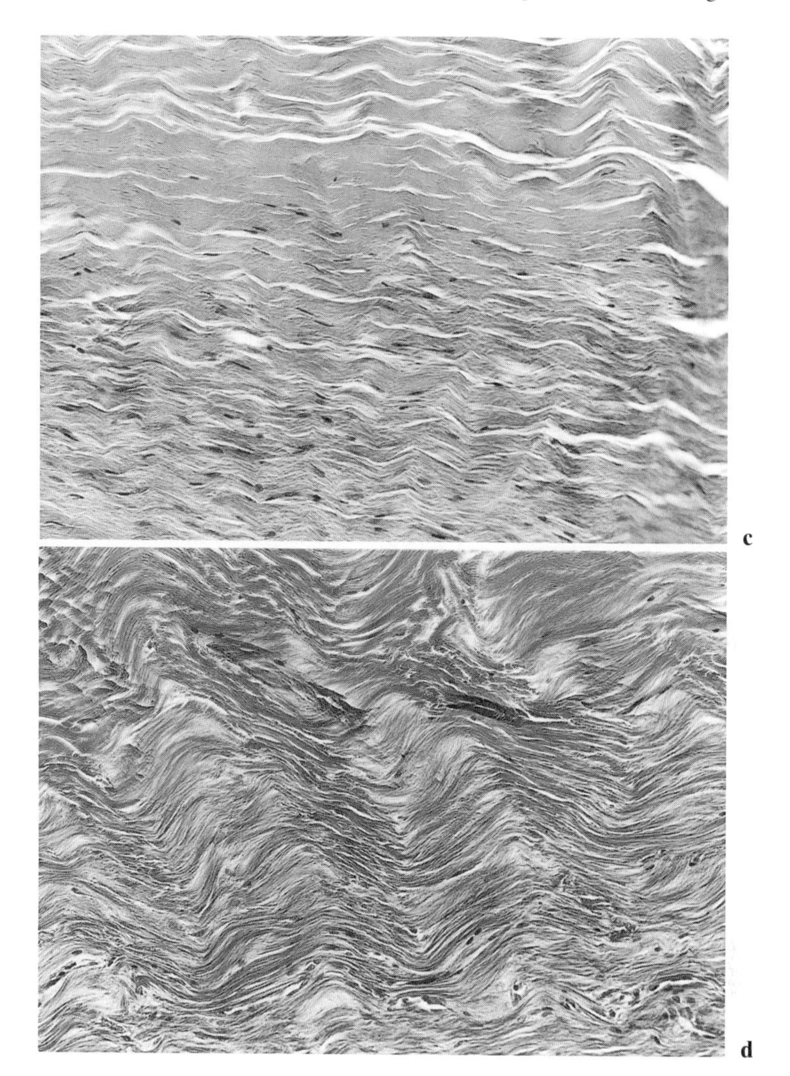

FIG. 5. *Continued*

seems to be compensating for the decrease of tensile strength and keeping the maximum failure load at a normal level. The maximum load in the CSS group was much lower in comparison with that in the sham group throughout the experimental period. Between the PSS and CSS groups, the maximum load was significantly larger in the former group than in the latter group at 3 and 6 weeks. These results imply that there is some limit in the compensation effect brought about by the increase of the cross-sectional area, depending on the extent of stress shielding.

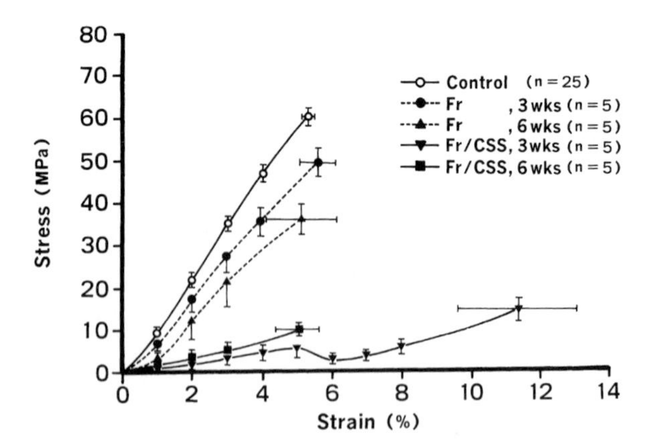

FIG. 6. Stress–strain curves (mean ± SE) for the control, sham (*Fr*, frozen in situ), and CSS (*Fr/CSS*, completely shielded from stress after freezing) groups. The slope of the Fr group was significantly steeper than that of the Fr/CSS group at each period. (From [100], with permission)

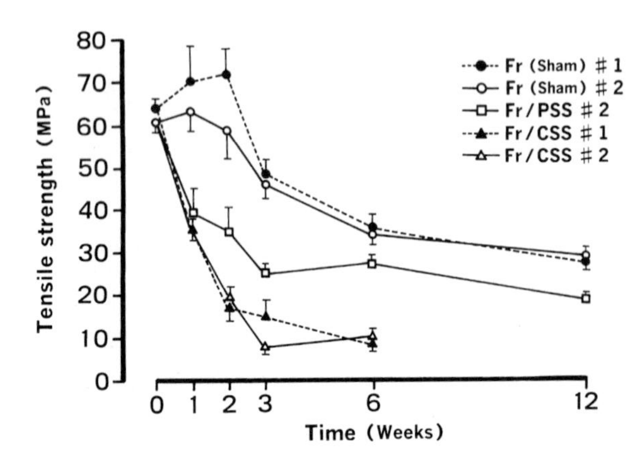

FIG. 7. Reduction of tensile strength (mean ± SE) in proportion to the extent of stress shielding. *Fr*, frozen; *Fr/PSS*, partially stress-shielded after freezing; *Fr/CSS*, completely stress-shielded after freezing. *#1*, Data reported by Ohno et al. [67]; *#2*, data reported by Majima et al. [66]. (From [100], with permission)

4.2.1.3 Ultrastructural Aspect of the Effects of Stress Deprivation

To clarify the effects of complete stress shielding on the ultrastructure of the in situ frozen patellar tendons, transmission electron micrographs obtained from the patellar tendons completely shielded from stress (Fr/SS tendons) for 3 and 6 weeks after the freeze–thaw treatment were compared with those obtained from the patellar tendons to which was applied normal stress for the same periods after the identical treatment (sham-operated tendons) [47,48]. Diameters of all col-

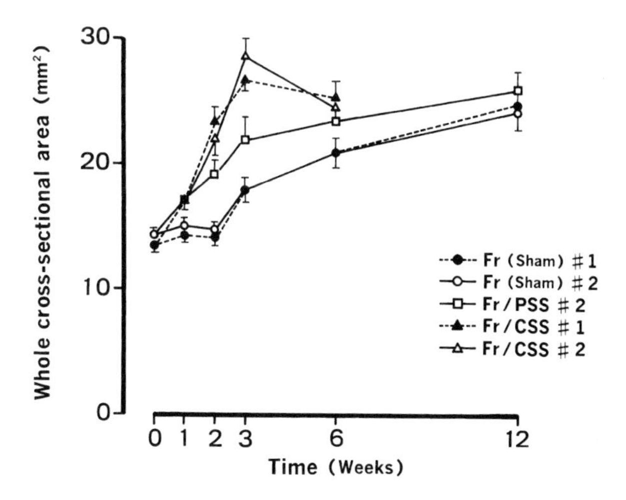

FIG. 8. Increase of cross-sectional area (mean ± SE) in proportion to the extent of stress shielding. *Fr*, Frozen; *Fr/PSS*, partially stress-shielded after freezing; *Fr/CSS*, completely stress-shielded after freezing. *#1*, Data reported by Ohno et al. [67]; *#2*, data reported by Majima et al. [66]. (From [100], with permission)

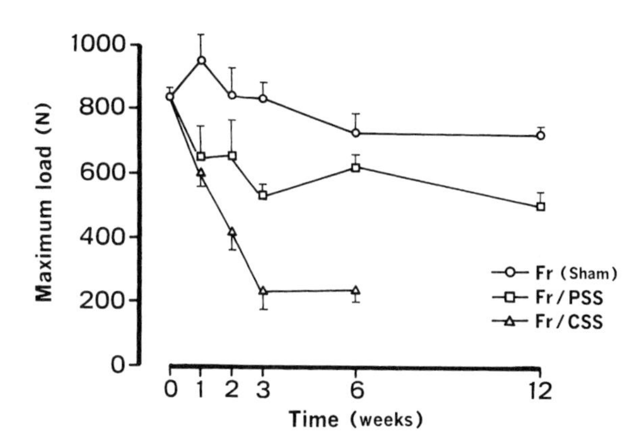

FIG. 9. Reduction of maximum load (mean ± SE) in proportion to the extent of stress shielding. There were no significant differences between the PSS and sham groups at 1, 2, and 6 weeks. *Fr*, Frozen; *Fr/PSS*, partially stress-shielded after freezing; *Fr/CSS*, completely stress-shielded after freezing. (From [100], with permission)

lagen fibrils in a 9-μm^2 area were measured and histograms of the diameters were determined. In addition, the average of the collagen fibril numbers in a 1-μm^2 area and the ratio of the total area of the collagen fibrils to the whole visualized area (%CF) were calculated.

This study clearly demonstrated that complete stress shielding had a strong influence on the ultrastructure of the frozen patellar tendon. In the Fr/SS tendons at 3 weeks, the number of thin fibrils decreased (Fig. 10), reducing the total

FIG. 10. Ultrastructure of the previously frozen and 3-week completely stress-shielded patellar tendon. *Bars*, 1 μm. Compare to Fig. 3. (From [100], with permission)

TABLE 5. Number of collagen fibrils per square in micrometer previously frozen patellar tendon completely shielded from stress for 0, 3, and 6 weeks.

Group	Period of stress shielding (weeks)		
	0	3	6
Control	24.4 ± 1.4		
Frozen (Fr)	24.6 ± 3.1	24.2 ± 2.1	27.3 ± 3.7
Frozen/stress-shielded (Fr/SS)		16.6 ± 0.9*	10.6 ± 1.1*,**

$n = 5$; values are mean ± SD.
* Significantly different in comparison with the data in the Fr group measured at each period ($P < 0.05$).
** Significantly different in comparison with the data in the Fr/SS group measured at 3 weeks ($P < 0.05$).

number of fibrils (Table 5). Thick fibrils having a diameter from 360 to 420 nm were observed, and the histogram showed a bimodal distribution. In contrast, no significant changes were observed between the sham-operated and control tendons at this period. At 6 weeks, the features observed at 3 weeks were more evident in the Fr/SS tendons, while the number of thin fibrils less than 90 nm in diameter increased in the sham tendons (see Fig. 3b). With respect to %CF, complete stress shielding reduced this parameter in the in situ frozen patellar tendon, while freezing did not change it (Fig. 11). These observations provide useful information for better understanding the cause of reduced tensile strength in the patellar tendon caused by freezing and stress shielding. Reduction of the

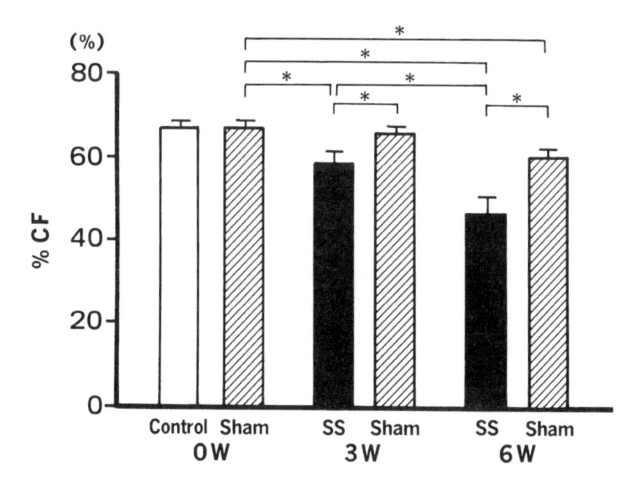

FIG. 11. Ratio of the total area of fibrils to the whole visualized area (*%CF*, percent collagen fibers). Complete stress shielding (*SS*) reduced this ratio in the in situ frozen patellar tendon; freezing (sham) did not change it. *, Significant difference between the two groups (ANOVA, $P < 0.01$; $n = 5$). (From [100], with permission)

total area of fibrils in cross section, reduction of the frictional force developed between fibrils, and decreased tensile strength of the collagen fibril itself were considered to be the reasons why the tensile strength of the patellar tendon was reduced.

A series of our studies showed that complete stress shielding changes the ultrastructure and tensile strength of the previously frozen patellar tendon at 3 weeks. These changes occurred in the acellular tendon. If "remodeling" is defined as the changes generated by living cells, we may not regard these described changes as the result of remodeling, but rather as results of direct effects applied by stress shielding on collagen fibers. It is an interesting phenomenon that changes of physical environment acutely affect the structure and the mechanical properties of the biological tissue, although we have not yet clarified the mechanism of this phenomenon in detail.

4.2.1.4 Clinical Relevance for Augmentation Procedure in Ligament Reconstruction

In the clinical field, the early weakness of the graft is a serious problem because the risk of failure increases if a graft is exposed to high load during physical activities soon after surgery. To prevent failure of patellar tendon grafts, they have been often reinforced with synthetic prostheses [21,76,103,104,105,113,114]. It has been believed that these augmentation procedures would protect the grafts from excessive load. In this mechanical condition, however, the devices shield the stress exerted on the grafts. Our studies imply that complete stress shielding yields adverse effects on augmented biological grafts at a very early stage after augmentation surgery. Therefore, we suggest that clinicians should avoid complete stress shielding when they apply augmentation techniques to ligment reconstruction. Our studies, however, also demonstrated that reduction of the maximum load of the patellar tendon was minimal in the case of partial stress shielding, suggesting that the concept of augmentation itself should not be discarded.

4.2.2 Effects of Stress Deprivation on the ACL Removed from the Tibia

We clarified effects of stress deprivation on the mechanical properties of the canine ACL removed from the tibia [58,59]. Twenty-seven mature mongrel dogs were divided into two groups: relaxed group and sham group. The tibial insertion of the ACL in the right knee was made free from the tibia as a cylindrical bone block by cutting with a hollow reamer (Fig. 12). In the relaxed group, this block was dislocated proximally by 3 mm to anatomically relax the ACL. In the sham group, this block was anatomically reduced (sham operation). The left knee in each dog was given no treatment and was used to obtain control data. After 6 or 12 weeks of unrestricted activities in a cage, each animal was sacrificed under general anesthesia. A tensile test was carried out on the femur–ACL–tibia complex. The preparation was attached to a set of specially designed grips and then mounted onto a tensile tester at 135° of knee flexion and 90° of rotation. The effects of the treatments (relaxation of the ACL and sham operation) on the tensile strength and maximum load in each dog were expressed as the ratio of each value in the treated knee to that in the counterpart intact knee [treated/untreated (T/U) ratio]. This study demonstrated that stress deprivation in the ACL rapidly increases the cross-sectional area, although this effect disappears by 12 weeks (Table 6). The tensile strength in the relaxed group was significantly lower ($P < 0.05$) than that in the sham group at 12 weeks, but not at 6 weeks (Table 6).

In this study, significant differences were observed at 6 weeks between the treated and the non-treated (control) ACLs even in the sham group. However, the effects of the sham operation were temporary: the differences of the strength

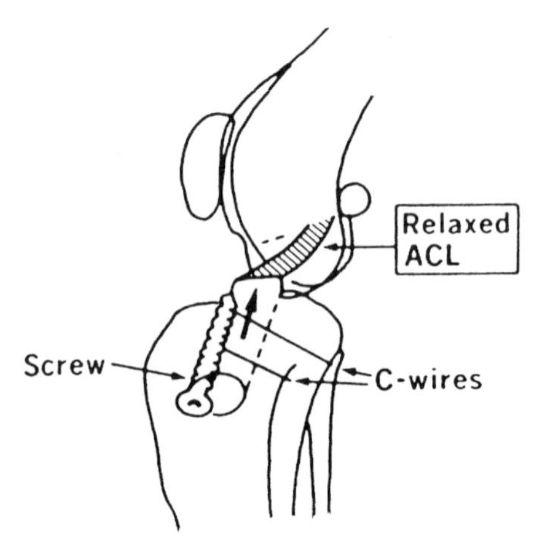

FIG. 12. Tibial insertion of the ACL was separated from the tibia with a cylindrical bone block. This block was anatomically reduced (sham group), or proximally dislocated by 3 mm to relax the ACL and then fixed with a screw (relaxed group). *C-wires*, stainless steel wire 1 mm in diameter

TABLE 6. Cross-sectional area (mm^2) and tensile strength (MPa) in canine ACLs reattached at the tibial side.

Groups	Cross-sectional area (mm^2)		Tensile strength (MPa)	
	6 weeks	12 weeks	6 weeks	12 weeks
Sham	8.5 ± 1.8**	9.1 ± 1.6**	793 ± 226*	891 ± 128
	(7.3 ± 1.4)	(7.7 ± 1.5)	(948 ± 2289)	(910 ± 98)
Relaxed	9.6 ± 1.6**	10.5 ± 2.0**	779 ± 193*	611 ± 139*
	(7.1 ± 1.5)	(8.2 ± 1.3)	(1325 ± 245)	(1020 ± 254)
Comparison	[$P < 0.01$]	[NS]	[NS]	[$P < 0.01$]

Values are mean ± SD.
( ), Normal control value (obtained from opposite side).
*, $P < 0.05$; **, $P < 0.01$ (treated *vs* normal control).
[ ], Sham *vs* relaxed; NS, not significant.
In the relaxed group, attachment of the ACL was made free from the tibia. After this treatment, the ACL attachment was reduced to the anatomical position in the sham group. Animals were sacrificed at 6 or 12 weeks after treatment.

and the modulus were not significant at 12 weeks, although the cross-sectional area remained increased. The invasive sham operation induces interruption of blood flow, intraarticular hemorrhage, and synovitis in the early phase after surgery, although these effects disappear within a few weeks postoperatively. Therefore, the significant differences in the mechanical properties between the sham-operated ACLs and control ACLs observed at 6 weeks are attributable to the early intraarticular pathological changes.

This result obtained from the canine ACL study is very interesting in comparison with the effects of stress deprivation on the rabbit patellar tendon reported by the authors and their colleagues [109]: the mechanical properties of the patellar tendon are dramatically reduced by stress deprivation. The effects of stress deprivation on the patellar tendon appeared within 1 week after treatment. On the other hand, tensile strength reduction of the relaxed ACL appeared more slowly, becoming apparent at 6 or more weeks after operation. Amiel et al. [22–24] revealed that there are many differences in morphology and biochemistry between ligament and tendon tissues. Moreover, these tissues are located in different environments, that is, intraarticularly and extraarticularly. Mechanical conditions to which these tissues are exposed are also different: for example, a markedly large load, exceeding the maximum load of the ACL, is frequently applied to the patellar tendon [98]. Therefore, mechanisms of remodeling induced by stress deprivation may differ between the tendon and the ligament. A series of our studies [66,67,114] have demonstrated in the rabbit patellar tendon that stress deprivation directly affects collagen fibers without mediation by cells. In the ACL, however, a period of stress deprivation of 6 weeks or longer was needed to affect the mechanical properties. Such a marked difference in the response to stress deprivation between the ACL and the patellar tendon is an interesting phenomenon that should be studied further.

4.2.3 Effects of Restressing on the Previously Frozen and Completely Stress-Shielded Patellar Tendon

On the basis of the results obtained from the studies we have described, this question was asked: Can early resumption of loading (restressing) to a stress-shielded autograft maintain the mechanical properties of the graft? Or, can this restressing provide recovery of the mechanical properties deteriorated by stress shielding? The following study was conducted to answer this question [115,116]. After complete stress-shielding treatment, rabbits were divided into the following four groups, CSS-1W, CSS-2W, CSS-3W, and CSS-6W, in which the tension in the patellar tendon was completely shielded for 1, 2, 3, and 6 weeks, respectively. The wire was then cut through a small skin incision to restress the patellar tendon. This procedure allowed the patellar tendon to be subjected again to normal stress.

As mentioned, the tensile strength measured immediately before restressing were significantly reduced in proportion to the duration of stress shielding, although there was no significant difference between the CSS-3W and CSS-6W groups (Fig. 13). At 3 weeks after starting restressing, the significant differences in the strength among the CSS-1W, CSS-2W, and CSS-3W groups observed immediately before restressing disappeared; however, change patterns of the tensile strength after restressing were different among the three groups (Fig. 13). In the CSS-6W group, the strength obtained immediately before restressing was not different from that after 3 weeks of restressing. Between 3 and 6 weeks after restressing, the strength did not change with time. However, this value significantly increased at 12 weeks. In the CSS-1W, CSS-2W, and CSS-3W groups, strength at 12 weeks recovered to the levels observed at corresponding periods in

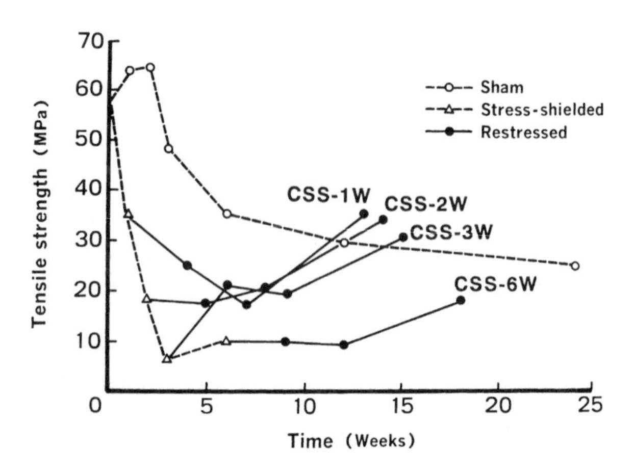

Fig. 13. Changes of tensile strength caused by restressing after complete stress shielding (*CSS*). Restressing cannot completely recover tensile strength that has been reduced by complete stress shielding. *Open circles*, sham; *triangles*, stress-shielded; *solid circles*, restressed

the sham group, although strength at 12 weeks in the CSS-6W group was significantly smaller than the values in the other three groups.

This study clearly answered the foregoing questions. When a patellar tendon autograft is completely shielded from stress for less than 3 weeks, first, restressing completely recovers the reduced mechanical strength to the normal level of the autograft, and second, 12 weeks of restressing (fourfold or more the period for stress shielding) is required for recovery. Even if stress is completely shielded for 6 weeks, it is expected that restressing may recover the mechanical strength in a period of much more than 12 weeks, because in the CSS-6W group restressing was gradually increasing the strength between 6 and 12 weeks. This study demonstrated that stressing essentially recovers the mechanical strength of patellar tendon autografts even if strength has been much reduced by complete stress shielding. However, the effects of restressing may depend on the period of stress shielding applied before restressing.

From the clinical aspect, early resumption of loading (restressing) to autografts has been expected to prevent adverse effects caused by stress shielding in augmentation procedures. Our studies demonstrated that this strategy is not completely wrong if we try to reduce the adverse effects of continuous stress shielding. However, it is also shown that it takes much time to recover the tensile strength and maximum load of the frozen patellar tendon, which were once reduced by complete stress shielding, to the values of the frozen patellar tendon without stress shielding. Therefore, we must conclude that the clinical utility of early resumption of loading is extremely low in augmentation procedures. As we have reported [66], partial stress shielding may have some possibility in searching for the solution. More suitable stress conditions for the remodeling of biological grafts, that is, magnitude of applied stress, duration of stress shielding, and timing of reloading, must be determined for the development of clinically acceptable and useful augmentation techniques for ligament reconstruction surgery.

4.2.4 Significance of Mechanical Environment for Remodeling of Autografts

A series of our studies suggested that mechanical environment plays a significant role in the remodeling of autografts. Specifically, stress deprivation has a large influence on the mechanical properties, histology, and ultrastructure of the previously frozen patellar tendon, and that the effects are dependent on the degree of stress shielding. This remodeling occurs independently of cells regardless of the extent of stress shielding. These are extremely interesting phenomena; however, the mechanisms are not sufficiently understood. Those studies implied that the lax graft may not apply enough mechanical loading to graft fibroblasts to alter the glycosaminoglycan-collagen biosynthesis, which determine the formation of large-diameter fibrils. This may explain one reason that the results of the pure ACL autograft reported by Nikolaou et al. [85] are superior to those of the patellar tendon autograft commonly used in ACL reconstruction; that is, because many or most collagen fibers in the patellar tendon graft, particularly in animal experiments, are placed in nonanatomical or nonisometric orientation, and so stress in those fibers may be reduced, leading to graft deterioration.

The difference of the degree of stress deprivation between human and animal surgeries may also partly explain why the postoperative results of human ACL reconstruction using the patellar tendon autograft are superior to those of animal ACL reconstruction using the same autograft; that is because the isometric placement of graft fibers has been very precisely investigated and can be more successfully made in human than in animal ACL reconstruction. Therefore, we believe that the physiological orientation and tension of graft fibers constitute one of the most important keys to success in ACL reconstruction using autografts.

5 Future Directions of Research on Remodeling of Tendon Autografts

Previous studies have demonstrated that the tensile strength of the autografts is reduced after transplantation, being accompanied by changes of the ultrastructure of collagen fibrils. However, the strength data reported in those studies are extremely divergent. Biological and mechanical properties of the tendon and ligament tissues and the remodeling process of these grafts after transplantation are considered to be different not only among species but also between surrounding mechanical environments. Each of these models has specific implications that should be interpreted with caution. Also, we must recognize that the data obtained from autograft studies are sensitively affected not only by many implantation variables but also by tensile test conditions. In future studies, each implantation variable should be well controlled, and standard tensile test conditions should be established for each tissue specimen.

Currently, it is necessary to reconstruct not only the ACL but also various ligaments in the knee, ankle, and elbow joints. Moreover, the semitendinosus and gracilis tendons have been commonly used as graft materials as an alternative to the patellar tendon [55,77–80]. Future studies therefore should focus not only on the patellar tendon and ACL but also on various tendon and ligament tissues including the semitendinosus tendon, the posterior cruciate ligament, and the medial collateral ligament, because these are located in biologically and physically different surrounding conditions.

The changes in the number as well as the diameter of autograft collagen fibrils are an important phenomenon, although the mechanism has never been clarified. A series of our studies have demonstrated that the mechanical condition exerted on autografts plays a significant role in the remodeling. Specifically, it is noted that stress deprivation acutely changes the mechanical properties, histology, and ultrastructure of the patellar tendon autograft model regardless of the existence of cells. This mechanism should be studied from biological, biochemical, biomechanical, and ultrastructural points of view. Moreover, newly infiltrated cells seem to play an important role in this mechanism [117,118]. It is also shown in our studies that the cells modify the remodeling process caused by stress deprivation. The role of cells in autografts also should be investigated. Development of a new technique to control cell infiltration into autografts may clarify this issue. Furthermore, effects of overloading on autografts should be a focus in future studies.

Finally, to develop more successful reconstruction procedures of ligament tissue, how to reduce or treat deterioration of the material and structural properties of tendon autografts after transplantation must be studied. This will lead to the establishment of "ideal" ACL reconstruction using tendon autografts.

Acknowledgments. Most of our research work described here was done in the Department of Biomedical Control, Research Institute of Applied Electricity (currently Department of Scientific Instrumentation and Control, Research Institute for Electronic Science), Hokkaido University. Animal experiments were carried out in the Institute of Animal Experimentation, Hokkaido University School of Medicine, under the Rules and Regulations of the Animal Care and Use Committee, Hokkaido University School of Medicine. Observation of the ultrastructure of the patellar tendon was performed in the Central Laboratory for Research and Education, Asahikawa Medical College. The authors appreciate their collaborators, Drs. Kiyoshi Kaneda, Noritaka Yamanoto, Kazunori Ohno, Harukazu Toyama, Motoharu Keira, Tokifumi Majima, Takamasa Tsuchida, Hirohide Ishida, Kunio Tanaka, Mr. Kiyoshi Miyakawa, Takahide Ishizaka, and Ms. Yoshie Tanabe, as well as their former students, Hiroyuki Kuriyama, Fumihiro Hayashi, and Takashi Fujii. This work was supported financially in part by the Grant-in-Aid for Scientific Research of Priority Areas, Biomechanics (nos. 04237102 and 04237104), and the Grant-in-Aid for General Scientific Research C (no. 06671433) from the Ministry of Education, Science and Culture in Japan, with K. Hayashi, H. Abe, and K. Yasuda as the principal investigators.

References

1. Clayton ML, Miles JS, Abdulla M (1968) Experimental investigations of ligamentous healing. Clin Orthop 61:146–153
2. Frank C, Woo SL-Y, Amiel D, et al (1983) Medial collateral ligament healing: a multidisciplinary assessment in rabbits. Am J Sports Med 11:379–389
3. Yasuda K, Erickson AR, Johnson JR, Pope MH (1993) Dynamic elongation behavior in the medial collateral and anterior cruciate ligaments during lateral impact loading. J Orthop Res 11:190–198
4. Ogata K, Whiteside LA, Andersen DA (1980) The intra-articular effect of various postoperative managements following knee ligament repair: an experimental study in dogs. Clin Orthop 150:271–276
5. Warren RF, Marshall JL (1978) Injuries of the anterior cruciate and medial collateral ligaments of the knee: a long-term follow-up of 86 cases. Part II. Clin Orthop 136:198–211

6. Clancy WG (1985) Advances in biologic substitution for cruciate deficiency. In: Finerman G (ed) American Academy of Orthopaedic Surgeons symposium on sports medicine: the knee. Mosby, St Louis, pp 222–229
7. Noyes FR, McGinniss GH (1985) Controversy about treatment of the knee with anterior cruciate laxity. Clin Orthop 198:61–76
8. Odensten M, Hamberg P, Nordin M, et al (1985) Surgical or conservative treatment of the acutely torn anterior cruciate ligament: a randomized study with short-term follow-up observations. Clin Orthop 198:87–93
9. O'Donoghue DH, Rockwood CA Jr, Frank GR, et al (1966) Repair of the anterior cruciate ligament in dogs. J Bone Joint Surg 48A:503–519
10. O'Donoghue DH, Frank GR, Jeter GL, et al (1971) Repair and reconstruction of the anterior cruciate ligament in dogs: factors influencing long-term results. J Bone Joint Surg 53A:710–718
11. Barfred T (1971) Experimental rupture of the Achilles tendon: comparison of experimental sutures in rats of different ages and living under different conditions. Acta Orthop Scand 42:406–428
12. Inoue M, McGurk-Burleson E, Hollis JM, et al (1987) Treatment of the medial collateral ligament injury: I. The importance of anterior cruciate ligament on the varus-valgus knee laxity. Am J Sports Med 15:15–21
13. Hey Groves EW (1917) Operation of the repair of the cruciate ligaments. Lancet 2:674
14. Ballock RT, Woo SL-Y, Lyon RM, Hokkis JM, Akeson WH (1989) Use of patellar tendon autograft for anterior cruciate ligament reconstruction in the rabbit: a long-term histologic and biomechanical study. J Orthop Res 7:474–485
15. Clancy WG, Narechania RG, Rosenberg TD, Gmeiner JG, Wisnefske DD, Lange TA (1981) Anterior and posterior cruciate ligament reconstruction in rhesus monkeys. J Bone Joint Surg 63A:1270–1284
16. Hurley PB, Andrish JT, Yoshiya S, Manley M, Kurosaka M (1987) Tensile strength of the reconstructed canine anterior cruciate ligament: a long-term evaluation of the modified Jones technique. Am J Sports Med 15:393
17. Jones KG (1963) Reconstruction of the anterior cruciate ligament: a technique using the central one-third of the patellar ligament. J Bone Joint Surg 45A:925–932
18. McPherson GK, Mendenhall HV, Gibbons DF, Plenk H, Rottmann W, Sanford JB, Kennedy JC, Roth JH (1985) Experimental mechanical and histologic evaluation of the Kennedy ligament augmentation device. Clin Orthop 196:186–195
19. Ryan JR, Grompp BW (1966) Evaluation of tensile strength of reconstructions of the anterior cruciate ligament using the patellar tendon in dogs: a preliminary report. South Med J 59:129–134
20. Yasuda K, Ohkoshi Y, Tanabe Y, Kaneda K (1992) Quantitative evaluation of the knee instability and muscle strength after anterior cruciate ligament reconstruction using patellar and quadriceps tendons. Am J Sports Med 20:471–475
21. Yoshiya S, Andrish JT, Manley MT, et al (1986) Augmentation of anterior cruciate ligament reconstruction in dogs with prostheses of different stiffnesses. J Orthop Res 4:475–485
22. Amiel D, Frank C, Harwood F, Fronek J, Akeson W (1984) Tendons and ligaments: a morphological and biochemical comparison. J Orthop Res 1:257–265
23. Amiel D, Kleiner JB, Roux RD, Harwood FL, Akeson WH (1986) The phenomenon of "ligamentization": anterior cruciate ligament reconstruction with autogenous patellar tendon. J Orthop Res 4:162–172

24. Amiel D, Kleiner JB, Akeson WH (1986) The natural history of the anterior cruciate ligament autograft of patellar tendon origin. Am J Sports Med 14:449–462
25. Amiel D, Kleiner JB (1988) Collagen. In: Nimni ME, Olsen B (eds) Biochemistry of tendon and ligament, vol 3. Biotechnology. CRC Press, Cleveland, pp 223–251
26. Amiel D, Billings E Jr, Akeson WH (1990) Ligament structure, chemistry, and physiology. In: Daniel D, Akeson W, O'Connor J (eds) Knee ligaments, structure, injury and repair. Raven, New York, pp 77–91
27. Alm A, Stromberg B (1974) Transposed medial third of patellar ligament in reconstruction of the anterior cruciate ligament: a surgical and morphologic study in dogs. Acta Chir Scand 445(suppl):37–49
28. Alm A, Liljedahl SO, Stronberg B (1976) Clinical and experimental experience in reconstruction of the anterior cruciate ligament. Orthop Clin North Am 7:181–189
29. Arnoczky SP, Tarvin GB, Marshall JL (1982) Anterior cruciate ligament replacement using patellar tendon: an evaluation of graft revascularization in the dog. J Bone Joint Surg 64A:217–224
30. Arnoczky SP, Warren RF, Ashlock MA (1986) Replacement of the anterior cruciate ligament using a patellar allograft. J Bone Joint Surg 68A:376–385
31. Arnoczky SP (1991) Basic science of anterior cruciate ligament repair and reconstruction. In: Tullos HS (ed) Instructional course lectures, vol XL. American Academy of Orthopaedic Surgeons, Park Ridge, IL, pp 201–212
32. Butler DL, Grood ES, Noyes FR, Olmstead ML, Hohn RB, Arnoczky SP, Siegel MG (1989) Mechanical properties of primate vascularized versus nonvascularized patellar tendon grafts; changes over time. J Orthop Res 7:68–79
33. Cabaud HE, Feagin JA, Rodkey WG (1980) Acute anterior cruciate ligament injury and augmented repair: experimental studies. Am J Sports Med 8:395–401
34. Chiroff RT (1975) Experimental replacement of the anterior cruciate ligament: a histological and microradiographic study. J Bone Joint Surg 57A:1124–1127
35. Curtis RJ, Delee JC, Drez DJ Jr (1985) Reconstruction of the anterior cruciate ligament with freeze-dried fascia lata allografts in dogs: a preliminary report. Am J Sports Med 13:408–414
36. Higgins RW, Steadman JR (1987) Anterior cruciate ligament repairs in world class skiers. Am J Sports Med 15:439–447
37. Holden JP, Grood ES, Butler DL, et al (1988) Biomechanics of fascia lata ligament replacements: early postoperative changes in the goat. J Orthop Res 6:639–647
38. Kleiner JB, Amiel D, Harwood FL, et al (1989) Early histologic, metabolic, and vascular assessment of anterior cruciate ligament autografts. J Orthop Res 7: 235–242
39. Kondo M (1979) An experimental study on reconstructive surgery of the anterior cruciate ligament. Nippon Seikeigeka Gakkai Zasshi (J Jpn Orthop Assoc) 53:521–533
40. van Rens TJ, van den Berg AF, Huiskes R, et al (1986) Substitution of the anterior cruciate ligament: a long-term histologic and biomechanical study with autogenous pedicled grafts of the iliotibial band in dogs. Arthroscopy 2:139–154
41 Oakes BW (1993) Collagen ultrastructure in the normal ACL and in ACL graft. In: Jackson DW (ed) The anterior cruciate ligament. Current and future concepts. Raven, New York, pp 209–217
42. Yasuda K, Tomiyama Y, Ohkoshi Y, Kaneda K (1989) Arthroscopic observations of autogenous quadriceps and patella tendon grafts after anterior cruciate ligament reconstruction of the knee. Clin Orthop 246:217–224

43. Goldberg VM, Burstein A, Dawson M (1982) The influence of an experimental immune synovitis on the failure mode and strength of the rabbit anterior cruciate ligament. J Bone Joint Surg 64A:900–906
44. Bray DF, Frank CB, Bray RC (1990) Cytochemical evidence for a proteoglycan-associated filamentous network in ligaments and the extracellular matrix. J Orthop Res 8:1–12
45. Bosch U, Decker B, Moller HD, Kasperczyk W, Oestern HJ (1995) Collagen fibril organization in the patellar tendon autograft after posterior cruciate ligament reconstruction. A quantitative evaluation in a sheep model. Am J Sports Med 23:196–202
46. Hart RA, Woo SL-Y, Newton PO (1992) Ultrastructural morphometry of the anterior cruciate and medial collateral ligaments: an experimental study in rabbits. J Orthop Res 10:96–103
47. Tsuchida T, Yasuda K, Hayashi K, Majima T, Yamamoto N, Kaneda K, Miyakawa K, Tanaka K (1994) Effects of stress shielding on the ultrastructure of the in situ frozen patellar tendon in the rabbit. Trans Orthop Res Soc 19:638
48. Tsuchida T, Yasuda K, Kaneda K, Hayashi K, Yamamoto N, Miyakawa K, Tanaka K (1995) Effects of stress shielding on the ultrastructure of normal and in situ frozen rabbit patellar tendons. A role of fibroblasts. Trans Orthop Res Soc 20:613
49. Kastelic J, Galeski A, Baer E (1978) The multicomposite structure of tendon. Connect Tissue Res 6:11–23
50. Clark JM, Sidkes JA (1990) The interrelation of fiber bundles in the anterior cruciate ligament. J Orthop Res 8:180–188
51. Yahia L-H, Drouin G (1988) Collagen structure in human anterior cruciate ligament and patellar tendon. J Mater Sci 23:3750–3755
52. Jackson DW, Grood ES, Goldstein J, Rosen MA, Kurzweil PR, Cummings JF, Simon TM (1993) A comparison of patellar tendon autograft and allograft used for anterior cruciate ligament reconstruction in goats. Am J Sports Med 21:176–185
53. Brand RA (1986) Knee ligaments: a new view. J Biomech Eng 108:106–110
54. Butler DL, Kay MD, Stouffer DC (1986) Comparison of material properties in fascicle-bone units from human patellar tendon and knee ligaments. J Biomech 19:425–432
55. Noyes FR, Butler DL, Grood ES, Zernicke RF, Hefzy MS (1984) Biomechanical analysis of human ligament grafts used in knee-ligament repairs and reconstructions. J Bone Joint Surg 66A:344–352
56. Woo SL-Y, Hollis JM, Adams DJ, Lyon RM, Takai S (1991) Tensile properties of the human femur-anterior cruciate ligament-tibia complex: the effects of specimen age and orientation. Am J Sports Med 19:217–225
57. Butler DL (1989) Anterior cruciate ligament: its normal response and replacement. J Orthop Res 7:910–921
58. Keira M, Yasuda K, Hayashi K, Yamamoto N, Kaneda K (1992) Mechanical properties of the canine anterior cruciate ligament chronically relaxed by elevation of the tibial insertion. Trans Orthop Res Soc 17:661
59. Keira M, Yasuda K, Kaneda K, Yamamoto N, Hayashi K (1996) Mechanical properties of the anterior cruciate ligament chronically relaxed by elevation of the tibial insertion. J Orthop Res (in press)
60. Rowe RWD (1985) The structure of rat tail tendon. Connect Tissue Res 14:9–20
61. Cetta G, Tenni R, Zanaboni G, et al (1982) Biomechanical and morphological modifications in rabbit Achilles tendon during maturation and ageing. Biomech J 204:61–67

62. Jimenez SA, Yankowski R, Bashey RI (1978) Identification of two new collagen alpha-chains in extracts of lathyritis chick embryo tendons. Biochem Biophys Res Commun 81:1298–1306
63. Piez KA, Miller EJ, Lane JM, et al (1969) The order of the CNBr peptides from the alpha-1 chain of collagen. Biochem Biophys Res Commum 37:801–805
64. Burks RT, Haut RC, Lancaster RL (1990) Biomechanical and histological observations of the dog patellar tendon after removal of its central one-third. Am J Sports Med 18:146–153
65. Loitz BJ, Zernicke RF, Vailas AC, Kody MH, Meals RA (1989) Effects of short-term immobilization versus continuous passive motion on the biomechanical and biochemical properties of the rabbit tendon. Clin Orthop 244:265–271
66. Majima T, Yasuda K, Yamamoto N, Kaneda K, Hayashi K (1994) Deterioration of mechanical properties of the autograft in controlled stress-shielded augmentation procedures. An experimental study with rabbit patellar tendon. Am J Sports Med 22:821–829
67. Ohno K, Yasuda K, Yamamoto N, Hayashi K (1993) Effects of complete stress shielding on the mechanical properties and histology of in situ frozen patellar tendon. J Orthop Res 11:592–602
68. Robert JM, Goldstrohm GL, Brown TD, Mears DC (1983) Comparison of unrepaired, primarily repaired, and polyglactin mesh reinforced achilles tendon laceration in rabbits. Clin Orthop 181:244–249
69. Viidik A (1969) Tensile strength properties of Achilles tendon systems in trained and untrained rabbits. Acta Orthop Scand 40:261–272
70. Webster DA, Werner FW (1983) Freezed-dried flexor tendons in anterior cruciate ligament reconstruction. Clin Orthop 181:238–243
71. Woo SL-Y, Gromez MA, Inoue M, et al (1987) New experimental procedures to evaluate the biomechanical properties of healing canine medial collateral ligaments. J Orthop Res 5:425–432
72. Noyes F, Butler DL, Paulos LE, Grood ES (1983) Intra-articular cruciate reconstruction. I: Perspectives on graft strength, vascularization, and immediate motion after replacement. Clin Orthop 172:71–77
73. Kleiner JB, Amiel D, Roux RD, Akeson WH (1986) Origin of replacement cells for the anterior cruciate ligament autograft. J Orthop Res 4:466–474
74. Alm A, Ekstrom H, Gillquist J, Stromberg B (1974) The anterior cruciate ligament. A clinical and experimental study on tensile strength, morphology and replacement by patellar ligament. Acta Chir Scand 140(suppl 445):4–49
75. Shino K, Kawasaki T, Hirose H, Gotoh I, Inoue M, Ono K (1984) Replacement of the anterior cruciate ligament by an allogeneic tendon graft: an experimental study in the dog. J Bone Joint Surg 66B:672–681
76. Yoshiya S, Andrish JT, Manley MT, Bauer TW (1987) Graft tension in anterior cruciate ligament reconstruction. An in vivo study in dogs. Am J Sports Med 15:464–470
77. Cho KO (1975) Reconstruction of the anterior cruciate ligament by semitendinosus tenodesis. J Bone Joint Surg 57A:608–612
78. Lipscomb AB, Johnson RK, Synder RB, Brothers JC (1979) Secondary reconstruction of anterior cruciate ligament in athletes by using the semitendinosus tendon: preliminary report of 78 cases. Am J Sports Med 7:81–84
79. Mott HW (1983) Semitendinosus anatomic reconstruction for cruciate ligament insufficiency. Clin Orthop 172:90–92

80. Yasuda K, Tsujino J, Ohkoshi Y, Tanabe Y, Kaneda K (1995) Isolated autogenous semitendinosus and gracilis tendon graft site morbidity. Am J Sports Med 23:706–714
81. Scott JE, Hughes EW (1986) Proteoglycan-collagen relationships in developing chick and bovine tendons: influence of the physiological environment. Connect Tissue Res 14:267–268
82. Toole BP, Lowther DA (1968) The effect of chondroitin sulphate-protein on the formation of collagen fibrils in vitro. Biochem J 109:857–866
83. Clancy WG Jr (1983) Anterior cruciate ligament functional instability: a static intra-articular and dynamic extra-articular procedure. Clin Orthop 172:102–106
84. Paulos LE, Cherf J, Rosenberg TD, et al (1991) Anterior cruciate ligament reconstruction with autografts. Clin Sports Med 10:469–485
85. Nikolaou DK, Seaber AV, Glisson RR, Ribbeck BM, Bassett FH (1986) Anterior cruciate ligament allograft transplantation: long-term function, histology, revascularization, and operative technique. Am J Sports Med 14:348–360
86. Noyes FR, Grood ES (1976) The strength of the anterior cruciate ligament in human and rhesus monkeys. Age-related and species-related changes. J Bone Joint Surg 58A:1074–1082
87. Graf B (1987) Isometric placement of substitutes for the anterior cruciate ligament. In: Jackson DW, Drez D Jr (eds) The anterior cruciate deficient knee. New concepts in ligament repair. Mosby, St Louis, pp 102–113
88. Hefzy MS, Grood ES, Noyes FR (1989) Factors affecting the region of most isometric femoral attachments. Part II: The anterior cruciate ligament. Am J Sports Med 17:208–216
89. Penner DA, Daniel DM, Wood P, Mishra D (1988) An in vitro study of anterior cruciate ligament graft placement and isometry. Am J Sports Med 16:238–243
90. Sidles JA, Larson RV, Garbini JL, Downey DJ, Matsen FA III (1988) Ligament length relationships in the moving knee. J Orthop Res 6:593–610
91. Arms SW, Pope MH, Johnson RJ, Renstrom P, Fischer RA, Jarvinen M, Beynnon B (1990) Analysis of ACL failure strength and initial strains in the canine model. Trans Orthop Res Soc 15:524
92. Beynnon BD, Huston DR, Pope MH, Fleming BC, Johnson RJ, Nichols CE, Renstrom P (1992) The effect of ACL reconstruction tension on the knee and cruciate ligaments. Trans Orthop Res Soc 17:657
93. Fleming B, Beynnon B, Howe J, McLeod W, Pope M (1992) Effect of tension and placement of a prosthetic anterior cruciate ligament on the anteroposterior laxity of the knee. J Orthop Res 10:177–186
94. Graf BK, Ulm MJ, Rogalski RP, Vanderby R Jr (1992) Effect of preconditioning on the viscoelastic response of primate patella tendon. Trans Orthop Res Soc 17:147
95. Kurosaka M, Yoshiya S, Andrish JT (1987) A biomechanical comparison of different surgical techniques of graft fixation in anterior cruciate ligament reconstruction. Am J Sports Med 15:225–229
96. Robertson DB, Daniel DM, Biden E (1986) Soft tissue fixation to bone. Am J Sports Med 14:398–403
97. Ohno K, Yasuda K, Yamamoto N, Hayashi K (1991) Effects of stress shielding on the mechanical properties of normal and in situ frozen patellar tendon. Trans Orthop Res Soc 16:134
98. Yasuda K, Sasaki T (1987) Exercise after anterior cruciate ligament reconstruction. The force exerted on the tibia by the separate isometric contractions of the quadriceps or the hamstrings. Clin Orthop 215:275–283

99. Ohno K, Yasuda K, Yamamoto N, Hayashi K (1991) Effects of in situ freezing on the mechanical and histological properties of the patellar tendon. J Jpn Orthop Assoc 65:S1081
100. Yasuda K, Hayashi K (1994) Effects of stress shielding on autografts in augmentation procedures: experimental studies using the in situ frozen patellar tendon. In: Hirasawa Y, Sledge CB, Woo SL-Y (eds) Clinical biomechanics and related research. Springer, Tokyo, pp 363–379
101. Jackson DW, Grood ES, Cohn BT, Arnozcky SP, Simon TM, Cummings JF (1991) The effects of in situ freezing on the anterior cruciate ligament. J Bone Joint Surg 73A:201–213
102. Vasseur PB, Rodrigo JJ, Stevenson S, Clark G, Sharkey N (1987) Replacement of the anterior cruciate ligament with a bone-ligament-bone anterior cruciate ligament allograft in dogs. Clin Orthop 219:268–277
103. Fowler P, Amendola A (1990) Allograft ACL reconstruction in a sheep model: the effect of synthetic augmentation. Trans Orthop Res Soc 15:80
104. McCarthy JA, Steadman JR, Dunlap J, Shively RT, Stonebrook S (1990) A nonparallel, nonisometric synthetic graft augmentation of a patellar tendon anterior cruciate ligament reconstruction: a model for assessment of stress shielding. Am J Sports Med 18:43–49
105. Andrish JT, Woods LD (1984) Dacron augmentation in anterior cruciate ligament reconstruction in dogs. Clin Orthop 183:298–302
106. Majima T, Yasuda K, Yamamoto N, Hayashi K, Kaneda K (1992) Effects of stress shielding on the mechanical properties of in situ frozen patellar tendon augmented with polyester artificial ligament. Trans Orthop Res Soc 17:674
107. Majima T, Fujii T, Yasuda K, Yamamoto N (1993) Effects of controlled stress shielding on the mechanical properties of the rabbits patellar tendon. Role of fibroblasts. Trans Orthop Res Soc 18:365
108. Yamamoto N, Hayashi K, Kuriyama H, Yasuda K, Kaneda K (1992) Mechanical properties of the rabbit patellar tendon. Trans ASME J Biomech Eng 114:332–337
109. Yamamoto N, Ohno K, Hayashi K, Kuriyama H, Yasuda K, Kaneda K (1993) Effects of stress shielding on the mechanical properties of rabbit patellar tendon. Trans ASME J Biomech Eng 115:23–28
110. Lewis JL, Lew WD, Schmidt J (1982) A note on the application and evaluation of the buckle transducer for knee ligament force measurement. Trans ASME J Biomech Eng 104:125–128
111. Toyama H, Ohno K, Yamamoto N, Hayashi K, Yasuda K, Kaneda K (1992) Stress-strain characteristics of in situ frozen and stress-shielded rabbit patellar tendon. Clin Biomech 7:226–230
112. Gomez MA (1988) The effect of tension on normal and healing medial collateral ligaments. PhD thesis, University of California, San Diego
113. Park JP, Grana WA, Chitwood JS (1985) A high-strength Dacron augmentation for cruciate ligament reconstruction: a two-year canine study. Clin Orthop 196:175–185
114. Roth JH, Kennedy JC, Lockstadt H, McCallum CL, Cunning LA (1985) Polypropylene braid augmented and nonaugmented intraarticular anterior cruciate ligament reconstruction. Am J Sports Med 13:321–336
115. Ishida H, Yasuda K, Majima T, Kaneda K, Yamamoto N, Hayashi K (1993) Effects of restressing on the mechanical properties of the in situ frozen patella tendon in the rabbit. Jpn J Orthop Assoc 67:S1514

116. Ishida H, Yasuda K, Kaneda K, Yamamoto N, Hayashi K (1995) Does restressing recover the mechanical strength of a patellar tendon autograft reduced by stress shielding? An experimental study. Trans Orthop Res Soc 20:611
117. Yamamoto N, Hayashi K (1995) Effects of stress shielding on the mechanical properties of rabbit patellar tendon: effects of inhibiting the invasion of fibroblasts. J Jpn Soc Clin Biomech 16: 119–273
118. Toyama H, Yasuda K, Kaneda, K (1996) Inhibition of extrinsic cell infiltration prohibits reduction of the mechanical properties of the patellar tendon following intrinsic cell necrosis. Trans Orthop Res Soc 21 (in press)

Instability of the Spinal System with Focus on Degeneration of the Intervertebral Disk

Sohei Ebara[1], Takeo Harada[1], Takenori Oda[1], Eiji Wada[1], Shimpei Miyamoto[1], Kazuo Yonenobu[1], Masao Tanaka[2], and Keiro Ono[3]

Summary. The intervertebral disk is a cartilaginous tissue. It is organized with a concentrated proteoglycan solution, which is the central nucleus pulposus, held within the strong collagen network, the outer anulus fibrosus. The disk exhibits a viscoelastic response when subjected to loads and deformations. Disk degeneration introduces a less stiff segment in the spinal column and is generally considered to be an age-dependent change. However, in the cervical spine of cerebral palsy patients who exhibit athetotic movements in the neck, there is a very early onset of disk degradation. Acceleration of disk degeneration occurs in the spines of animals subjected to excessive extension–flexion of the head and neck or to spinal instability induced by surgery. Repetitive torsion of the disk may lead to structural regression in in vitro studies using animal spines. Delamination or disruption of the anulus fibrosus is always recognized as the beginning stage of destruction of intervertebral disk structure. This disruption of the collagen network may result from fatigue failure by repetitive loadings, which cause high tensile stresses in the anulus fibrosus because large hydrostatic pressures develop within the nucleus pulposus. Loosening of the collagen network may be responsible for the loss of proteoglycans and water, and also may be a key factor leading to the development of disk degeneration. A "degenerated disk" can be induced through pure mechanical fatigue failure of the tissue as an age-independent change of the cartilaginous tissue.

Key words: Intervertebral disk—Nucleus pulposus—Anulus fibrosus—Fatigue failure—Disk degeneration

[1] Department of Orthopaedic Surgery, Osaka University School of Medicine, 2-2 Yamadaoka, Suita, Osaka, 565 Japan
[2] Faculty of Engineering Science, Osaka University, 1-3 Machikaneyama-cho, Toyonaka, Osaka, 560 Japan
[3] Osaka Koseinenkin Hospital, 4-2-78 Fukushima, Fukushima-ku, Osaka, 553 Japan

1 Introduction

The discomfort and pain associated with shoulder, neck, and lower back problems are major health concerns in the aging population. Physical impairment inevitably leads to a reduction in the quality of life. The impact in terms of workdays lost to illness from these health problems is also substantial [1,2]. Occasionally, in addition to pain, age-related spinal disorders cause physical disabilities such as difficulty in standing or walking, clumsiness of the hand, or intermittent limping. Hip fracture is recognized as a notorious killer of the aged, but a primary cause of this fracture might be a faltering gait secondary to a forerunning spine disorder.

Orthopedists, radiologists, and neurosurgeons have tried to identify the causes of "age-related spine disorder" by means of postmortem anatomy, pathology, and radiology. These experts have developed spine surgery to alleviate the pain and disability associated with this disorder. The surgical procedure consists of saving nerve tissue from impingement by bone–cartilage protrusions that grow around the intervertebral disk, which is the nonosseous linkage portion of the spine. Observation and analysis of this disorder have implicated structural degeneration of the disk as the initiating pathological event, followed by growth of the bone–cartilage. Various theories have been advanced to explain the primary involvement of the intervertebral disk. Because the disk itself has no inherent remodeling or regenerative capacity (the core structure of the disk lacks blood vessels), we have suspected, with little supporting evidence, that it must be the site most susceptible to the effects of aging and accumulation of trauma.

We have recently noticed, in the cervical spine of cerebral palsy (CP) patients who exhibit athetotic movements in the neck, that there is a very early onset of spondylotic age-related degradation of the disk with attendant bone–cartilage growth around this structure [3,4]. This structural degradation coupled with histological evidence suggested that degeneration of the disk was not solely a function of aging because surgical disk specimens from athetoid CP patients were comparable with spondylotic discs. An animal model of spondylosis also has been developed in rabbits and mice [5–7]. In these models, acceleration of disk degeneration took place in the spines of animals subjected to excessive extension–flexion of the head and neck, or to spinal instability induced by surgery. Repetitive torsion of the disk could lead to structural regression through fatigue failure in follow-up in vitro studies using the spines from these animals [8].

This chapter describes these studies and discusses a potential conflict between the biological properties and functional demands of the intervertebral disk.

2 What Conditions Accelerate Structural Degradation in the Intervertebral Disk?

Cervical spinal cord or nerve root involvement often develops in athetoid CP patients at a relatively young age (30–50 years) and, as a consequence, exacerbates muscular incoordination [9–11]. Neck–shoulder–arm pain, progressive muscle weakness, dysesthetic sensation on the extremities, and sustained athetoid movements in the neck constitute convincing evidence of lesions localized in the cervical spine. This deduction can be supported by plain X-ray films that show the following abnormalities: instability and malposture of the cervical spine, premature spondylosis, or disk protrusions.

One of the authors of this chapter established that there is much earlier onset of spondylosis (as judged from the marked narrowness of the intervertebral disk spaces on plain roentgenograms) in the cervical spine (Fig. 1). Putting the symptoms and treatment of the affliction aside, this would appear to be a good example in humans of structural degradation being secondary to excessive flexion–extension or torsion. The intervertebral disk is composed of a central core (a semiliquid gelatinous substance, that is, the nucleus pulposus) and an external envelope (a laminated collagen annulus, that is, the anulus fibrosus) (Fig. 2). To simplify the biomechanical analysis of the spine, the study was conducted in the vertebra–disk–vertebra complex or spinal functional unit (SFU). Biomechanical studies of the SFU have revealed that the disk has a superb ability to transform axial compression load into tensile force on the annulus and diffuse pressure on the endplate of the vertebra. During bending of

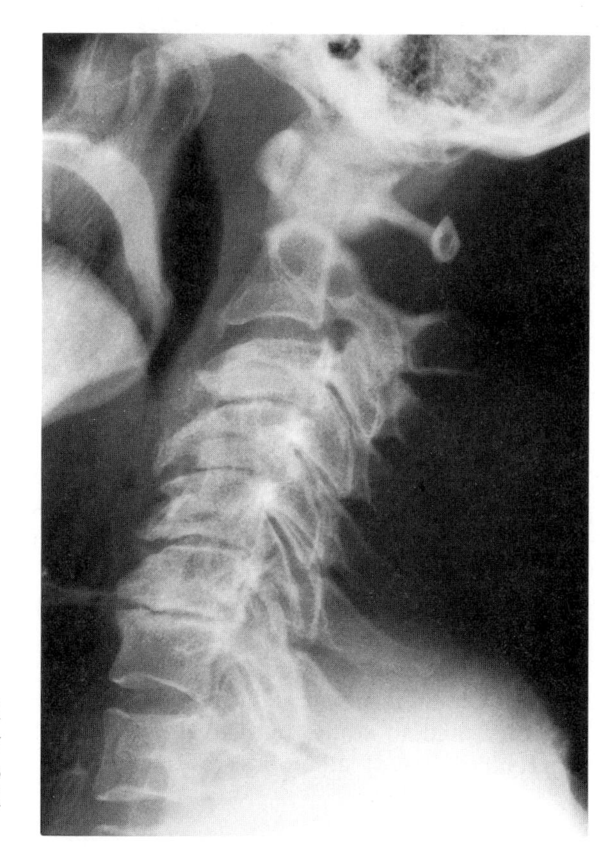

Fig. 1. Spondylosis of cervical spine in athetoid cerebral palsy (CP). Note marked narrowness of the intervertebral disk and osteophyte formation

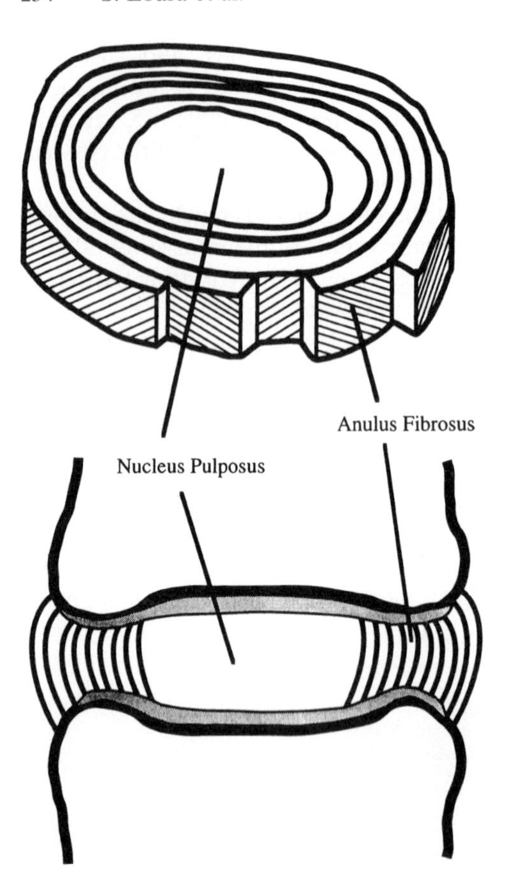

Fig. 2. The intervertebral disk is composed of a central core (a semiliquid gelatinous substance, the nucleus pulposus) and an external envelope (a laminated collagen annulus, the anulus fibrosus)

the spine, one side of the disk (the nucleus and the annulus) is subjected to compression and the other side to tension, events synchronized so as to provide an instantaneous axis of rotation. When the disk is subjected to torsion, there are shear stresses in the horizontal as well as the axial plane [12]. Thus, various patterns of loading and bending or torsion of the disk generate tensile stress in the annulus, friction between the laminae of the annulus, and increased pressure within the nucleus. Impact load absorption and impact energy dissipation characteristic to the disk are therefore explainable by tension and friction of the annulus and movements of free water in the intervertebral disk like cartilage [13].

As the spine ages, increases in collagen fiber content and fibrosis in the nucleus along with delamination or disruption of the annulus fiber can be observed. The intervertebral disk in cases of premature cervical spondylosis in athetoid CP has histological findings similar to those seen in surgical specimens. We attempted to study, by use of radiological and biomechanical analyses, the components of athetotic movements that accelerate spondylosis in the cervical spine.

2.1 Motion Analysis of the Cervical Spine in Athetoid CP Patients

We conducted a study of motion analysis of the cervical spine in athetoid CP [14]. Controlled flexion–extension motions of the neck were performed under instruction. Subjects were immobilized at the shoulders and chest, and the movements were recorded cineradiographically from the lateral direction. The motion records for each cervical vertebra were examined by use of image-analysis tech-

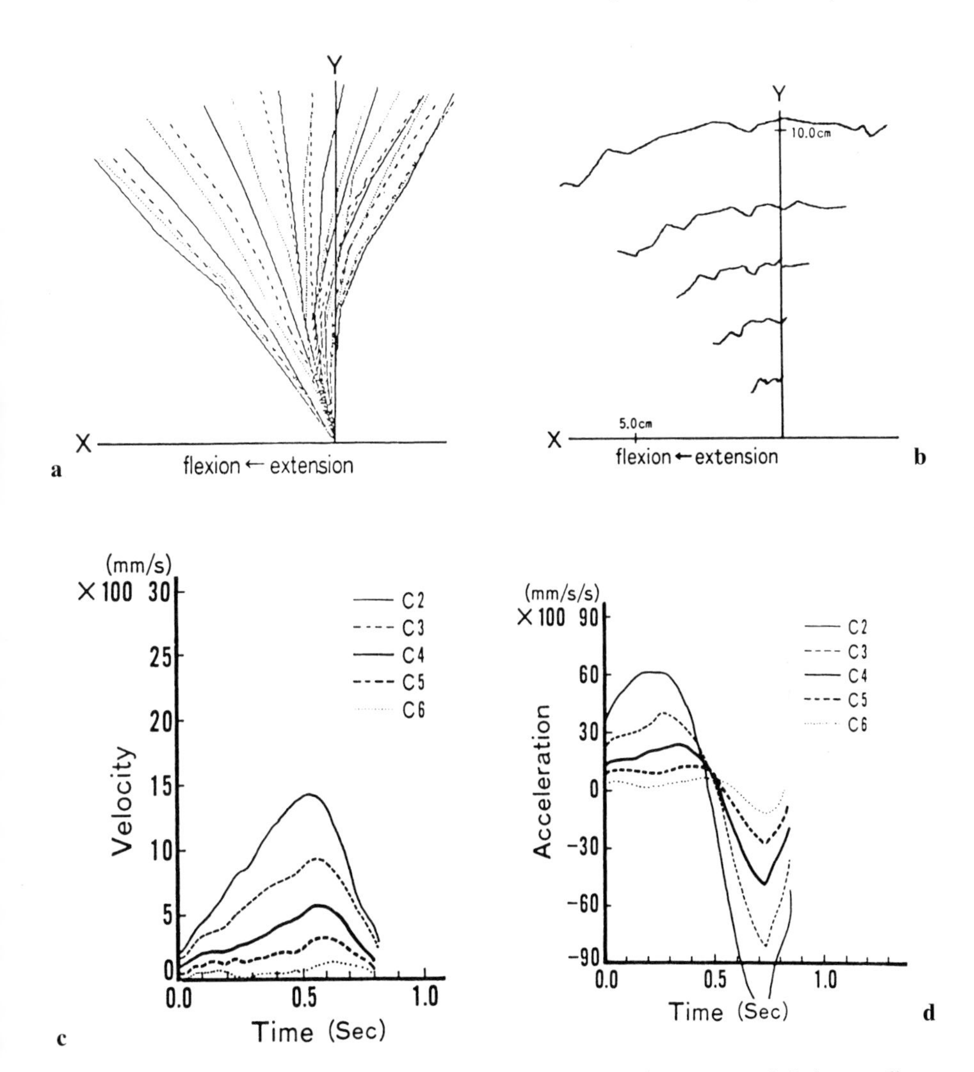

Fig. 3. **a,b** Trajectory of a normal case (case 3; Table 1) shows a definite gradient, craniocaudally, between the centers of each vertebra. **c,d** Velocity and acceleration. There was no sudden change or significant individual difference in velocity and acceleration in motions of the cervical spine of normal subjects. (From [14], with permission)

nology. The trajectory of the center point of each vertebral body could be traced and its velocity and acceleration calculated as a value relative to the C7 vertebra. Velocity and acceleration were measured along a line tangential to the tracings at intervals over the duration of the motion period. Technical details of the protocol used can be found in an earlier report [14].

Trajectories of the cervical spine during extension–flexion motion, as represented by the center of each vertebra, are illustrated in Figs. 3a,b, 4a,b, and 5a,b.

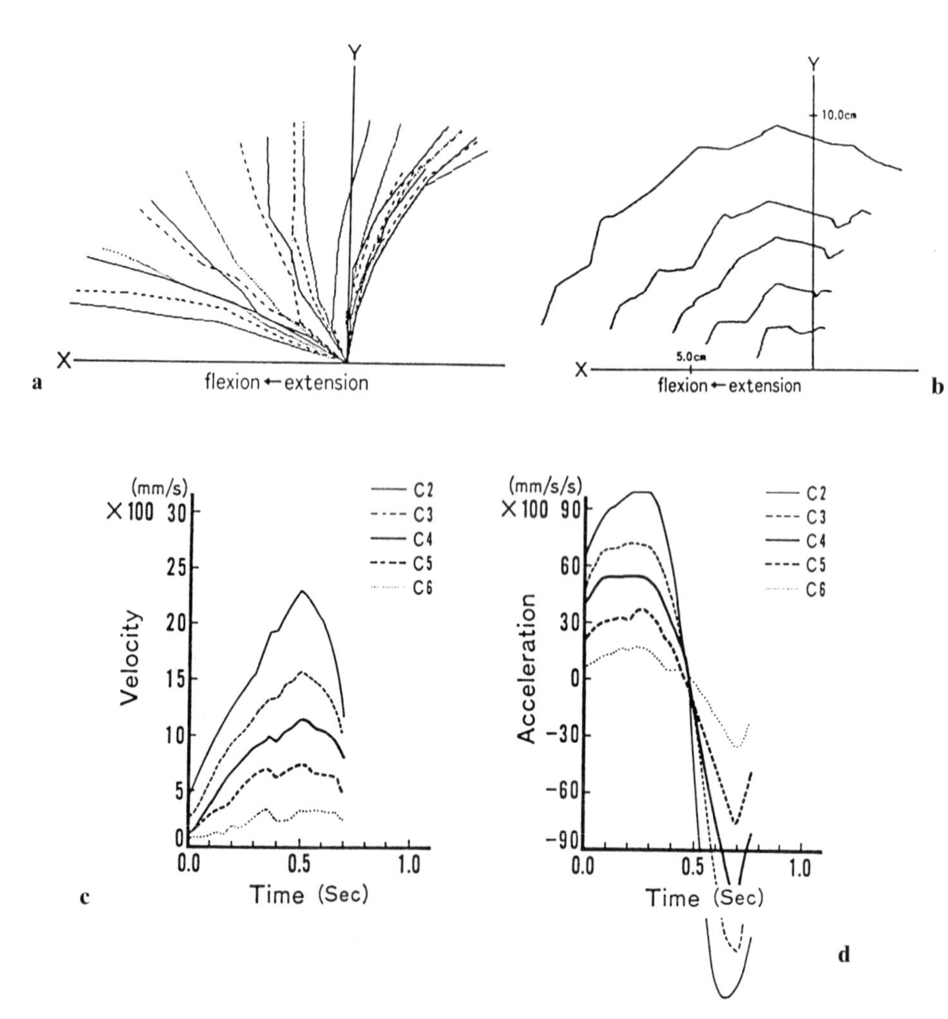

Fig. 4. **a,b** Trajectory of a CP case (case 5; Table 2) with a large range of motion. **c,d** Patient showed higher than normal velocity and acceleration. Maximum speed was 231.2 cm/s at C2; maximum acceleration was 988.2 cm/s^2; maximum deceleration was 2179.5 cm/s^2 at C2. The maximum velocity is about 1.7 fold and the maximum acceleration about 1.9 fold greater than the average of normal subjects at the C2 level. (From [14], with permission)

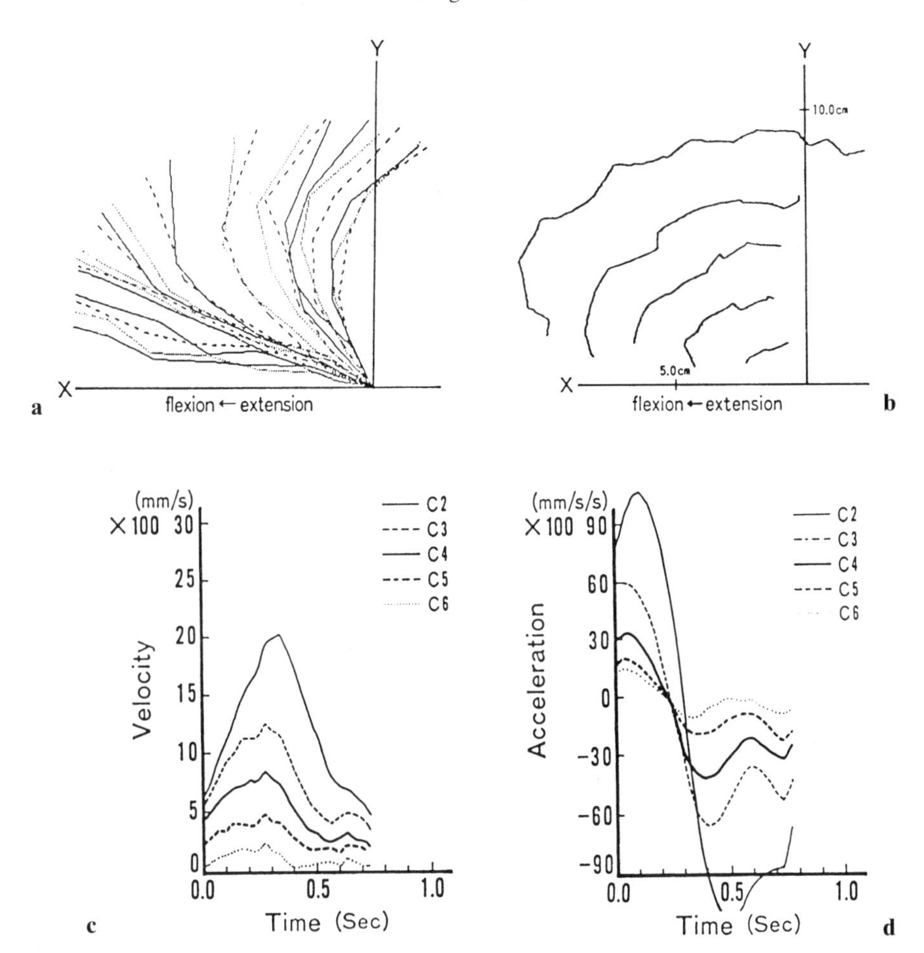

Fig. 5a–d. Segmental variation of velocity and acceleration at a certain disk level (case 2; Table 2). **a,b** The trajectories reveal inconsistent and changeable patterns in comparison with a uniform pattern in normal subjects. The characteristic feature of this cervical motion is "whip movement." **c,d** The difference in velocity between C4 and C5 is remarkable compared with that between the lower disk levels. This indicates that greater stress, that is, shear or bending stress, is generated at that particular location during motions of vertebral bodies. (From [14], with permission)

The distance from the center of each vertebra to the center point of the C7 vertebra is indicated on the *X*- and *Y*-axis, and was measured on a cineradiograph at intervals of 1/30s. Velocity and acceleration of each vertebral body, calculated from the displacement of the center at 1/30-s intervals during extension–flexion motion, are illustrated in Figs. 3c,d, 4c,d, and 5c,d. The velocity and acceleration were calculated along a line tangential to the trajectory of the vertebral body center at 1/30-s intervals. There were no significant differences between individuals in velocity or acceleration of the cervical spine among normal subjects (Table

1). There was, however, a definite gradient, descending craniocaudally, between the centers of each vertebra for both velocity and acceleration (Fig. 3).

Conversely, in patients with athetoid CP, velocity and acceleration showed a much greater range, and the pattern of movement varied from case to case. Trajectories of the cervical spine, as well as the velocity and acceleration of individual vertebrae during extension–flexion motion of representative cases, are illustrated in Figs. 4 and 5. The average maximum speed was 195.8 cm/s at the C2 and 84.5 cm/s at the C4 levels (Table 2). The average maximum acceleration in the acceleration phase was 746.1 cm/s^2 at the C2 and 335.9 cm/s^2 at the C4 levels. In the deceleration phase, the average maximum deceleration was 1422.5 cm/s^2 at the C2 and 642.6 cm/s^2 at the C4 levels (Table 2).

Kinematic analysis of the cervical spine of athetoid CP patients revealed certain characteristics: (1) a sudden increase in velocity and acceleration at certain levels during extension–flexion motion; (2) greater velocity and acceleration of the cervical spine, especially at the upper cervical levels, than those of normal subjects; and (3) a larger range of motion displayed by the cervical spine in association with (1) or (2).

It is notable that substantial differences in both acceleration and deceleration between two adjacent vertebrae might have clinical significance because of the generation of a substantial shear or bending moment in the intervening disk and facet. Figure 4 shows an example (case 5; table 2) of a large range of motion and high velocity and acceleration. The maximum speed was 231.2 cm/s at the C2 and 116.0 cm/s at the C4 levels, the maximum acceleration was 988.2 cm/s^2 at

TABLE 1. Maximum velocity, acceleration, and deceleration of each vertebra during extension–flexion motion in normal cases. (From [14], with permission).

Vertebra	C2	C3	C4	C5	C6
Maximum velocity (cm/s)					
Case 1	155.1	107.0	67.3	35.6	14.1
Case 2	135.5	70.5	47.6	30.3	16.8
Case 3	145.7	94.1	59.6	34.2	16.5
Case 4	123.1	72.4	46.9	34.0	20.9
Average	139.9	86.0	55.3	33.5	17.1
Maximum acceleration (cm/s^2)					
Case 1	586.1	342.8	218.4	107.8	28.5
Case 2	307.2	204.2	142.5	107.8	75.8
Case 3	605.0	399.9	219.2	106.6	54.4
Case 4	575.2	336.1	222.2	164.7	99.1
Average	518.4	320.8	200.6	121.6	69.5
Maximum deceleration (cm/s^2)					
Case 1	1203.0	894.1	573.1	333.4	127.4
Case 2	623.3	410.2	305.7	194.3	101.5
Case 3	1143.2	820.8	502.7	298.8	136.5
Case 4	980.1	561.6	379.5	259.7	131.8
Average	987.4	671.7	440.3	290.9	124.3

TABLE 2. Maximum velocity, acceleration, and deceleration of each vertebra during extension–flexion motion in athetoid cerebral palsy (CP) cases. (From [14], with permission).

Vertebra	C2	C3	C4	C5	C6
Maximum velocity (cm/s)					
Case 1	255.9	171.3	107.6	55.7	12.4
Case 2	264.2	128.6	86.1	49.0	26.6
Case 3	140.2	85.1	54.5	35.1	12.6
Case 4	139.1	94.5	70.6	48.5	25.4
Case 5	231.2	158.7	116.0	76.9	36.4
Case 6	144.1	95.6	72.6	43.1	17.4
Average	195.8	122.3	84.5	51.4	21.8
Maximum acceleration (cm/s^2)					
Case 1	496.9	340.1	282.9	151.3	44.6
Case 2	1044.4	596.1	316.4	178.3	130.1
Case 3	582.9	270.7	170.9	120.6	32.8
Case 4	620.8	436.5	358.7	248.1	125.8
Case 5	988.2	686.7	543.6	365.3	170.0
Case 6	744.1	498.3	331.0	220.3	78.4
Average	746.1	471.4	335.9	214.0	97.0
Maximum deceleration (cm/s^2)					
Case 1	2591.1	1732.7	1092.2	557.8	107.6
Case 2	1121.7	664.2	443.5	216.0	130.8
Case 3	979.2	580.4	359.3	233.5	97.4
Case 4	729.9	494.3	278.6	238.8	123.2
Case 5	2179.5	1600.3	1199.9	778.2	385.6
Case 6	933.4	648.4	482.1	270.8	140.3
Average	1422.5	953.4	642.6	382.5	164.2

the C2 and 543.6 cm/s^2 at the C4 levels, and the maximum deceleration was 2179.5 cm/s^2 at the C2 and 1199.9 cm/s^2 at the C4 levels. The velocity and acceleration were approximately 1.7 fold and 1.9 fold greater, respectively, than the average values recorded from normal subjects at the C2 level.

Athetotic neck motions in CP patients take place involuntarily and incessantly whenever the patients are awake and are often exaggerated in association with various daily activities. What are the changes one might expect to observe in a cervical spine that is moving with increased range, velocity, acceleration, and deceleration for a period of 30 or more years?

2.2 The Characteristic Features of the Cervical Spine in Athetoid CP Patients

The second- or third-decade group of athetoid CP patients often present with a malposture or segmental instability in their cervical spine [15]. These patients suffer from chronic pain and discomfort around the neck and shoulder, and while physicians might suspect spinal involvement, it is difficult to diagnose the malposture and segmental instability as characteristics preceding premature disk degeneration or spondylosis of the cervical spine. On a lateral roentgenogram, the cervical spine assumes a lordotic posture (a line connecting the anterior surfaces of the vertebrae has a convex curvature), which can vary in terms of curvature from nearly straight to pronounced lordosis. In our study, malpostures such as an S-shaped curve (lordokyphosis), a reverse S-shaped curve (kypholordosis), and kyphosis were observed in more than half the patients. By roentgenographical analysis, a segmental instability of the spine has been defined as follows: (1) forward or backward slipping by more than 3mm of adjacent vertebrae, and (2) a comparison of lateral roentgenograms of the cervical spine in extension and flexion that reveals an excessive rotation in the sagittal plane between two adjacent vertebrae beyond the normal range, as determined by Penning [16].

The incidence of the malpostures just listed was as follows: an S-shaped curve was seen in 35%, a reverse S-shaped curve in 14%, kyphosis in 9%, and a lordotic pattern in the remainder (42%). In athetoid CP, the incidence of the S-shaped curve was exceptionally high compared to normal subjects. The incidence of segmental instability of the cervical spine varied with the level of the segment (disk level) and posture of the cervical spine as a whole, as depicted in Table 3 and Fig. 6, respectively. Cervical instability was mainly found at the C3–C4, C4–C5, or, occasionally, at the C5–C6 disk levels. The malpostures of the cervical spine often resulted in segmental instability at a specific disk level when compared to normal or increased lordotic curvature of the cervical spine.

As mentioned earlier, large acceleration and deceleration differences between two adjacent vertebrae proved to be capable of inducing multiple segmental instability (either slipping or an excessive range of bending between two adjacent vertebrae) or malposture. Therefore, involuntary and incessant repetition of neck movement appeared to be the cause of the reduction in mechanical strength in the linkage structure of the cervical spine in young adult patients.

What are the consequences of these mechanical abnormalities? For this study, we carried out roentgenographic examinations in a larger number of patients with a broader age range than in previous studies [3,4]. We invited 180 patients who had attended our affiliated hospitals for diagnosis and treatment of athetoid CP to join our study project. The group consisted of 67 men and 49 women aged from 15 to 60 (mean, 31.9) years. The control group consisted of 103 subjects who attended the same hospitals for examination of ailments other than spinal prob-

Table 3. The range of motion versus malposture. (From [15], with permission).

Disk level	C2–C3	C3–C4	C4–C5	C5–C6	C6–C7
Lordokyphosis					
Transitional vertebra: C4	13.2 (7.2)	15.7 (8.5)	26.3 (14.1)	19.4 (6.0)	11.3 (9.6)
Transitional vertebra: C5	8.4 (4.3)	17.5 (7.3)	19.4 (10.5)	20.6 (9.1)	10.1 (5.1)
Lordosis	9.0 (4.7)	11.0 (4.6)	16.8 (6.5)	15.8 (6.7)	9.2 (5.9)

Data are in degrees (mean ± SD).

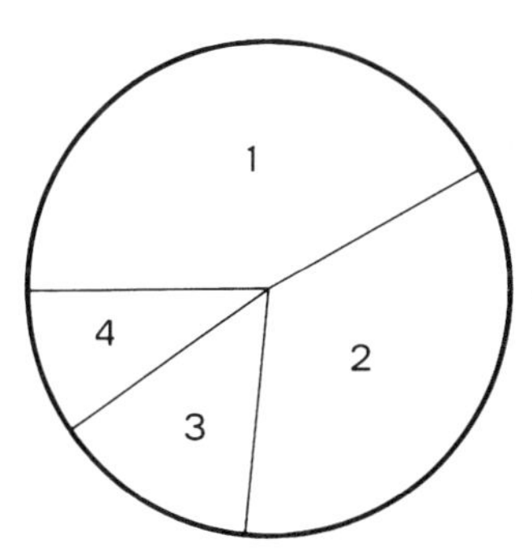

FIG. 6. Malposture of the cervical spine in athetoid CP. The incidence of malposture was S-shaped curve (lordokyphosis, *2*) in 35%, reversed S-shaped curve (kypholordosis, *3*) in 14%, and kyphosis (*4*) in 9%; the rest (42%) assumed a lordotic pattern (*1*). In athetoid CP, the incidence of the S-shaped curve was exceptionally high compared with normal subjects. (From [15], with permission)

lems. The age range and gender composition of the control group were comparable to the CP patient group. Again, meticulous care, sedation, and stabilization of the trunk were needed for taking plain and dynamic lateral roentgenographs of the cervical spine in the patients.

For each cervical spine (lateral view), the disk height at each level, bony spur formation at the margin of the vertebra, and segmental instability view in the control and CP groups were compared. Disk degeneration in a given case was assessed according to Lawrence [17]. Accordingly, grade 1 stood for slight anterior wear of the vertebral lip, grade 2 for the presence of anterior osteophytes, grade 3 for the presence of anterior osteophytes and narrowing of the disk space, and grade 4 for the presence of anterior osteophytes, disk space narrowing, and sclerosis of the vertebral plates. Epidemiological studies have repeatedly correlated these grades with the progression of spondylosis.

In male patients with athetoid CP, cervical disk degeneration irrespective of severity (grade 1 to 4) was found in 66% of patients between 15 and 24 years of age, in 96% of those 25 to 34 years of age, and in 100% of those more than 35 years of age. Advanced disk degeneration (grade 3 to 4) was not found in athetoid CP patients aged 15–24 years, but was detected in 36% of those 25–34 years old, in 85% of those aged 35–44 years, in 82% of those aged 45–54 years, and in 100% of those more than 55 years of age (Fig. 7).

The incidence and grade of disk degeneration increased with age in both male and female patients (Fig. 7). There was no correlation with gender with respect to prevalence of disk degeneration in patients with athetoid CP (Fig. 7). In the control group, the frequency of disk degeneration and the severity of disk degeneration increased with age (Fig. 8). Grade 1 and 2 cases were encountered predominantly in this group, but in the control subjects of 55–64 years of age, the frequency of grade 3 and 4 disks was 33% in men and 31% in women. No significant differences between males and females were noted (Fig. 8). In

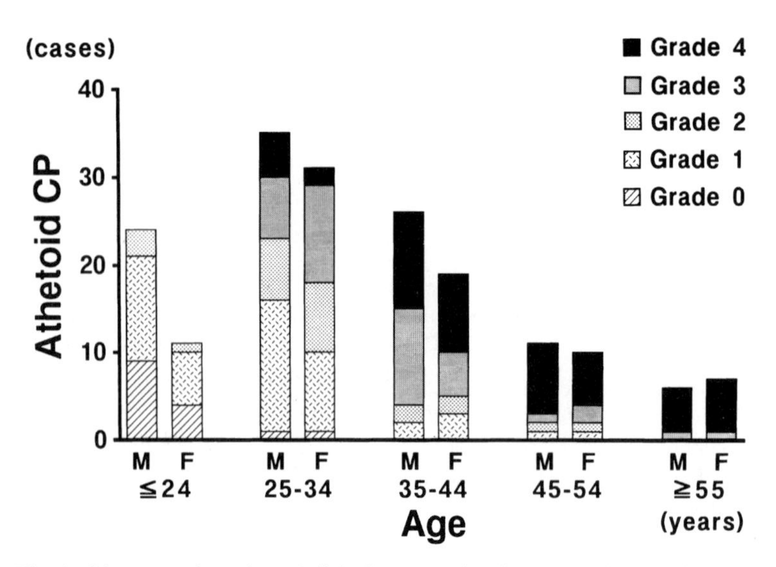

Fig. 7. The incidence and grades of disk degeneration increase with age in athetoid CP patients. *M*, Male; *F*, female

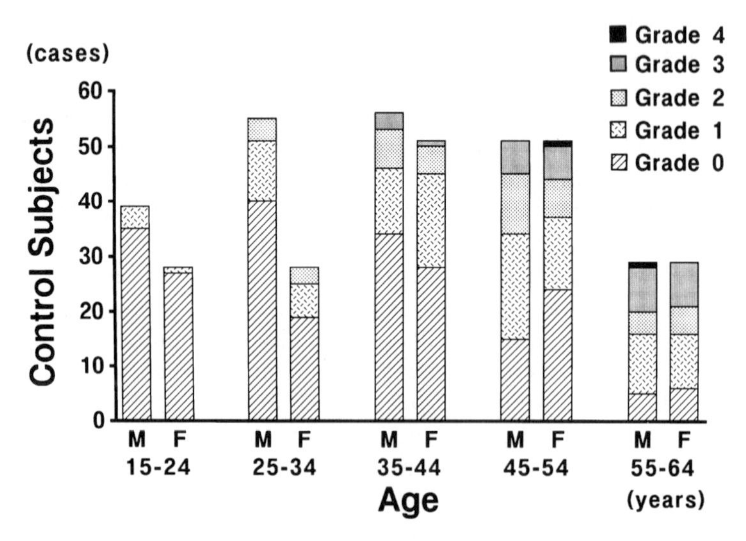

Fig. 8. The incidence and grades of disk degeneration also increase with age in the normal control group for both men and women

athetoid CP, disk degeneration was found to start at a younger age and progress more rapidly compared with the control group.

The incidence and grade of disk degeneration at each disk level increased with age. The levels most frequently affected were the C3–C4, C4–C5, C5–C6, and C6–C7 disks, either singularly or in combination. At ages of 34 years or less, the frequency of grades 1 through 4 peaked at C5–C6 and then increased parallel and multisegmentally from C3–C4 down to C6–C7 beyond this age. The most severe form of disk degeneration, grade 3 or 4, developed after the fifth decade and affected the disks diffusely from C3–C4 to C6–C7. Generally speaking, disk degeneration in athetoid CP started at the C5–C6 disk level and progressed into C3–C4, C4–C5, and C6–C7 levels with increasing severity. In the control group, disk degeneration affected C5–C6, C6–C7, and C4–C5 levels and tended not to progress beyond grade 1 or 2.

In summary, repeated neck motions of high velocity and acceleration proved to cause a measurable reduction of mechanical strength at the intervertebral disk, the medical sequelae of which are an early onset of structural degradation, disk degeneration, or spondylosis. Although our study was cross-sectional in nature and targeted particular disease patients, the findings point to an age-independent disk degeneration in the absence of systemic spine disorders.

2.3 Animal Experiments

The question therefore remains whether overuse or instability of the spine can really induce disk degeneration. We produced two different types of animal model to examine this hypothesis: (1) an overuse animal model in which repetitive extension–flexion of 5 cycles/min was applied to a rabbit cervical spine, and (2) an instability animal model in which the erectorspinae muscles were detached and the spinous processes were resected in mice.

To study accelerated structural degradation in the intervertebral discs of an overuse animal model (the Japanese white rabbit) [5,6], we succeeded in promoting repetitive extension–flexion movements in the neck through a cyclic electrostimulation with electrodes implanted into the trapezius muscles. The frequency of the stimulation was 5 cycles/min, the average time of loading was 10 h/day, and the total number of cyclic loads was approximately 3000 cycles/day. Histological examinations of the intervertebral disks were made after 15 through

TABLE 4. Description of the histological grades in rabbit intervertebral disk[a]. (From [5], with permission).

Grade	Nucleus	Anulus	Vertebral body
1	Gel-like and clear distinction from anulus fibrosus	Discrete fibrous lamellae	Margin rounded
2	Fibrous tissue extending from anulus fibrosus	Slightly delaminated fibrous lamellae	Margin pointed
3	Consolidated fibrous tissue and loss of anular–nuclear demarcation	Extensively delaminated fibrous lamellae and tear in peripheral rim	Early chondrophytes and osteophytes at margins
4	Horizontal clefts	Focal disruptions and separation from vertebral body	Marked osteophytes at margins

[a]Based on the Vernon-Roberts grading system [61].

70 days of cyclic loading, respectively. Structural degradation was found in the C6–C7 disk and the margin of the adjacent vertebra in those animals loaded with 200 000 cycles of flexion and extension. The anulus fibrosus of the disc was delaminated or focally disrupted. Early osteophyte formation occurred at the anterosuperior margin of the C7 vertebra. Loads of 50 000 or 100 000 cycles generated only mild changes in the anulus fibrosus with no other major differences between loaded and nonloaded groups.

The severity of structural degradation in the intervertebral disks was assessed in terms of the histological grades quoted from the scale system devised for

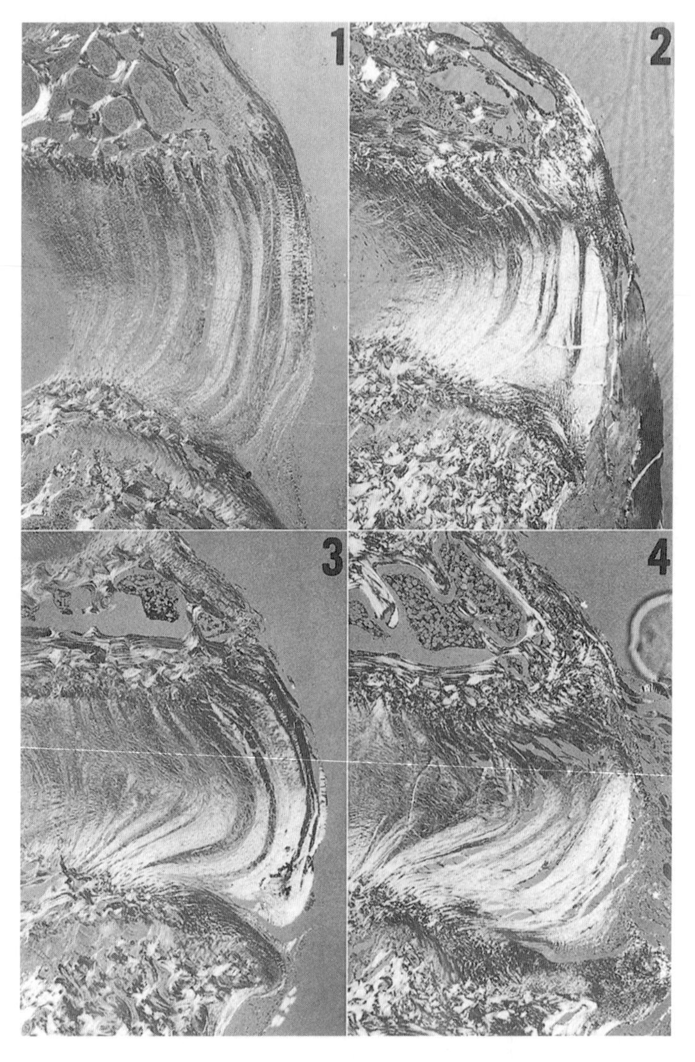

Fig. 9. Typical appearance of anulus fibrosus for each grade (*1–4*) (Congo red with polarized microscope, ×40). (From [5], with permission)

spondylosis for humans [18,61]. Increasing load cycles in the animals led to more advanced structural degradation as outlined by the scale (Table 4). In overuse animals loaded with a huge number of extension–flexion cycles, structural degradation took place in the disk and resembled disk degeneration in spines with spondylosis (Figs. 9 and 10). Age-independent disk degeneration (from histological comparisons) advanced with increasing load on the disk. Overuse in this experiment appeared to induce fatigue failure in the disk structure. Further study of the similarity between structural degradation from overuse and degeneration caused by spondylosis is needed, based on mechanical and biochemical analyses of these disks.

To study accelerated structural degradation in the intervertebral disks of a surgically induced unstable spine that leads to disk herniation or osteophyte formation at the discovertebral junction, we succeeding, using adult, 6-month-old, male ICR strain mice [7], in producing an unstable spine animal model as follows. The electorspinae muscles were detached from the spinous processes, laminae, and facets, followed by resection of the spinous processes including the supra- and interspinous ligaments in the whole spine. The animals were housed and fed in the same manner as the control group and killed after a 2- or 12-month period for histological examination of the disk. The structural changes in the disks incorporated into a chronic unstable spine were also contrasted with disks from normal animals of the same age.

Six months post surgery, roentgenographic examinations revealed pathological kyphosis of the spine in 50% of the mice, thus providing evidence that the surgical manipulations had induced instability. Disk space narrowing and anterior osteophytosis were observable predominantly at the C4–C5, C5–C6, and C6–C7 disk levels in all the treated animals at 12 months after surgery. Histological examination revealed marked degeneration of the disks at the same location,

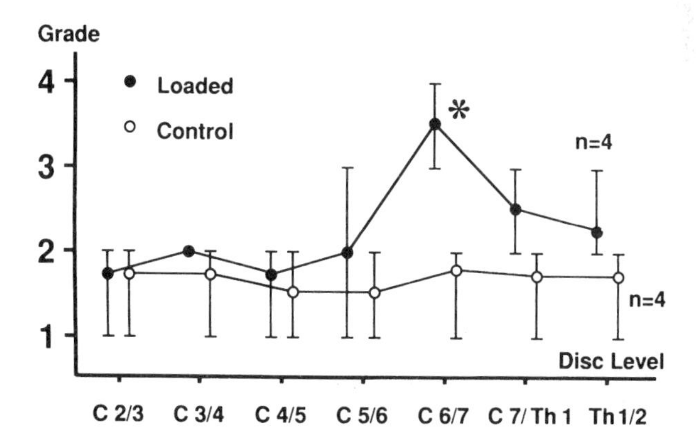

FIG. 10. Mean grade of degeneration of anulus fibrosus in the loaded group and control group at each disk level. Repetitive loading of 2×10^5 cycles generated more severe degeneration of anulus fibrosus at the lower cervical spine (*, $P < 0.05$, Mann–Whitney U test). *Bars* denote distributions of grade. (From [5], with permission)

particularly at more than 6 months after surgery. The lesions were characterized by shrinkage or disappearance of the nucleus pulposus, disruption of the anulus fibrosus, and chondrocyte proliferation. The latter often displaced and protruded beyond the outer margin of the anulus, with evidence of partial ossification indicating osteophyte formation (Figs. 11 and 12). Further degenerative changes in the disks were visible as time elapsed after surgery, but the overall appearance was similar to that of disks of the same grade in an untreated group of animals aged 12 months or more. For this reason, surgically induced instability was considered to accelerate disk degeneration, intensify structural degradation, and promote the onset of spondylosis.

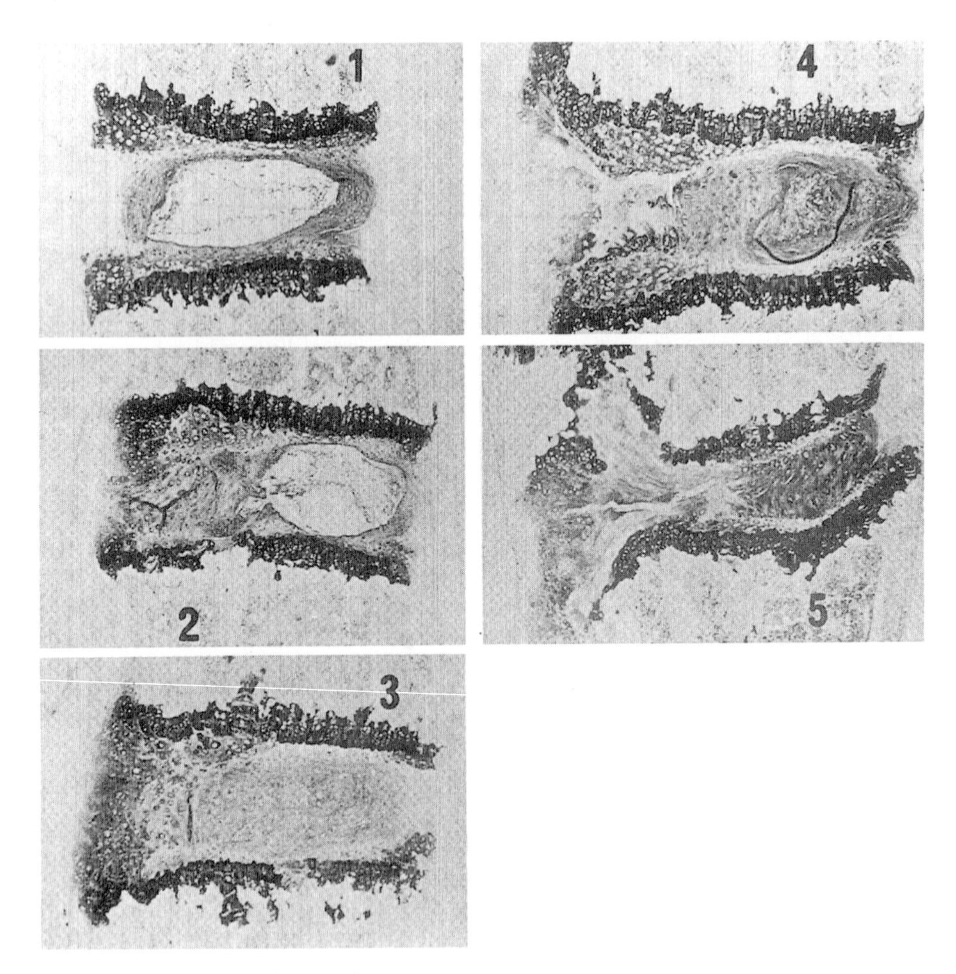

Fig. 11. The appearance of disks in mice typical of each grade (*1–5*) (toluidine blue stain). (×40) (From [7], with permission)

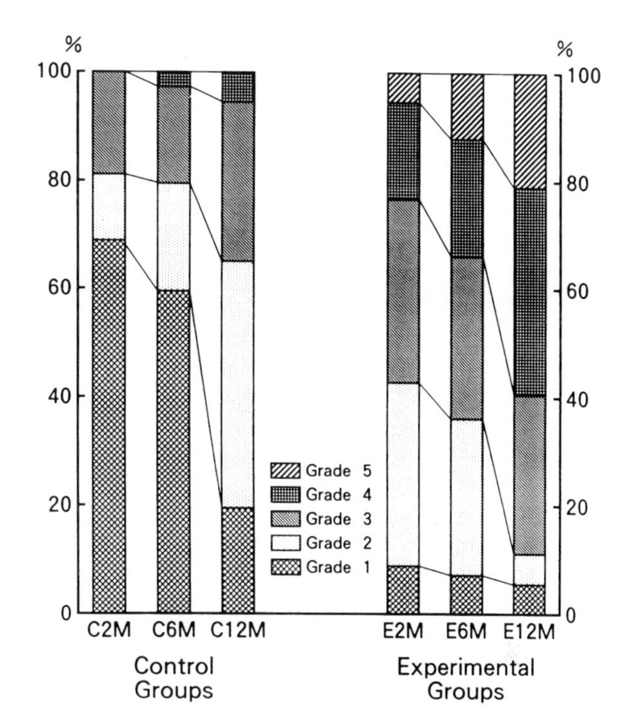

FIG. 12. The percentage of disks in mice in each morphological grade (1–5) for each control (*left*, $n = 81$/group) and each experimental (*right*, $n = 90$/group) group at 2, 6, or 12 months of age (*2 M*, *6 M*, *12 M*). (From [7], with permission)

3 Measurement of Axial Stability in the Lumbar Segment

In the preceding sections, we drew attention to the structural degradation or degeneration, in medical terms, of the intervertebral disk and the mechanism responsible for its degeneration in humans and animal models. However, to what extent does disk degeneration affect spinal stability, and what is the most sensitive test for spinal instability caused by disk degeneration? How can we measure the degree of instability accurately in vivo? Although the damaged lumbar disks are usually excised as part of spinal fusion surgery, it is not known how this affects stability in the spine.

As a preliminary procedure, we were able to measure tensile stiffness of the respective segment of the lumbar spine during surgery [19,20]. Although the definition and quantitation of spine stability are still controversial issues, we viewed tensile stiffness of a particular motion segment as an indicator of its stability and correlated it with disc degeneration as visualized by roentgenography and magnetic resonance imaging (MRI). Tensile stiffness of a motion segment (vertebra–disk–vertebra) was measured with a spinal distractor during spinal surgery (Fig. 13). Stiffness (comparison of load and displacement) was assessed by suspending a vertebral spreader equipped with a load strain gauge and a displacement transducer between two adjacent spinous processes (Fig. 14).

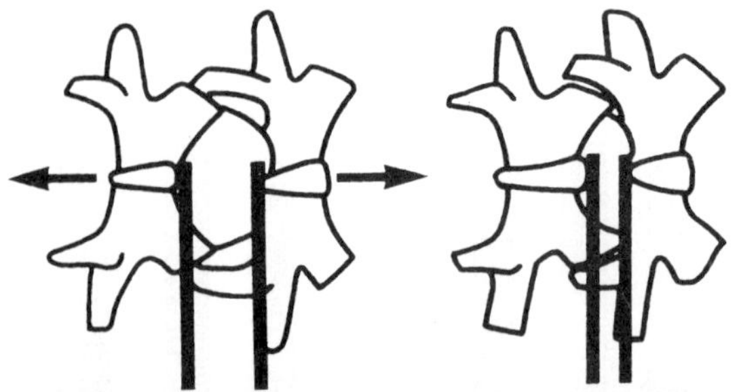

FIG. 13. To quantify spinal instability, the tensile stiffness of a motion segment (vertebra–disk–vertebra) was measured with a spinal distractor during spinal decompression surgery. (From [19], with permission)

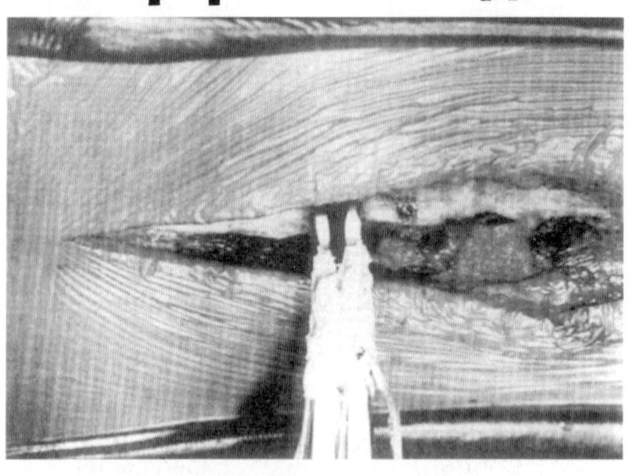

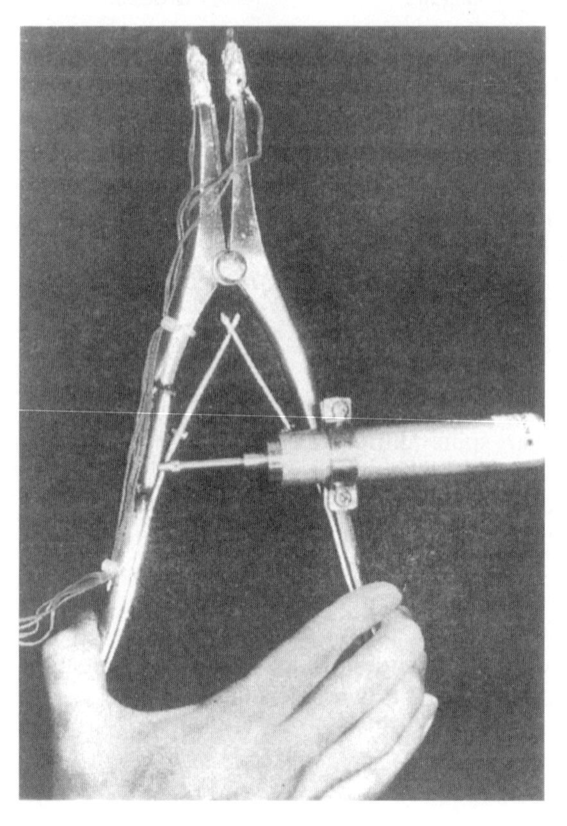

FIG. 14. Device for measuring stiffness of spinal motion segment, a lumbar spinal spreader equipped with two measurement devices: one is a load strain gauge attached to the distal legs of the spreader to measure load during distraction of two spinous processes; the other is a displacement transducer attached between two proximal legs of the spreader to measure displacement between two adjacent spinous processes. (From [19], with permission)

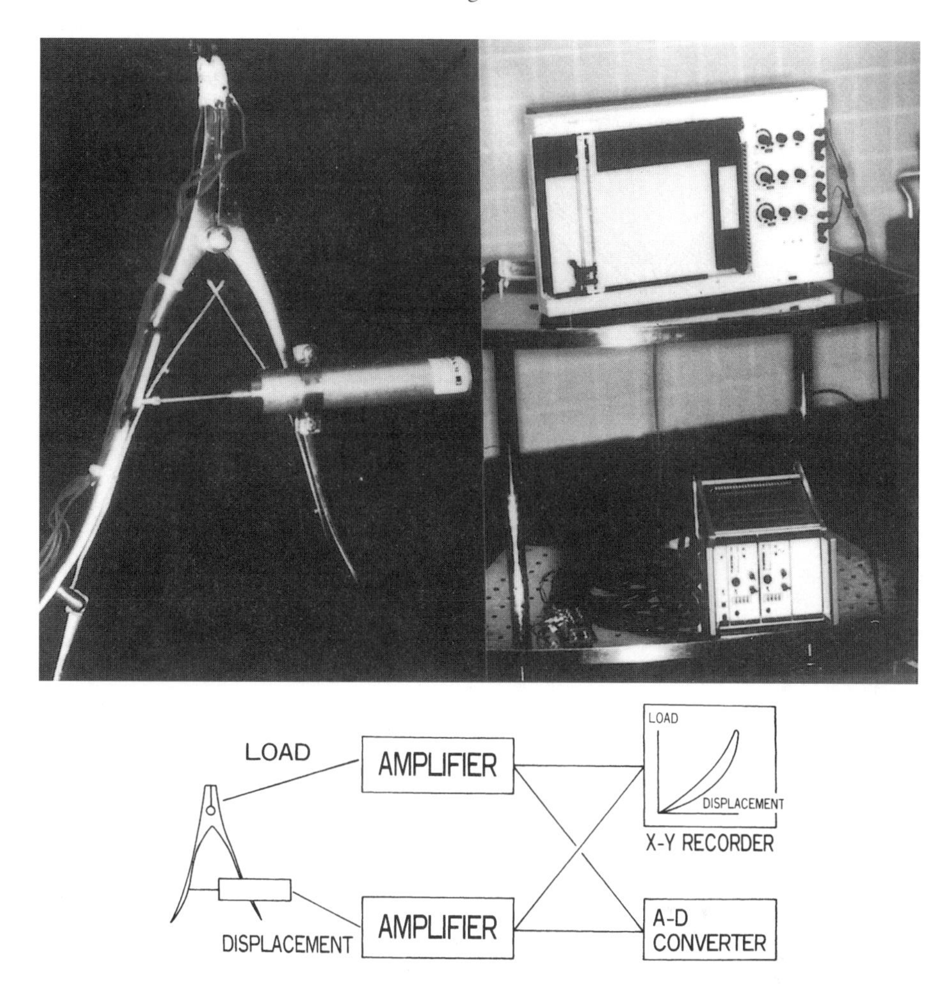

FIG. 15. A load-displacement curve was recorded while the motion segment was being distracted at a constant speed. (From [19], with permission)

A load-displacement curve was recorded while the motion segment was being distracted at a constant speed (Fig. 15). The measurements gave load-displacement curves of a high reproducibility and showed hysteresis loops in all cases (Fig. 16). The average stiffness of the spinal motion segments was 14.7 N/mm, unless the intervening disk showed signs of degeneration. The stiffness of the motion segment diminished markedly with age in this series (Fig. 17). Disks with greater evidence of degeneration showed a reduction in stiffness (Fig. 18). In patients suffering from slipping of the vertebra either forward or backward and disk degeneration, the lowest stiffness (5.4 N/mm, on average) was recorded. The average value for stiffness in spines with diffuse disk degeneration was 8.5 N/mm. Therefore, disk degeneration, either single or multiple, caused a marked lower-

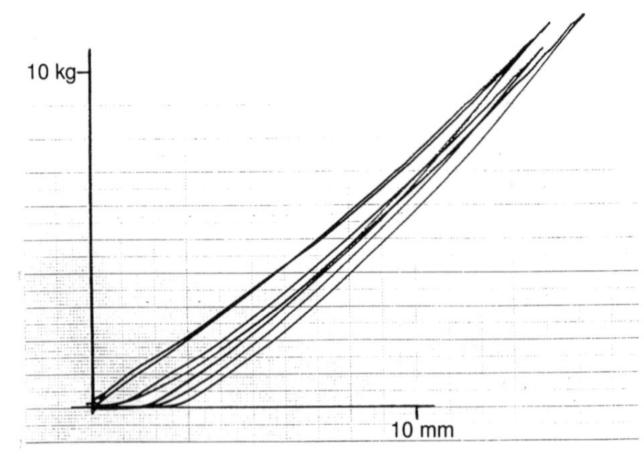

Fig. 16. Measurement gave load-displacement curves of high reproducibility that show hysteresis loops in all cases. (From [19], with permission)

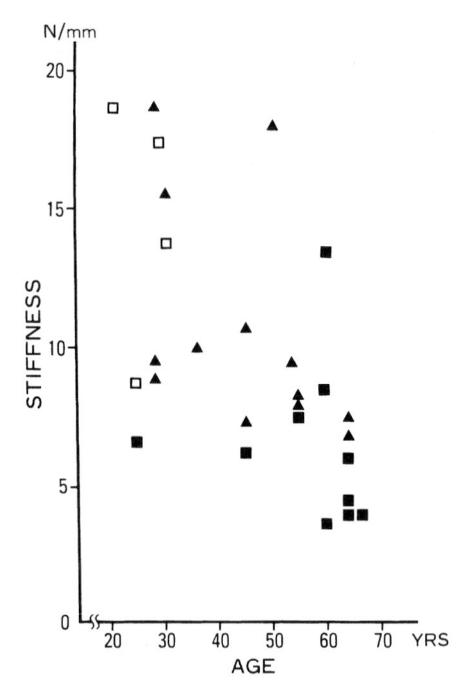

Fig. 17. Variation in stiffness in relation to both aging and disk degeneration. Stiffness of motion segments markedly diminished with age in this series. The *open squares*, *solid triangles*, and *solid squares* represent magnetic resonance imaging (MRI) findings of normal, moderate, and severe disk degeneration, respectively. (From [19], with permission)

ing of stiffness in a motion segment of the spine and as a result rendered the spine unstable.

Is it possible for stiffness or stability to improve in a motion segment shown to be unstable because of disk degeneration? Intraoperative measurements of tensile stiffness could not be repeated in the same subject over a certain interval. Thus, we could not perform a longitudinal study of disk degeneration in terms of the development of tensile stiffness or segmental stability.

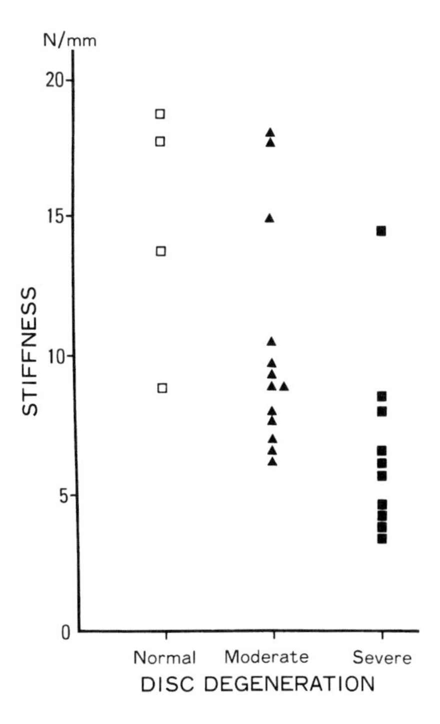

FIG. 18. Stiffness of spinal motion segment diminished as disk degeneration developed. Average stiffness of spinal motion segments was 14.7, 10.7, and 6.8 N/mm in groups of normal, moderate, and severe degeneration, respectively. The *open boxes*, *solid triangles*, and *solid boxes* represent MRI findings of normal, moderate, and severe disk degeneration, respectively. (From [19], with permission)

It is difficult to accept that recovery of stiffness occurs in the clinical population on the basis of the data from our cross-sectional study. Advanced disk degeneration was correlated with lowered tensile stiffness of the disk. Previous reports have described recovery of motor segment stiffness, which was believed to be restabilization of the segment with advancing disk degeneration [21]. This conclusion was drawn from roentgenographic findings, which showed a reduction in the flexion–extension range of the motor segment as the intervening disk degenerated. In the lateral view, the motor segment lost its expansion and compression properties with progressive disk degeneration. However, this phenomenon was the result of marginal osteophyte formation or stiffening of the intervertebral joints with osteoarthrosis. Measurement of tensile stiffness by our approach and apparatus did not add support to the restabilization theory, at least for axial stability. MRI observation also failed to provide evidence to support recovery of disk structure, once degeneration had started.

The fundamental mechanisms underlying disk degeneration remain to be determined. Histological, biochemical, and imaging studies have implicated a variety of causes. Methods capable of quantifying disk properties, e.g., measurement of tensile stiffness of the disk, can be used to assess disk degeneration and therefore monitor this process more precisely. Employing stiffness as a quantitative indicator of structural degradation of the disk in in vivo animal experiments will also help to unravel the pathobiology of disk degeneration.

4 Quantification of the Structural Degradation of the Rabbit Spinal Unit Under Cyclic Torsional Loading

In previous sections, we described the principal factors that could accelerate structural degradation compatible with degeneration in the disk, both in a clinical setting and in animal experiments. Repetitive application of extension–flexion motion, bending or torsion, and chronic instability of the spine were found to cause structural degradation in the disk. What is the significance of repetition and chronic malfunction in this process? We hypothesized that repetition and chronicity elicited a certain condition that accumulated and led toward the resulting degradation of the disk structure. What condition was initiated and accumulated through repetition and chronicity?

We further reasoned that this must not be a simple trauma severe enough to impart detectable damage to the structure. Rather, it must be a microtrauma, which at first is undetectable but is manifested after sufficient accumulation of its effects. We selected first a cyclic torsional loading as an appropriate modality for the rabbit spinal unit (motion segment, vertebra–disc–vertebra), with the expectation of causing a cumulative effect on the disc [8]. This approach was based on deduction and clinical experience, which suggested that a high frequency of disk degeneration might result from the cumulative effects of cyclic torsion on the lower lumbar segment generated during daily walking.

Fresh cadaveric lumbar spinal functional units of L2–L3 and L4–L5 were removed from Japanese white rabbits (age, 8–9 weeks; weight, 1.9 kg on average). After removing the muscles and posterior elements, the disk vertebra units were used in experiments. The specified angular displacement was applied to the specimen in a cyclic manner. The loading apparatus converts the translational movement of the piston head of the electrohydraulic universal testing machine

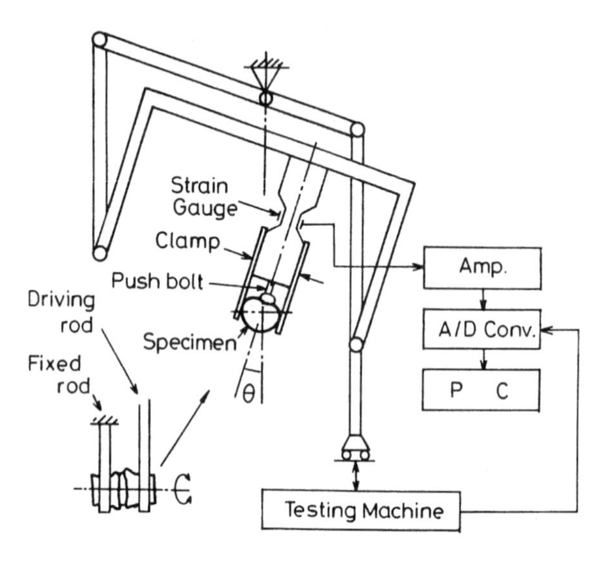

Fig. 19. The loading apparatus converts the translational movement of the piston head of the electrohydraulic universal testing machine into angular movement on the specimen by using the parallelogram mechanism. (From [8], with permission)

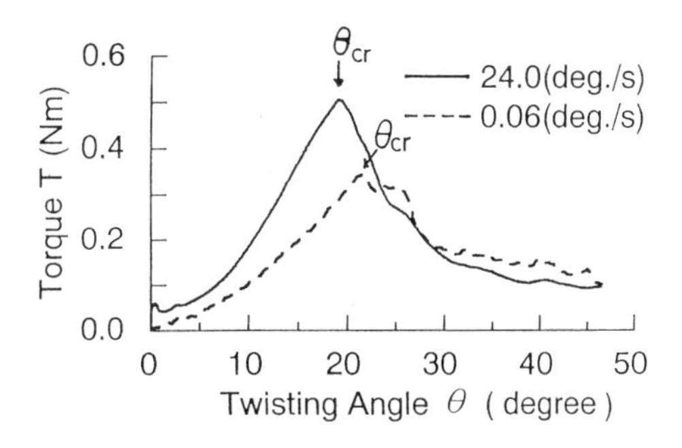

FIG. 20. Typical torque angle curves of the L2–L3 disk body unit when the specimens were twisted until failure at two different but constant loading rates. (From [8], with permission)

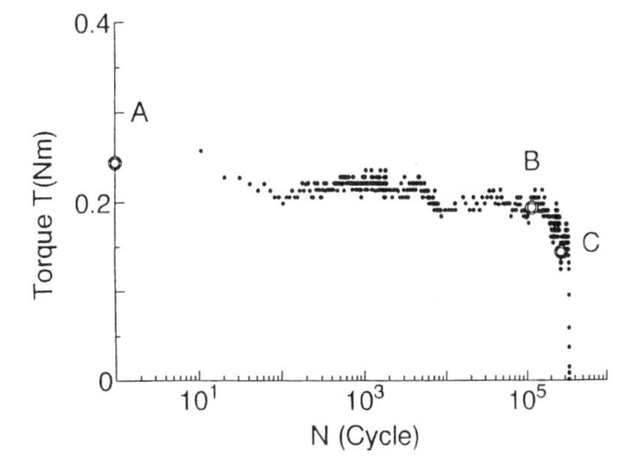

FIG. 21. Typical peak torque decrease under cyclin torsional loading. The peak torque decreased as the number of torsional cycles increased, displaying two phases: an initial gradual decrease phase followed by a sudden decrease phase. (From [8], with permission)

into an angular movement on the specimen by using a parallelogram mechanism (Fig. 19). The center of the angular movement of the specimen was the geometric center of the intervertebral disk. Upper and lower vertebral bodies were firmly held by custom-designed clamps. A pair of strain gauges were bonded to the driving rods to measure the torque acting on the specimen.

First, we determined the maximum torque (critical angle, Θ_{cr}) of the disk–vertebra unit against twist. Typical torque angle curves of the L2–L3 disk–vertebra unit, depicted in Fig. 20, were recorded when the specimens were twisted until failure at two different but constant loading rates. To avoid a one-time twisting failure, the amplitude of axial rotation (Θ_a) during cyclic testing was regulated below the critical angle. For example, when the amplitude of axial rotation was set at 12°, the frequency of one-way sinusoidal loading was 1 Hz. The specimens were placed in a physiological saline bath at a constant temperature of 20°C. The peak torque decreased as the number of torsional cycles increased (Fig. 21). The decrease of peak torque displayed two phases, an initial gradual decrease phase followed by a sudden decrease phase. After a sudden decrease of peak torque, macroscopic failure of the specimen occurred without exception. Figure 22 demonstrates the change of tangential stiffness at the three different cycles. Stiffness decreased as the number of torsional cycles increased. Lengthening of the initial toe part in the torque angle curve suggested an increase of the

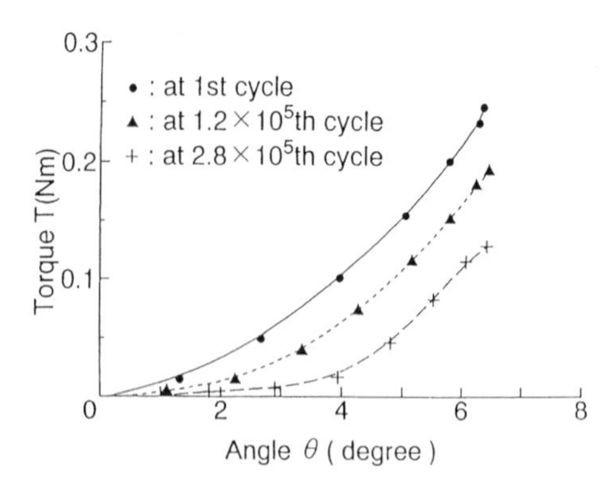

Fig. 22. Torque–angle behavior during cyclic torsional loading shows the change of tangential stiffness at the three different cycles. (From [8], with permission)

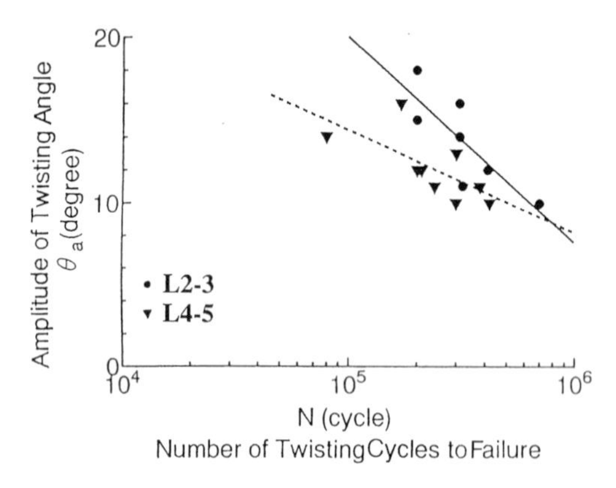

Fig. 23. A strong correlation between the number of cycles to failure (N_f) and the amplitude of rotation (Θ_a), is expressed in the following equation: $\Theta_a = a + b\log_{10} N_f$. (From [8], with permission)

Table 5. The coefficients *a* and *b* and the critical angle, Θ_{cr}. (From [8], with permission).

Level	a (degrees)	b (degrees)	θ_{cr} (degrees)
L2–L3	82	−12	22.1 ± 2.4
L4–L5	45	−6	18.3 ± 2.6

neutral zone, which was defined by Pope and Panjabi [22] and is also regarded as a good parameter of instability. A strong correlation between the number of cycle loads to failure (N_f) and the amplitude of rotation (Θ_a) was verified, as expressed in the following equation and in Fig. 23 (coefficients *a* and *b* are shown in Table 5):

$$\Theta_a = a + b\log_{10} N_f$$

The negative value of the coefficient *b* means that the number of cycles to failure decreases exponentially as the amplitude of rotation increases. This exponential decrease is a typical feature encountered in the fatigue phenomenon and is not

FIG. 24. Histological findings at the intervertebral disk: delamination and disruption of the anulus fiber (×40). (From [8], with permission)

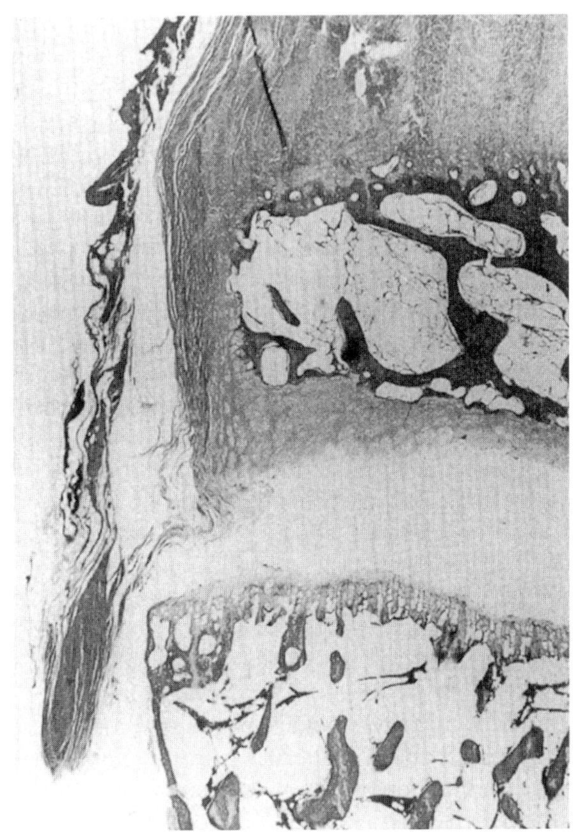

FIG. 25. Separation at the growth plate of the vertebra. (×40) (From [8], with permission)

found in one-time failure responses. Also, significant differences occurred between the values of critical angle (Θ_{cr}) and the values of coefficient a ($\Theta_a = a$ when $N_f = 1$) as shown in Table 5. A decrease of stiffness caused by repetitive loading suggested an increase in instability and structural weakening of the motion unit.

Histological examination revealed corresponding damage (Figs. 24 and 25). Delamination or disruption of the anulus fiber and separation at the growth plate of the vertebrae were observed. The latter finding suggests that the rabbit may be an inappropriate model for this experiment. We are planning to continue the experiments in another animal, e.g., the sheep.

5 Discussion

The intervertebral disk is a cartilaginous tissue with properties and macromolecular constituents similar to those found in articular cartilage. However, the structure of the intervertebral disk is characterized by the outer anulus fibrosus and the central nucleus pulposus. The disk is comprised mostly of water (70%), collagens, and proteoglycans in an extracellular matrix. In the nucleus pulposus, collagen accounts for only 10%–20% of the nucleus dry weight with proteoglycans and other matrix proteins accounting for the remainder [23]. Collagens dominate the composition of the anulus fibrosus, accounting for approximately 60%–70% of the dry weight of this structure.

The fibrous framework of the disk is built from type I and type II collagen fibrils [24]. These fibrils are distributed radially in opposing concentration gradients with fine type II collagen exclusively in the nucleus pulposus and strong type I increasingly concentrated toward the outer edge of the anulus fibrosus. The latter forms the tough lamellar sheets that anchor the disk into the bone of the vertebral bodies [24–26]. This network of collagen fibers, in which polyanionic proteoglycans are embedded, is able to distribute load and provide tensile properties. Because of extreme charge density, proteoglycans have an extended structure and try to occupy as large a volume as possible. Their expansion is prevented by the tensile properties of the collagen fibers. Thus, the organization of disk structure is such that a concentrated proteoglycan solution, the nucleus pulposus, is held within the strong collagen network of the anulus fibrosus.

In a manner similar to articular cartilage, the intervertebral disk exhibits a viscoelastic response when subjected to loads and deformations. It creeps under a constant applied load and stress-relaxes under a constant applied deformation [27,28]. The viscoelastic response of articular cartilage depends on two different physical mechanisms, according to Mow and associates [27,28]. One is the intrinsic viscoelastic properties of the macromolecules, mainly collagens and proteoglycans, that form the organic solid matrix, and the other is the frictional drag arising from the flow of the interstitial fluid through the porous, permeable solid matrix composed of collagens and proteoglycans.

The mechanical properties of cartilaginous tissues depend not only on differences in the proportions of collagens and proteoglycans but also on the organization of these molecules [29]. The sulfate and carboxylate groups closely spaced on the glycosaminoglycan disaccharide repeating units along the proteoglycans create strong intra- and inter-molecular repulsive forces. This charge repulsion extends and stiffens proteoglycan aggregates in the interfibrillar space formed by the collagen network. Motile cations such as Na^+ and Ca^{2+} in the solution are attracted to the fixed negatively charged groups on the glycosaminoglycans [30–32] and this leads to an inbibition of water and creation of an osmotic swelling pressure (Donnan osmotic effect) [31]. This swelling pressure is resisted and balanced by tension in the collagen network [31,33] thus confining the proteoglycans to about 20% of their free solution domain. This pressure exposes the surrounding collagen meshwork to a state of "pre-stress" even in the absence of external loads [33].

Most of the water, excluding that strongly associated with the collagen fibrils or in the intracellular compartment, occupies the intermolecular space and is free to move when a load is applied to the tissue [27,30,31,34,35]. Approximately 70% of the water within the cartilaginous tissue may be moved upon loading. This movement is essential in controlling tissue deformation and mechanical behavior [27,28,30,32,34–38]. The collagen and proteoglycans that make up the solid matrix are highly dispersed in the interstitial fluid, resulting in a microporous material with extremely low permeability. The flow of the interstitial fluid and the transport of solutes through this microporous solid matrix are resisted by a high frictional drag.

When the cartilaginous tissue is loaded, the interstitial fluid pressure in the porous solid matrix rises. The interstitial fluid begins to flow and exudation takes place. The load applied is balanced by the compressive stress developed within the collagen–proteoglycan solid matrix and by the frictional drag generated by the flow of the interstitial fluid during exudation. As the interstitial fluid is depleted, the proteoglycan concentration within the solid matrix increases, which increases charge–charge repulsive force and the Donnan osmotic swelling pressure until they are in equilibrium with the applied external load. Through this mechanism, the proteoglycan gel trapped within the collagen meshwork enables the cartilaginous tissue to resist compression.

As discussed, the intervertebral disk behaves as a viscoelastic material and exhibits a cartilage-like response to mechanical loading [38–40]. However, because of its special structure, high tensile stresses are generated in the anulus fibrosus due to the development of large hydrostatic pressures within the nucleus pulposus [39,41] when the disk is loaded through compression, bending, and torsion. This tensile stress in the outer region of intervertebral disk may be the reason for the abundant type I collagens (the main macromolecules of tendons) and the tight fibrous organization of the anulus fibrosus. Variations in the magnitude of tensile stress and strain generated in the anulus fibrosus may also occur from the inner to outer regions due to differences in tissue structure [42,43], biochemical composition [25,44,45], and material properties [38].

As the results of this study suggest, the repetition and chronicity of loading elicited a cumulative condition that led to degradation of the intervertebral disk structure. It is interesting to speculate as to the nature of the events engendered by repetitive loading of the disk. From a structural perspective, high tensile stresses take place in the anulus fibrosus because large hydrostatic pressures develop within the nucleus pulposus [39,41] under various mechanical conditions. The collagen fibers in the anulus fibrosus are constantly exposed to repetitive tensile stresses during daily mechanical loading. Furthermore, various patterns of loading cause friction between the laminae of the anulus that may cause microtrauma in the network of collagen fibers and delamination in the anulus fibrosus. These events may be undetectable initially but after prolonged repetition, the cumulative damage leads to a loss of structural integrity within the disk. Data from clinical studies and experimental models of overuse, instability, and in vitro cyclic loading indicate that delamination of the anulus fibrosus is always present at the initial stage of destruction in the intervertebral disk. Disruption of the collagen network may be a result of fatigue failure by repetitive loadings, and also may be a key factor leading to the development of disk degeneration. Loosening of the collagen network may be responsible for the increased swelling and water content at the initial stage and loss of proteoglycans and water in the advanced stage of cartilaginous tissue degradation [46–48]. Because of these changes, the disk is unable to produce viscoelastic response when subjected to loads and deformation.

The presence of a damaged disc introduces a less stiff segment in the spine and initiates a marginal reaction, or spondylosis, around the affected cartilage. Segmental instability will further accelerate structural degradation and increase disintegration or disruption of the anular fibers. Disk degeneration is generally considered to be an age-dependent change [49–55,60] but it can also be induced through pure mechanical fatigue failure as an age-independent change in the cartilaginous tissue.

The next question is whether the disk is capable of undergoing repair or remodeling. Explant culture studies have shown that the metabolic activity of cartilage will respond to applied hydrostatic and osmotic pressures [56,57]. Chondrocytes most likely respond to mechanical forces transmitted and modulated by the extracellular matrix through signal transduction at the matrix–cell membrane interface which leads to regulation of the biological remodeling process. However, at the tissue level, the repair or remodeling processes do not appear to be well organized. Although less information is available on the potential of the disk for repair, the sequence of events after disk herniation has been described in a rabbit model in which a small stab incision was made through the anterior anulus [18,58,59]. In this model, there was rapid loss of nuclear material, an ingrowth of fibrotic replacement tissue, scar formation, and little or no repair or replacement with normal disk tissue. There is no available evidence that the anulus fibrosus can repair the intrinsic geometry of the collagen network once it is destroyed. At the cell level, cartilaginous tissue reacts to produce collagen fibers and proteoglycans to restore the cell–matrix environment. However, at the tissue level, these collagen fibers may fail to contain the compressed proteoglycan core which spreads and may impede reassembly of the collagen fibril network.

The occurrence of age-unrelated spondylosis in athetoid CP patients prompted the initiation of our studies on the structural degradation of the intervertebral disk in human subjects and animal models. Clearly, the data suggests that the intervertebral disk has a propensity for fatigue failure under loading conditions. Structural degradation appears to start in the anulus fibrosus region of the intervertebral disk and is unrelated to aging. The disk cartilage, through its collagen-and proteoglycan-rich composition can sequester water and dissipate the energy from a loading challenge. Unfortunately, the intrinsic hydrodynamic properties of the cartilage that permit the dissipation of energy may be responsible for the enhanced fatigue nature of the intervertebral disk. This tendency of the cartilage to undergo cumulative fatigue damage, coupled with poor reparative and remodeling capacities, leads to structural degradation of the intervertebral disk.

References

1. Kelsey JL (1980) Epidemiology: natural course of the disease. In: White AA III, Gordon SL (eds) AAOS symposium on idiopathic low back pain. Mosby, St Louis, pp 3–22
2. Praemer A, Furner S, Rice DP (1991) Musculoskeletal conditions in the United States. American Association of Orthopedic Surgeons, Park Ridge, pp 1–20
3. Ebara S, Harada T (1992) Cervical spondylosis in athetoid cerebral palsy (in Japanese). Sagyo Ryoho Janaru 26:80–84
4. Harada T, Ebara S, Kajiura I, Ohkawa A, Hiroshima K, Anwar MM, Ono K (1992) Cervical disc degeneration in athetoid cerebral palsy. Orthop Trans 16:790
5. Wada E, Ebara S, Saito S, Ono K (1992) Experimental spondylosis in rabbit spine. Overuse could accelerate the spondylosis. Spine 17:S1–S6
6. Wada E, Ebara S (1993) Experimental study of cervical facet degeneration in rabbits. Orthop Trans 17:756–757
7. Miyamato S, Yonenobu K, Ono K (1991) Experimental cervical spondylosis in the mouse. Spine 16:S495–S500
8. Oda T, Ebara S, Tanaka M, Kuraya T, Ono K (1994) Biomechanical properties of the disc body unit under cyclic torsional loading. In: Hirasawa Y, Sledge CB, Woo SLY (eds) Clinical biomechanics and related research. Springer-Verlag, Tokyo, pp 326–334
9. Harada T, Ebara S, Kajiura I, Hiroshima K, Ono K (1994) Surgical treatment for cervical spondylotic radiculopathy and myelopathy in patients with athetoid cerebral palsy (in Japanese). Cent Jpn J Orthop Traumat 37:555–556
10. Nishihara N, Tanabe G, Nakahara S, Imai T, Murakawa H (1984) Surgical treatment of cervical spondylotic myelopathy complicating athetoid cerebral palsy. J Bone Joint Surg 66B:504–508
11. Fuji T, Yonenobu K, Fujiwara K, Yamashita K, Ebara S, Ono K (1987) Cervical radiculopathy or myelopathy secondary to athetoid cerebral palsy. J Bone Joint Surg 69A:815–821
12. White AA III, Panjabi MM (1990) Clinical biomechanics of the spine, 2nd edn. Lippincott, Philadelphia, pp 1–83

13. Mow VC, Kuei SC, Lai WM, Armstrong CG (1980) Biphasic indentation of articular cartilage in compression. Theory and experiment. J Biomech Eng 102:73–84
14. Ebara S, Yamazaki Y, Harada T, Hosono N, Morimoto Y, Tang Li, Seguchi Y, Ono K (1990) Motion analysis of the cervical spine in athetoid cerebral palsy. Extension-flexion motion, Spine 15:1097–1103
15. Ebara S, Harada T, Yamazaki Y, Hosono N, Yonenobu K, Hiroshima K, Ono K (1989) Unstable cervical spine in athetoid cerebral palsy. Spine 14:1154–1159
16. Penning L (1978) Normal movement of the cervical spine. AJR 130:317–326
17. Lawrence JS (1969) Disc degeneration. Ann Rheum Dis 28:121–137
18. Vernon-Roberts B, Pirie CJ (1977) Degenerative changes in the intervertebral discs of the lumbar spine and their sequelae. Rheumatol Rehabil 16:13–21
19. Ebara S, Harada T, Hosono N, Inoue M, Tanaka M, Morimoto Y, Ono K (1992) Intraoperative measurement of lumbar spinal instability. Spine 17:S44–S50
20. Ebara S, Tanaka M, Morimoto Y, Harada T, Hosono N, Yonenobu K, Ono K (1993) Intraoperative measurement of lumbar spinal stiffness. In: Yonenobu K, Ono K, Takemitsu Y (eds) Lumbar fusion and stabilization. Springer-Verlag, Tokyo, pp 45–53
21. Kirkaldy-Willis W, Farfan HF (1982) Instability of the lumbar spine. Clin Orthop 165:110–123
22. Pope MH, Panjabi MM (1985) Biomechanical definitions of spinal instability. Spine 10:255–256
23. Heinegard D, Sommarin Y (1987) Proteoglycans. An overview. In: Cunningham LW (ed) Methods of enzymology, vol 144. Structural and contractile proteins. Part D, Extracellular matrix. Academic Press, Orlando, pp 305–319
24. Eyre DR (1979) Biochemistry of the intervertebral disc. Int Rev Connect Tissue Res 8:227–291
25. Beard HK, Ryvar R, Brown R, et al (1980) Immunochemical localization of collagen types and proteoglycan in pig intervertebral discs. Immunology 41:491–501
26. Beard HK, Roberts S, O'Brien JP (1981) Immunofluorescent staining for collagen and proteoglycan in normal and scoliotic intervertebral discs. J Bone Joint Surg 63B:529–534
27. Mow VC, Holmes MH, Lai WM (1984) Fluid transport and mechanical properties of articular cartilage: a review. J Biomech 17:377–394
28. Mow VC, Kuei SC, Lai WM, Armstrong CG (1980) Biphasic indentation of articular cartilage in compression. Theory and experiment. J Biomech Eng 102:73–84
29. Mow VC, Mak AF, Lai WM, et al (1984) Viscoelastic properties of proteoglycan subunits and aggregates in varying solution concentrations. J Biomech 17:325–338
30. Maroudas A (1975) Biophysical chemistry of cartilaginous tissues with special reference to solute and fluid transport. Biorheology 12:223–248
31. Maroudas A (1979) Physicochemical properties of articular cartilage. In: Freeman MAR (ed) Adult articular cartilage, 2nd edn. Pitman Medical, Tunbridge Wells, pp 215–290
32. Maroudas A (1968) Physicochemical properties of cartilage in the light of ion exchange theory. Biophys J 8:575–595
33. Myers ER, Armstrong CG, Mow VC (1984) Swelling pressure and collagen tension. In: Hukins DWL (ed) Connective tissue matrix. MacMillan, London, pp 161–186
34. Armstrong CG, Mow VC (1982) Variations in the intrinsic mechanical properties of human articular cartilage with age, degeneration, and water content, J Bone Joint Surg 64A:88–94
35. McCutchen CW (1962) The frictional properties of animal joints. Wear 5:1

36. Holmes MH (1985) A theoretical analysis for determining the nonlinear hydraulic permeability of a soft tissue from experiment. Bull Math Biol 47:669–683
37. Holmes MH (1986) Finite deformation of soft tissue: analysis of mixture model in uniaxial compression. J Biomech Eng 108:372–381
38. Best BA, Guilak F, Setton LA, Zhu WB, Saed-Nejad F, Ratcliffe A, Weidenbaum M, Mow VC (1994) Compressive mechanical properties of the human annulus fibrosus and their relationship to biochemical composition. Spine 19:212–221
39. Galante JO (1967) Tensile properties of the human lumbar annulus fibrosus. Acta Orthop Scand Suppl 100:4–91
40. Panagiotacopulos ND, Knauss WG, Bloch R (1979) On the mechanical properties of human intervertebral disc material. Biorheology 16:317–330
41. Nachemson A (1960) Lumbar intradiscal pressure. Acta Orthop Scand Suppl 43: 6–104
42. Marchand F, Ahmed AM (1990) Investigation of the laminate structure of lumbar disc annulus fibrosus. Spine 15:402–410
43. Tsuji H, Hirano N, Ohshima H, Ishihara H, Terahate N, Motoe T (1993) Structural variation of the anterior and posterior annulus fibrosus in the development of human lumbar intervertebral disc. Spine 18:204–210
44. Brickley-Parsons D, Glimcher MJ (1984) Is the chemistry of collagen in intervertebral disc an expression of Wolff's law? Spine 9:148–163
45. Eyre DR, Benya P, Buckwalter J, Caterson B, Heinegard D, Oegema T, Pearce R, Pope M, Urban J (1989) Intervertebral disk: basic science perspectives. In: Frymoyer JW, Gordon SL (eds) New perspectives on low back pain. American Association of Orthopedic Surgeons, Park Ridge, pp 147–207
46. Bollet AJ, Nance JL (1966) Biochemical findings in normal and osteoarthritic articular cartilage. II. Chondroitin sulfate concentration and chain length, water, and ash content. J Clin Invest 45:1170–1177
47. Mankin HJ, Thrasher AZ (1975) Water content and binding in normal osteoarthritic human cartilage. J Bone Joint Surg 57A:76–80
48. Marodas A (1976) Balance between swelling pressure and collagen tension in normal and degenerate cartilage. Nature 260:808–809
49. Brain L (1963) Some unsolved problems of cervical spondylosis. Br Med J 23: 771–777
50. Friedenberg ZB, Miller WT (1963) Degenerative disc disease of the cervical spine. A comparative study of asymptomatic and symptomatic patients. J Bone Joint Surg 45A:1171–1178
51. Hayashi H, Okada K, Hamada M, Tada K, Ueno R (1987) Etiologic factors of myelopathy. A radiographic evaluation of the aging changes in the cervical spine. Clin Orthop 214:200–209
52. Kelsey JL (1982) Epidemiology of musculoskeletal disorders. In: Lilienfeld AM (ed) Monographs in epidemiology and biostatistics, vol 3. Oxford University Press, Oxford, pp 155–158
53. Lawrence JS (1969) Disc degeneration. Its frequency and relationship to symptoms. Ann Rheum Dis 28:121–37
54. Payne EE, Spillane JD (1957) The cervical spine. Anatomico-pathological study of 70 specimens (using a special technique) with particular reference to the problem of cervical spondylosis. Brain 80:571–596
55. Sasaki A (1980) Radiology of normal cervical spine (in Japanese). J Jpn Orthop Assoc 54:615–631

56. Gray ML, Pizzanelli AM, Grodzinsky AJ, Lee RC (1988) Mechanical and physicochemical determinants of the chondrocyte biosynthetic response. J Orthop Res 6:777–792
57. Sah RLY, Kim YJ, Doong JYH, Grodzinsky AJ, Plaas AHK, Sandy JD (1988) Biosynthetic response of cartilage explants to dynamic compression. J Orthop Res 7:619–636
58. Lipson SJ, Muir H (1980) Vertebral osteophyte formation in experimental intervertebral disc degeneration. Morphologic and proteoglycan changes over time. Arthritis Rheum 23:319–324
59. Lipson SJ, Muir H (1981) Volvo award in basic science. Proteoglycans in experimental intervertebral disc degeneration. Spine 6:194–210
60. Lyons G, Eisenstein SM, Sweet MBE (1981) Biochemical changes in intervertebral disc degeneration. Biochim Biophys Acta 673:443–453
61. Vernon-Roberts B (1987) The pathology of intervertebral discs and apophyseal joints. In: Jayson MIV (ed) The lumbar spine and back pain. Chvrchill-Livingstone, Edinburgh, pp 37–55

Biological Response in Orthodontics

SHINJI NAKAMURA[1], HIROYUKI ISHIKAWA[1], YOSHIAKI SATOH[1], TOMOO KANEKO[1], NAOYUKI TAKAHASHI[1], and MINORU WAKITA[2]

Summary. This study was conducted to provide a biomechanical description of periodontal tissue changes caused by orthodontic forces. The maxillary canine of a cat was moved distally in the alveolar bone with an initial force of 100 g. First, the tissue accommodation process on the pressure side was observed by means of three-dimensionally reconstructed images of microscopic images. The degenerated tissue in the compressed periodontium was reduced as the tissue accommodation process continued. During the process, bone resorption occurred in the alveolar bone area adjacent to the degenerated tissue. This enlarged the periodontal space, loosening the compressed periodontium and releasing the internal stress. The process of accommodation to orthodontic force appeared to have a cycle of almost 4 weeks. A second experiment was designed to investigate the process of formation of degenerated tissue using light and electron microscopy. The intradegeneration zone (IZ) was observed in the compressed periodontium at 12 h after the initial force was applied. At 4 days, a cell-free zone (CFZ), free of pyknotic cells or debris, had appeared at the peripheral area of the degenerated tissue. The results suggest that the CFZ is formed where cells and debris are carried away by the tissue fluid flow through interfiber gaps remaining under conditions less compressed than in the IZ.

Key words: Orthodontic tooth movement—Bone resorption—Intradegeneration zone—Cell-free zone—Accommodation process

1 Introduction

The purpose of orthodontic treatment is to obtain occlusions with desirable functional and esthetic features. The treatment procedure is characterized by the application of force to move the teeth. When force is applied to a tooth crown,

[1] Department of Orthodontics, School of Dentistry, Hokkaido University, N13 W7, Kita-ku, Sapporo, 060 Japan
[2] Department of Oral Anatomy II, School of Dentistry, Hokkaido University, N13 W7, Kita-ku, Sapporo, 060 Japan

the tooth tips in the alveolar socket with the center of rotation apically at one-third of the root [1]. The periodontal membrane is compressed on one side of the cervical area and, in the opposite direction, the apical area; on the other side, the periodontal fibers are stretched (Fig. 1). Since Sandstedt [2] observed hyalinization on the compressed periodontal membrane and its surrounding bone resorption in animal experiments, many investigators have reported periodontal tissue responses during orthodontic tooth movement [3–20]. When force is applied to teeth, they move in the alveolar bone accompanied by histological changes involving degeneration of the periodontal membrane and osteoclastic bone resorption on the pressure side, with new bone formation on the tension side. However, the relation between histological changes and force distribution is not well understood, mainly because of the complexity in tissue response that is influenced by anatomical structures. At present, the application of orthodontic force is largely based on the orthodontist's experience.

To achieve a better understanding of the phenomena involved, a new approach incorporating biological and mechanical considerations is required. Biomechanics attempts to clarify the mechanism of accommodation to external forces in living substances. Force applied to teeth is transmitted to the periodontium and the alveolar bone, and this induces various tissue responses. This is exactly the biomechanical phenomenon. Our histological experiments were designed to describe the tissue changes caused by orthodontic force from a biomechanical point of view.

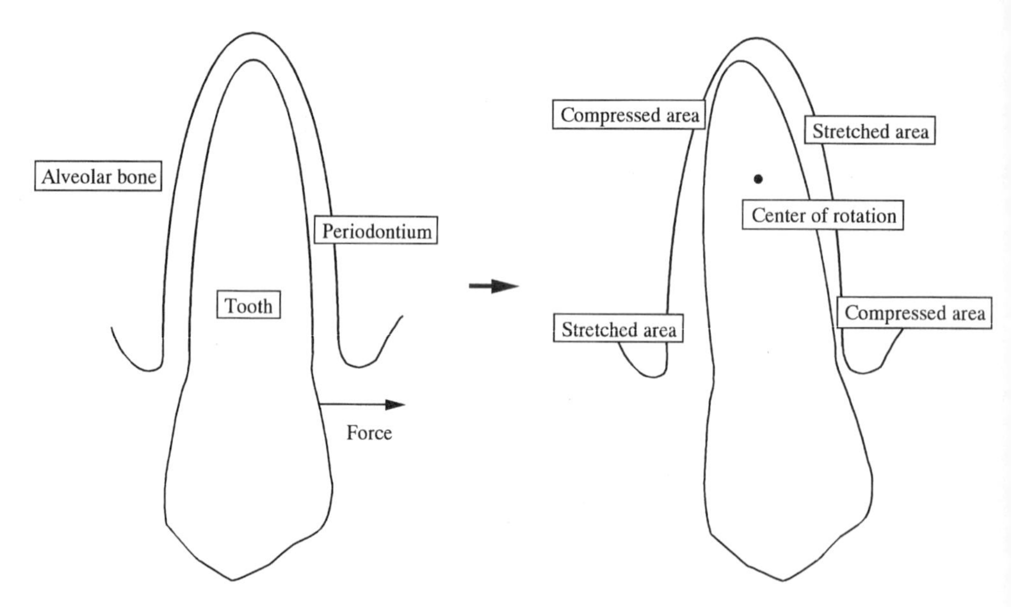

Fig. 1. Displacement of a tooth in the alveolar bone socket by orthodontic force. Compression and stretch of periodontal fibers occur in the opposite sides of the periodontium

Two experiments were performed. The first was designed to observe the tissue accommodation process on the pressure side by means of three-dimensional (3-D) reconstruction of microscopic images [21]. The second experiment was conducted to obtain histological evidence to show that stress was distributed with a certain range in the periodontal membrane [22]. Our target was the area of degeneration in the compressed periodontium. The process of formation of degenerated tissue was observed with light and electron microscopes.

1.1 Definitions of Terms Relating to the Histological Response Used in this Study

Before describing the present study, we should define some of the technical terms related to the histological events. There are numerous terms for degeneration and types of bone resorption on the pressure side of periodontal tissue during orthodontic tooth movement [9,16,19,20]. Our usage and definitions are based on the results of animal experiments conducted by us and reported previously [23]. Here, the maxillary canines of cats were moved distally, that is, posteriorly in the alveolar bone. Figure 2 shows the appliance used for experimental tooth movement. The canine was continuously retracted by a closed coil spring with an initial force of 100g for 4 or 7 days. In this experiment, the compression of the periodontium occurs at the cervical area of the distal side and at the apical area of the mesial side (Fig. 3). At several stages of tooth movement, histological changes on the pressure side were investigated.

Degenerated tissue had appeared at the cervical area of the compressed periodontium 4 days after the application of force. It showed different features in the peripheral and central parts, which were named the cell-free zone and the intradegeneration zone. The definitions of these zones are as follow: (1) cell-free zone (CFZ), a homogeneous region stained by eosin in no nuclei or other debris of periodontal cells can be seen; and (2) intradegeneration zone (IZ), an inner

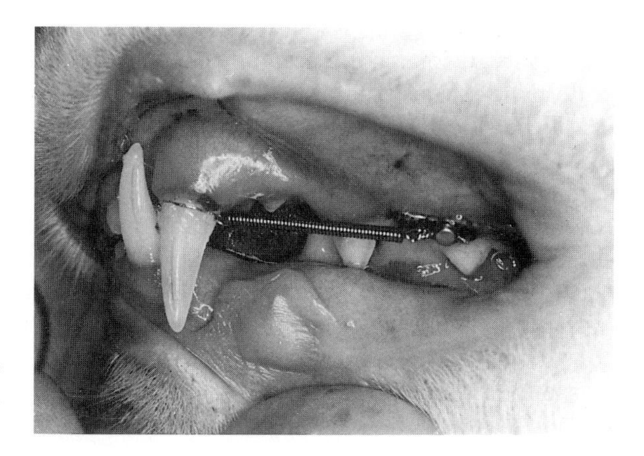

FIG. 2. An experimental appliance for tooth movement. (From [23], with permission)

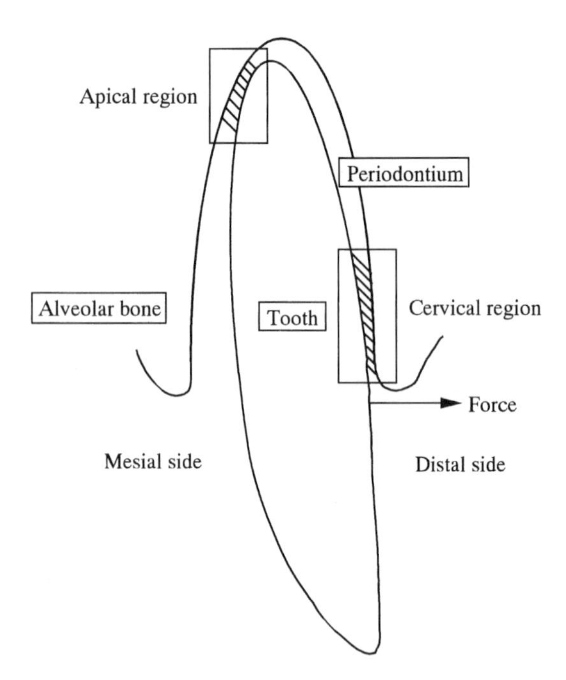

FIG. 3. Compression of the periodontium by force for distal tooth movement

region surrounded by cell-free zones in which pyknotic nuclei of periodontal cells can be seen.

The terms related to bone resorption, direct bone resorption, undermining bone resorption, and rear bone resorption were newly defined. The conventional usage of these terms is based on the morphological features and locations of osteoclast distributions. The problems are that there is no consideration for differences in stress conditions among areas, which influence osteoclast appearance. Our basis of the definitions adds more accuracy to the positional relations, considering that the locations of histological events are closely relate to the manner in which osteoclasts are subject to stress in the periodontal tissue. The definitions are as follow. (1) In undermining bone resorption, osteoclasts distribute in the alveolar bone area adjacent to the degenerated tissue, appearing to undermine it. This type of bone resorption is considered to relate to the manner in which osteoclasts are subject to the stress directly in the periodontium. (2) In direct bone resorption, osteoclasts distribute directly along the alveolar bone surface in the area adjacent to the undermining resorption area. This type of bone resorption is also considered to relate to the manner in which osteoclasts are subject to stress directly in the periodontium. (3) In rear bone resorption, osteoclasts distribute in the alveolar bone marrow spaces, canals, and openings communicating with the degenerated periodontium. This type of bone resorption is considered to relate to the manner in which osteoclasts are subject to the stress transmitted through the periodontium and the alveolar bone.

2 Experiment 1

2.1 Materials and Methods

Male cats of 4–6kg body weight were used. The maxillary canine was continuously retracted distally for 4, 7, 14, and 28 days. An initial force of 100g was applied with a closed coil spring by the method described in the previous study. After the tooth movement, the cats were sacrificed and the maxillary jaw of each cat was cut into a block containing canine and periodontal tissue. After decalcification with Plank-Rychlo, these blocks were embedded in celloidin. The celloidin blocks were given two datum points with an 0.8-mm-diameter drill and sectioned serially at 30-μm-thick slices in the coronary direction. The sections were stained with hematoxylin and eosin. Microscopic photographic images at 20× magnification were made of every fifth section. The structural information of the images was traced and fed into a computer for 3-D reconstruction of the structures. The tracings showed the alveolar bone surface facing the periodontium, the root surface of the canine, the degenerated tissue area of the periodontium, the distributional regions of osteoclasts, and the two datum points. The distal and the lateral (buccal or lingual) views of the 3-D images were used for ease of observation of the pressure side.

2.2 Results

2.2.1 Control

In the cervical region, the periodontium was of uniform width. Bone marrow, canals, and openings were observed on the distal side of the alveolar bone. The direction of periodontal fibers was not disrupted (Fig. 4).

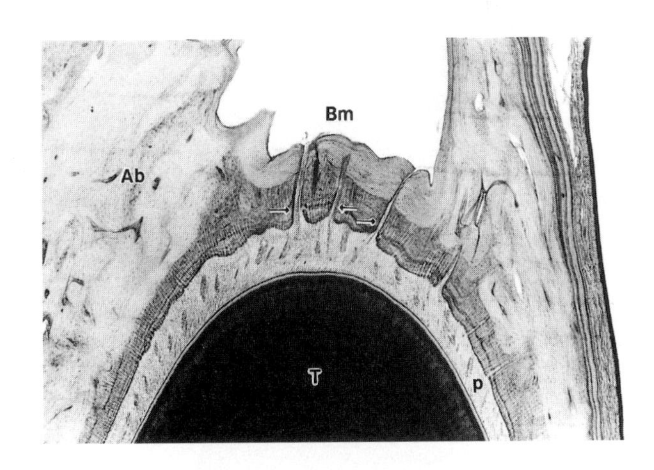

Fig. 4. Control. Histological images on the distal side of the cervical region. *Ab*, Alveolar bone; *T*, tooth root; *P*, periodontium; *Bm*, bone marrow. *Arrows* indicate Bm canals and openings. (×10)

2.2.2 Experiment

2.2.2.1 Experiment after 4 Days

Histological Study. A sandwiched intradegeneration zone (IZ) was observed between cell-free zones (CFZ) on the distal side of the cervical region. Capillaries in the IZ were compressed and blocked. Some osteoclasts of undermining bone resorption were observed in the alveolar bone area adjacent to the CFZ. There was a CFZ but no IZ on the mesial side of the apical region (Fig. 5).

3-D Study. There was much IZ surrounded by circular CFZ on the distal side of the cervical region. On the mesial side of the apical region, only a CFZ was observed in a narrow belt extending from the apical region in the vertical direction. Some osteoclasts were scattered near the CFZ in both the cervical and apical regions (Fig. 6).

2.2.2.2 Experiment after 7 Days

Histological Study. An IZ was sandwiched between CFZs on the distal side of the cervical region, the same as at 4 days. Osteoclasts involved in undermining bone resorption were found lateral to the CFZ, and direct bone resorption was

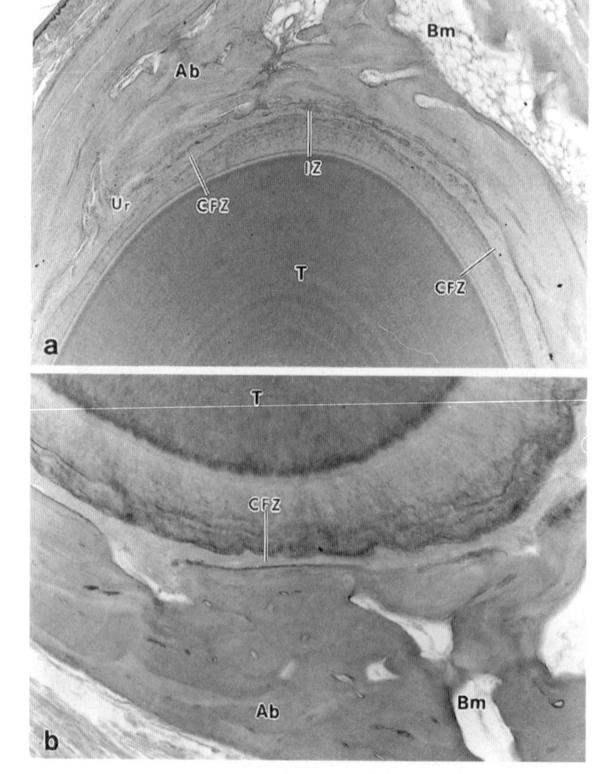

Fig. 5a,b. Experiment after 4 days. Histological images on the distal side of the cervical region (**a**) and on the mesial side of the apical region (**b**). *IZ*, Intradegeneration zone; *CFZ*, cell-free zone; *Ur*, undermining bone resorption. (**a** ×10, **b** ×20). (From [23], with permission)

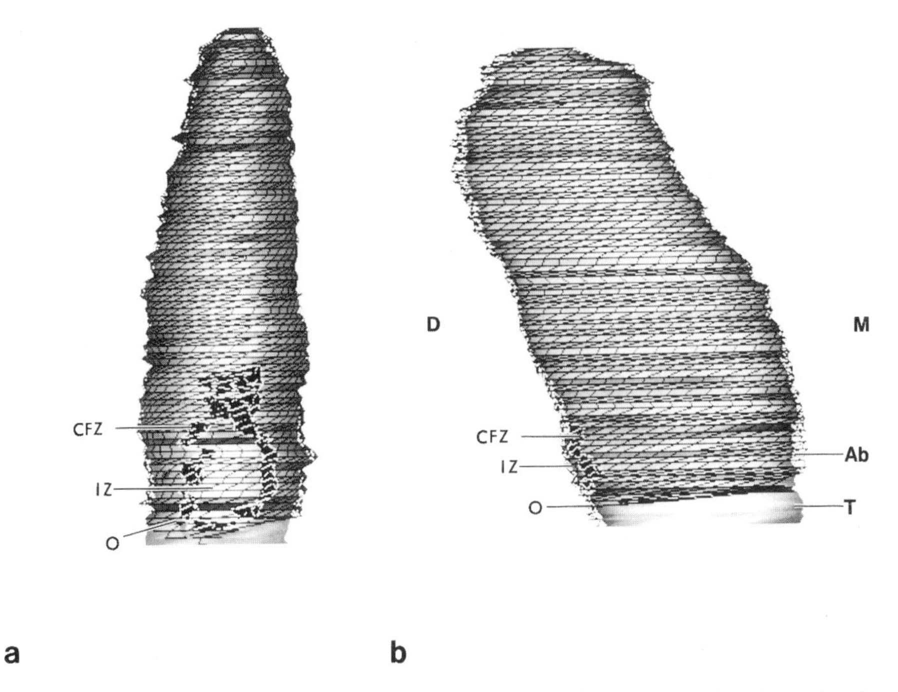

FIG. 6a,b. Experiment after 4 days. Three-dimensional (3-D) images in the distal view (**a**) and the lateral view (**b**). *D*, Distal side; *M*, mesial side; *Ab*, alveolar bone surface; *T*, tooth root surface; *IZ*, intradegeneration zone; *CFZ*, cell-free zone (indicated by *dark gray*); *O*, distributions of osteoclasts (indicated by *light gray*)

observed along the bone surface adjacent to undermining bone resorption. Some rear bone resorption in the alveolar bone marrow space and opening had taken place behind the IZ. Only a CFZ was observed in the degenerated area on the mesial side of the apical region and with undermining and direct bone resorption lateral to this (Fig. 7).

3-D Study. In the cervical and the apical regions, the area of CFZ and IZ remained as large as at 4 days. The osteoclasts had changed to a large extent from the area around the CFZ in the lateral direction. For the rear bone resorption, osteoclasts were observed to overlap with the IZ in the distal view and separated at a distance from the IZ in the lateral view (Fig. 8).

2.2.2.3 Experiment after 14 Days

Histological Study. Both CFZ and IZ were still observed on the distal side of the cervical region. Undermining and direct bone resorption as well as rear bone resorption were clearly observed. Rear bone resorption mainly appeared at the openings connecting with the degenerated periodontium. On the mesial side of the apical region, there was a CFZ and the undermining bone resorption surrounding it (Fig. 9).

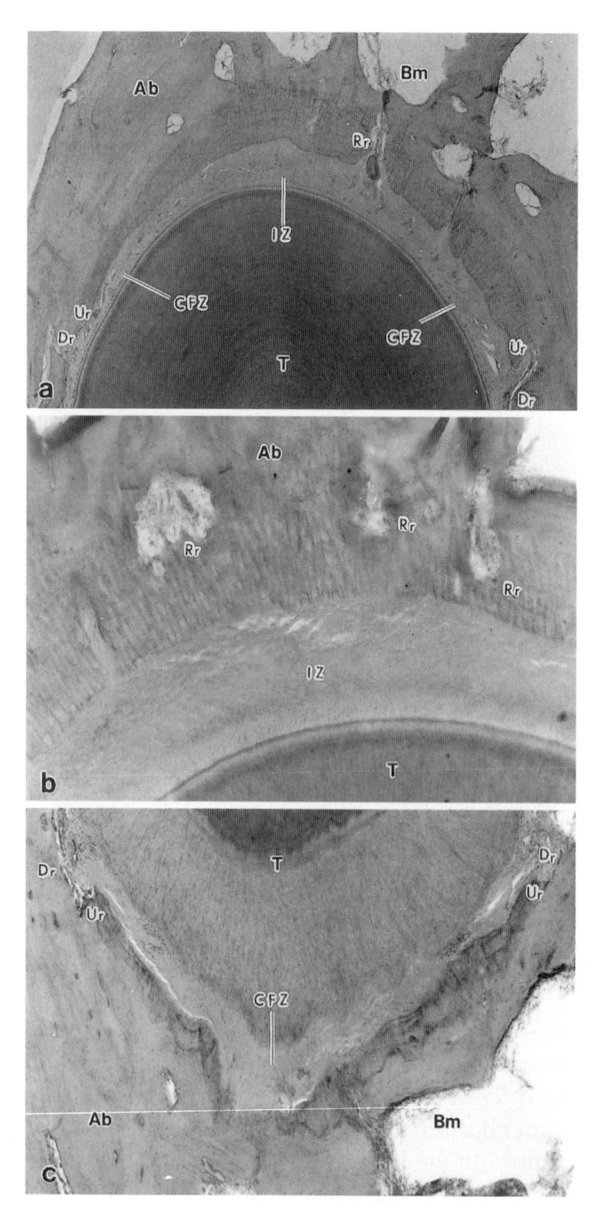

FIG. 7a–c. Experiment after 7 days. Histological images on the distal side of the cervical region (**a,b**) and on the mesial side of the apical region (**c**). *Dr*, Direct bone resorption; *Rr*, rear bone resorption. (**a** ×10, **b** ×25, **c** ×20)

3-D Study. On the distal side of the cervical region, there was less degenerated periodontium than at 7 days, especially in the IZ. There were fewer osteoclasts in the area lateral to the CFZ, but more in the region facing the IZ, the rear bone resorption area. On the mesial side of the apical region, the CFZ was very small, and the distribution of osteoclasts was limited to the area near the CFZ (Fig. 10).

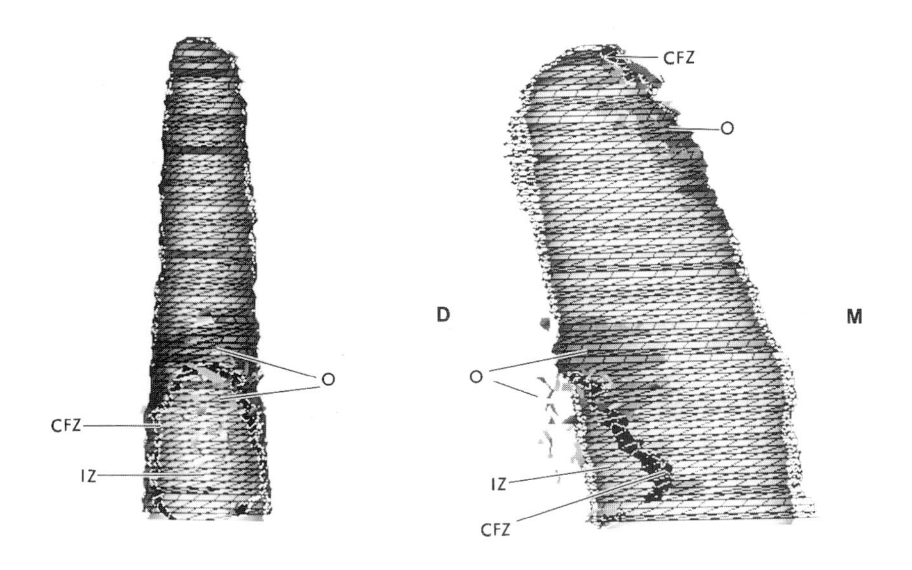

FIG. 8a,b. Experiment after 7 days. 3-D images in the distal view (**a**) and the lateral view (**b**)

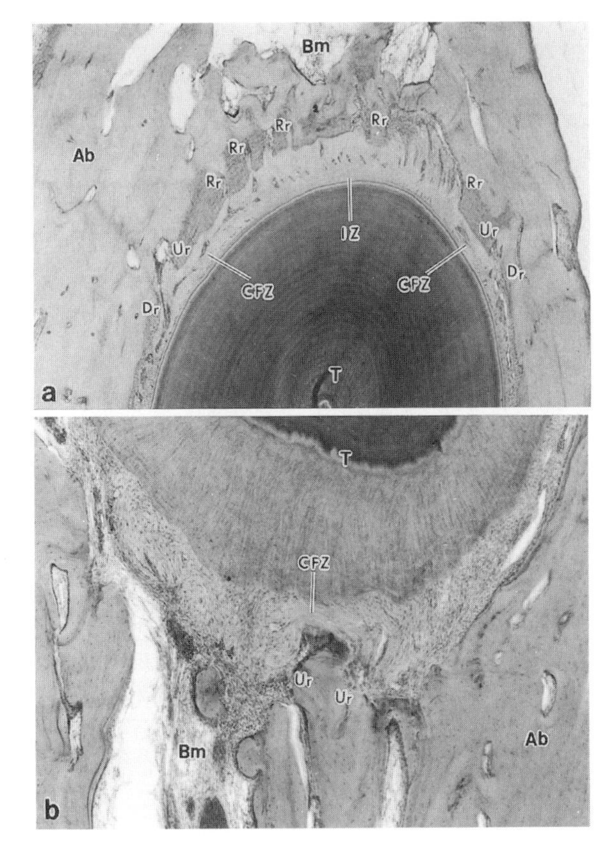

FIG. 9a,b. Experiment after 14 days. Histological images on the distal side of the cervical region (**a**) and on the mesial side of the apical region (**b**). (**a** ×10, **b** ×20)

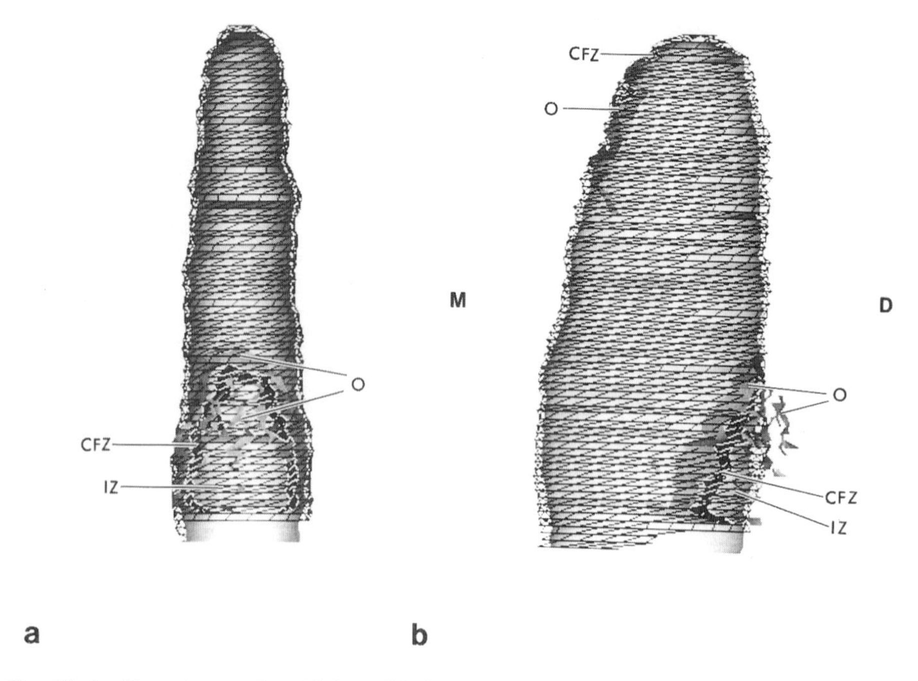

Fig. 10a,b. Experiment after 14 days. 3-D images in the distal view (**a**) and the lateral view (**b**)

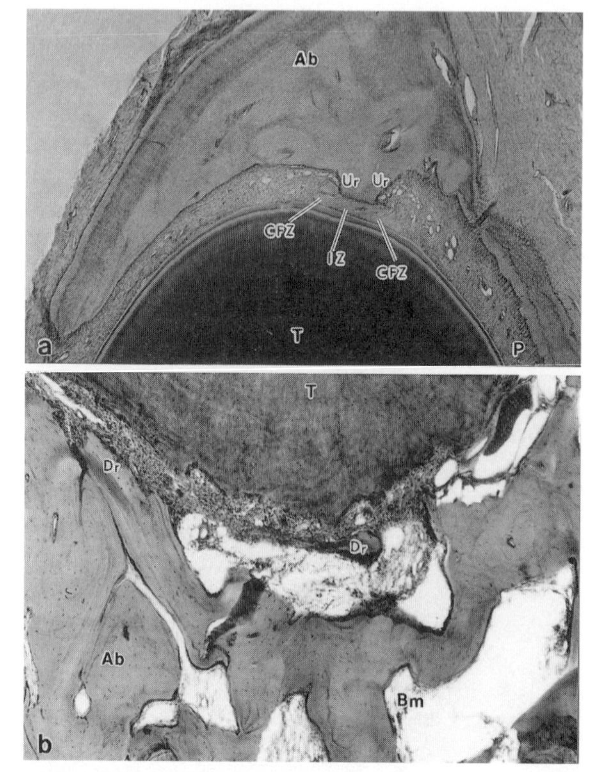

Fig. 11a,b. Experiment after 28 days. Histological images on the distal side of the cervical region (**a**) and on the mesial side of the apical region (**b**). (**a** ×10, **b** ×20)

2.2.2.4 Experiment after 28 Days

Histological Study. On the distal side of the alveolar bone crest, the CFZ and IZ could be discriminated, but shared only a very narrow region. A small region of alveolar bone remained behind this degenerated periodontium. There were no osteoclasts on the surface of the alveolar bone except at the lateral side of this remaining bone. On the mesial side of the apical region, the CFZ had already disappeared. Osteoclasts, which cause direct bone resorption were observed widely (Fig. 11).

3-D Study. The degenerated periodontium had almost entirely disappeared, remaining only on the distal side of the cervical area near the alveolar bone crest in a narrow region. On the distal side, the distribution of osteoclasts was limited to the area close to the degenerated periodontium near the alveolar bone crest. On the mesial side, osteoclasts were observed with a wide extent at the apical region.

3 Experiment 2

3.1 Materials and Methods

To establish the process of formation of degenerated tissue, the upper canines of male cats were moved distally by the same method as in experiment 1. The initial force of 100 g was used for 6 h, 12 h, 1 day, or 4 days.

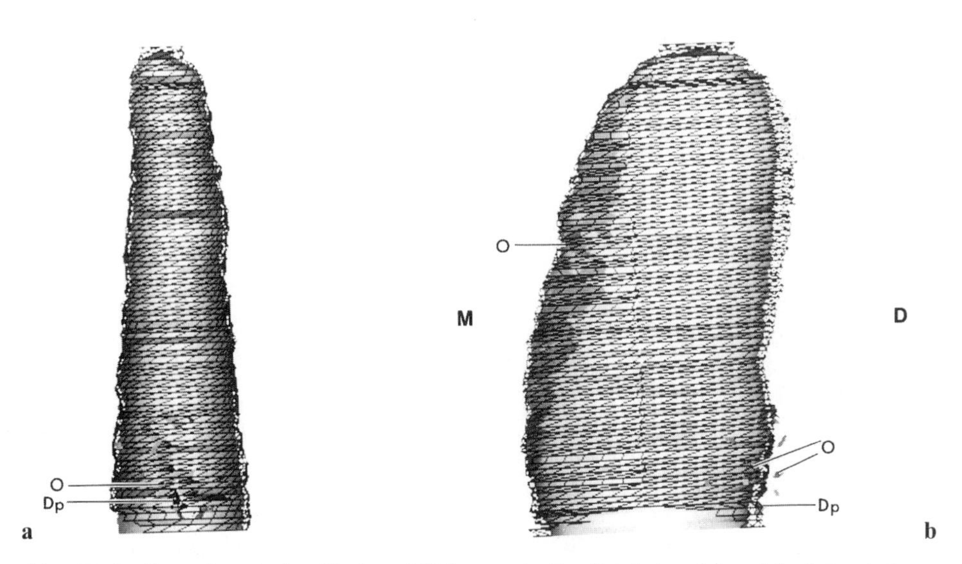

Fig. 12a,b. Experiment after 28 days. 3-D images in the distal view (**a**) and the lateral view (**b**). *Dp*, Degenerated periodontium

Anesthetized cats were perfused with a mixture of 2% paraformaldehyde and 1.25% glutaraldehyde in 0.05 *M* cacodylate buffer. Then, the maxillary jaw of each cat was excised and cut into a small block that contained upper canine and periodontal tissue. After decalcification with cold 5% ethylenediamine tetraacetate (EDTA) solution, these blocks were cut into halves and postfixed in 1.0% osmium tetroxide in 0.1 *M* cacodylate buffer (pH7.2), and then embedded in Epon 812. A part of the specimens were semithin-sectioned for methylene blue Azure II staining to observe the degenerated area by the light microscope. For transmission electron microscopy, specimens were cut into ultrathin sections about 80 nm thick and stained with uranylacetate and lead citrate.

3.2 Result

3.2.1 Control

3.2.1.1 Light Microscopic Study

The periodontal fibers Normally run straight between the alveolar bone and the cementum, but around blood vessels they assume a curved shape. In the interfi-

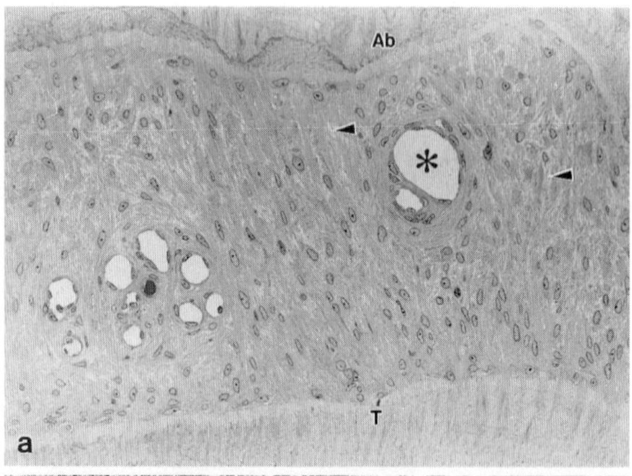

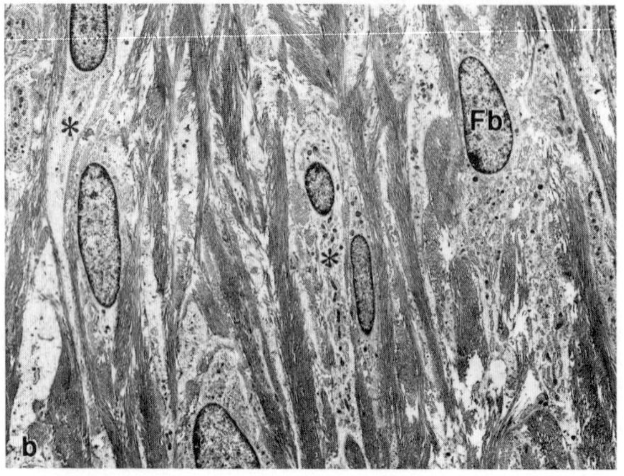

Fig. 13a,b. Control. Light microscopic image (**a**) and electron microscopic images (**b**) (from [22], with permission). *Ab*, Alveolar bone; *T*, tooth root; *Fb*, fibroblast. *Asterisk* in (**a**) indicates blood vessels; *arrowhead* indicates unstained areas. *Asterisk* in (**b**) indicates elongated processes of fibroblast. (**a** ×250, **b** ×3000)

ber space, there were unstained areas and clearly stained nuclei and nucleoli (Fig. 13a).

3.2.1.2 Electron Microscopic Study

The periodontal fibers were observed as bundles of interlacing collagen filaments. Many fibroblasts with oval nuclei and chromatin could be seen in the interfiber space. Most of the unstained area seen by light microscope had fibroblasts with elongated processes (Fig. 13b).

3.2.2 Experiment

The IZ in the central area of the compressed periodontium showed little morphological change during the experimental period. To investigate the process of formation of the CFZ, observations were made in the relatively peripheral area of the compressed periodontium. Both IZ and CFZ were observed only in the 4-day sample.

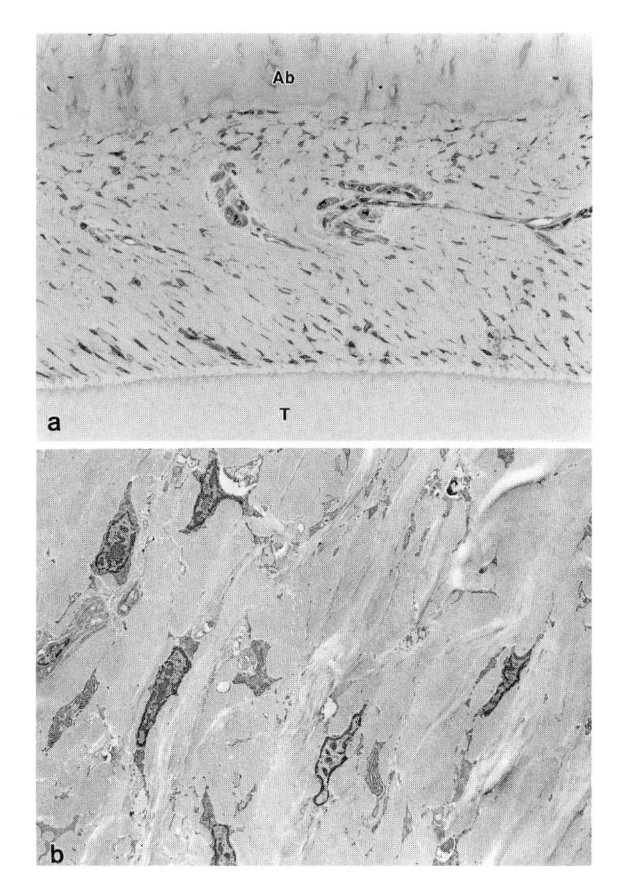

FIG. 14a,b. Experiment after 6h. Light microscopic image (**a**) and electron microscopic image (**b**). (**a** ×250, **b** ×3000). (From [22], with permission)

3.2.2.1 Experiment After 6 Hours

Light Microscopic Study. The blood vessels were pressed flat, but there was no CFZ or IZ. The periodontal fibers waved, and only deformed nuclei of the periodontal cells were seen in the interfiber space (Fig. 14a).

Electron Microscopic Study. Collagen fibers were compressed closely and the cytoplasm and chromatin of the periodontal cells adhered closely, but the unclear membrane and a part of the cytoplasm were still observed, particularly around the nuclei (Fig. 14b).

3.2.2.2 Experiment After 12 Hours

Light Microscopic Study. Compression of the periodontal space had progressed and the nuclei of the periodontal cells were much more deformed. The IZ was characterized by heavily degenerated cells, and compressed fibers had appeared in the central area of the compressed periodontal space, but no CFZ had appeared yet. The nucleolus of the periodontal cells had disappeared, and many

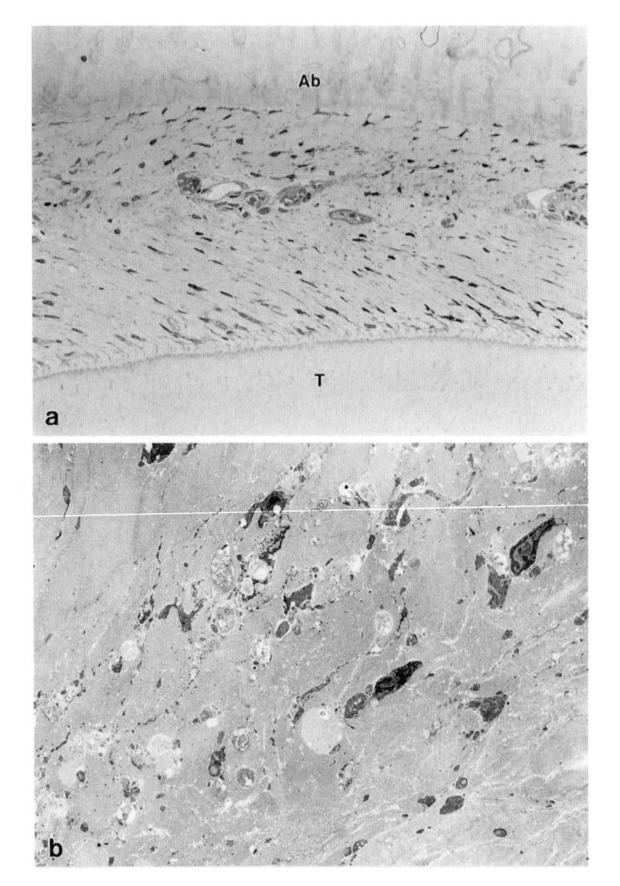

Fig. 15a,b. Experiment after 12 h. Light microscopic image (**a**) and electron microscopic image (**b**). (**a** ×250, **b** ×3000). (From [22], with permission)

small fragments darkly stained by methylene blue Azur II were observed around the cells (Fig. 15a).

Electron Microscopic Study. The interfiber space was narrower than at 6h, and the periodontal cells were further deformed. The cell membrane became indistinct and the cytoplasm could not be clearly identified. There were scattered narrow gaps throughout the area (Fig. 15b).

3.2.2.3 Experiment After 1 Day

Light Microscopic Study. The periodontal space was narrower than at 12h. The area of the IZ became larger but no CFZ had appeared. The nuclei of the periodontal cells and the small fragments were smaller than at 12h, while large gaps had appeared around the cells (Fig. 16a).

Electron Microscopic Study. The periodontal cells had collapsed to pyknotic cells, and there was deformation of the cytoplasm. The collapsed cells were much smaller than at 12h, but the gaps were larger and some of the gaps were filled with amorphous substances (Fig. 16b).

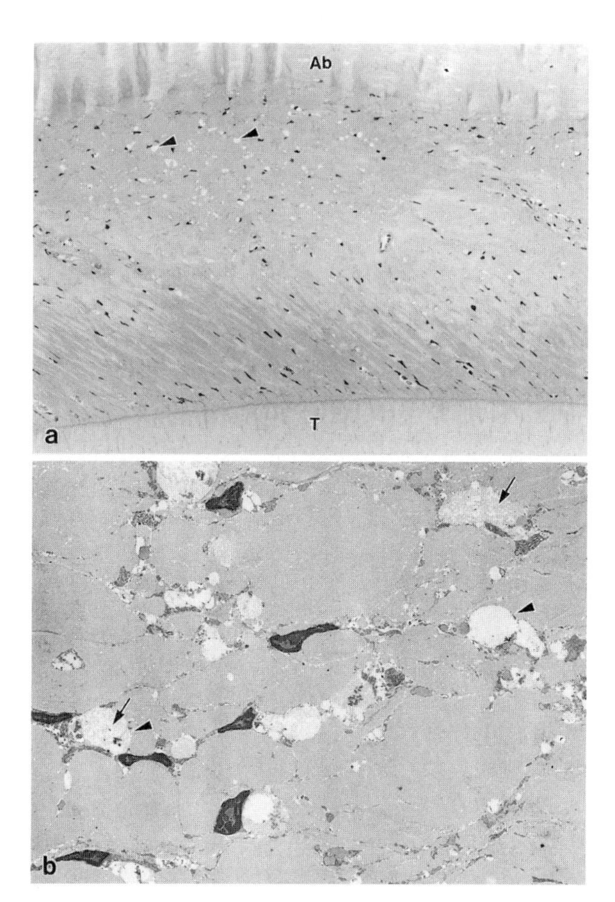

Fig. 16a,b. Experiment after 1 day. Light microscopic image (**a**) and electron microscopic image (**b**) (from [22], with permission). *Arrowheads* indicate gaps; *arrows* in (**b**) indicate gaps filled with amorphous substance. (**a** ×250, **b** ×3000)

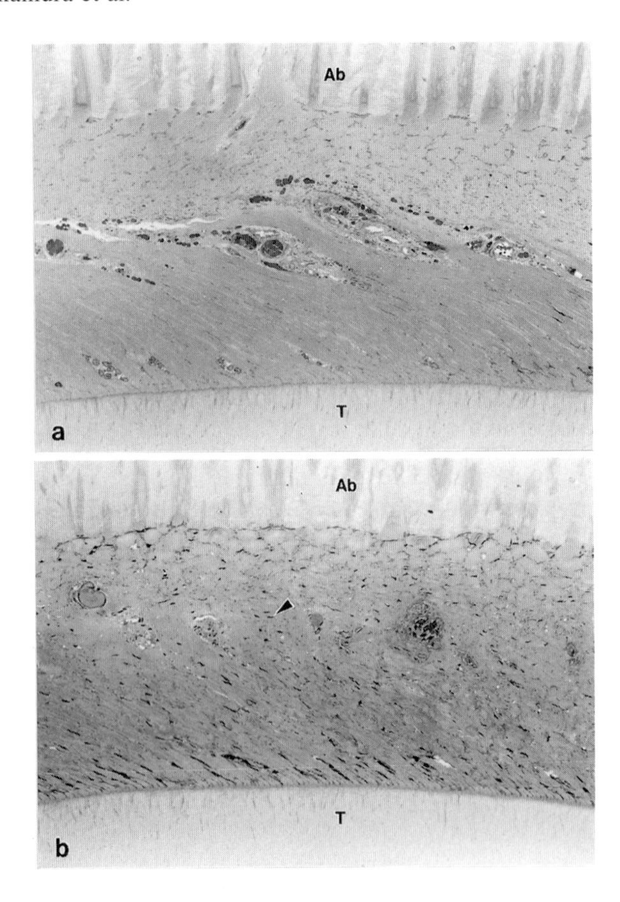

Fig. 17a–d. Experiment after 4 days. Light microscopic image (**a,b**) and electron microscopic image (**c,d**) of CFZ and IZ (from [22], with permission). CFZ is shown in (**a,c**); IZ is in (**b,d**). *Arrows* in (**c**) indicate the gaps in CFZ. In the images of IZ, *arrowhead* in (**b**) indicates deformed nuclei of cells in the interfiber space; *arrowhead* in (**d**) indicates large electron-dense substances, and *arrow* indicates small electron-dense substances. (**a**, **b** ×250, **c**, **d** ×3000)

3.2.2.4 Experiment After 4 Days

Light Microscopic Study. The CFZ appeared as a homogeneous area as in experiment 1. In the CFZ, these were no nuclei, and many small substances could be observed by semithin section. The IZ showed severely deformed nuclei of cells in the interfiber space, and the fiber stained darkly as at 6h, 12h, or 1 day (Fig. 17a,b).

Electron Microscopic Study. The electron-dense nuclei of cells observed after 1 day had disappeared in the CFZ, but an electron-lucent substance was visible. Some narrow gaps were scattered throughout the area. In the IZ, the interfiber

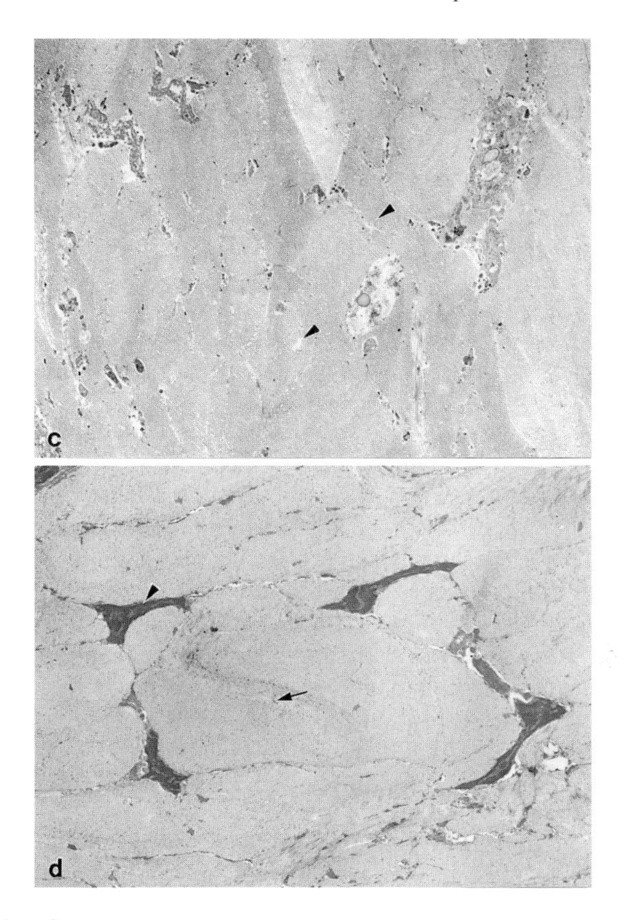

FIG. 17. *Continued*

space was filled with large or small electron-dense substances, and narrow gaps as seen in the CFZ were not observed (Fig. 17c,d).

4 Discussion

Histological changes in the periodontal tissue during orthodontic tooth movement have been investigated extensively [1–20]. In these investigations, the relationship between force and histological events in tissue response at different forces has been studied experimentally. Tissue response is influenced by force distribution and also anatomical and physiological factors, and force distribution is further influenced by anatomical structures. Because of this complexity in tissue response, there are numerous unexplained aspects of the mechanism of orthodontic tooth movement.

The first experiment (experiment 1) was conducted to investigate accommodation to orthodontic force in the periodontal tissue [21]. It is difficult to understand the process of tissue response by conventional observation of sectional images. The histological changes vary at different positions in a specimen, and it is also difficult to identify anatomically similar positions of sections of different specimens. To compare tissue response at different time points more easily, 3-D images of periodontal tissue were constructed from sectional images. The tissue accommodation process was investigated in terms of both histological appearance and the area it affects. The experimental period was a maximum of 28 days after the application of force, as orthodontic appliances are generally adjusted every 4 weeks during tooth movement, based on experiential knowledge of orthodontic treatment.

Degenerated tissue was widely distributed at the cervical and apical areas on the pressure side 4 days after application of the initial force to the tooth. There was still much degenerated tissue at 7 days, and restoration of the tissue damage had not yet begun. On the contrary, the distribution of osteoclasts changed to a large extent. This was the initial stage of accommodation to the external force in which undermining and direct bone resorption enlarges the periodontal space to loosen compressed periodontium and to release the internal stress. At 14 days, the degenerated region was reduced. Granulation tissue took the place of degenerated tissue. The tissue accommodation process was proceeding from 7 days to 14 days, and rear bone resorption, rather than undermining and direct bone resorption, was dominant. This plays a role in eliminating the remaining bone behind the degenerated periodontium, which results in tooth movement in the alveolar bone.

Because osteoclasts appear mainly at the openings of the bone marrow communicating with the degenerated periodontium, the effects of rear bone resorption depend on anatomical structures, the existence of large marrow spaces. This seems to be the major factor influencing the speed of tooth movement. At 28 days, there was much less degenerated tissue, now seen only at the cervical area near the alveolar crest, and the periodontal space was filled with fibrous tissue; the accommodation process was nearly complete. The distribution of osteoclasts had reduced at the cervical area but increased at the apical area.

Reitan [7,8,10] observed that the tooth was at a standstill while the degenerated tissue and the alveolar bone behind it remained, and that it started to move when these were removed. Therefore, the large extent of osteoclastic region at the apical area indicates that this portion of the tooth root has already started to move in the alveolar bone. This occurs because the continuous force exerted pressure on the apical area, with the fulcrum at the cervical area where a small amount of alveolar bone remained behind the degenerated tissue. It was expected that the remaining bone and degenerated tissue would soon disappear and osteoclasts would appear also at the cervical area. This second bone resorption would result in a dramatic reduction of the localized force. The results have allowed us to conclude that the process of accommodation to orthodontic force has a cycle of about 4 weeks.

Observations of the initial tissue changes showed that the degenerated region and the osteoclastic area were positionally interrelated. The IZ, CFZ, and bone resorption were located, in that order, from the center of compression of the periodontium in the lateral direction (Fig. 18). It is well established that blood perfusion is required for bone resorption [24–28]; however, the formation of degenerated tissue results from blocking of the blood supply [11,29,30]. Therefore, when applying a force of 100g to a cat's canine, the degree of compression of the periodontium has a wider range of level from blocking blood flow and tissue fluid flow to maintaining them. Also, it tends to reduce dramatically from the central area of compression to the lateral. Considering these points, it is suggested that the histological differences between the IZ and CFZ result from the degree of pressure on the periodontal space. To prove this, the second experiment (experiment 2) was designed to investigate the process of formation of the degenerated tissue by light and electron microscopy [22].

From 6h to 1 day after the force was applied, compression of the periodontal space gradually increased, indicating viscoelasticity of the periodontal fibers [31]. In the electron microscopic study, the fiber bundles were closely compressed and recognition of their structures became difficult. Deformation and collapse of the periodontal cells were increasing as the interfiber space was reduced. Light microscopy showed an IZ at the central area of the compressed periodontium at 12h, and this was enlarged at 1 day. However, the CFZ was not observed at this time.

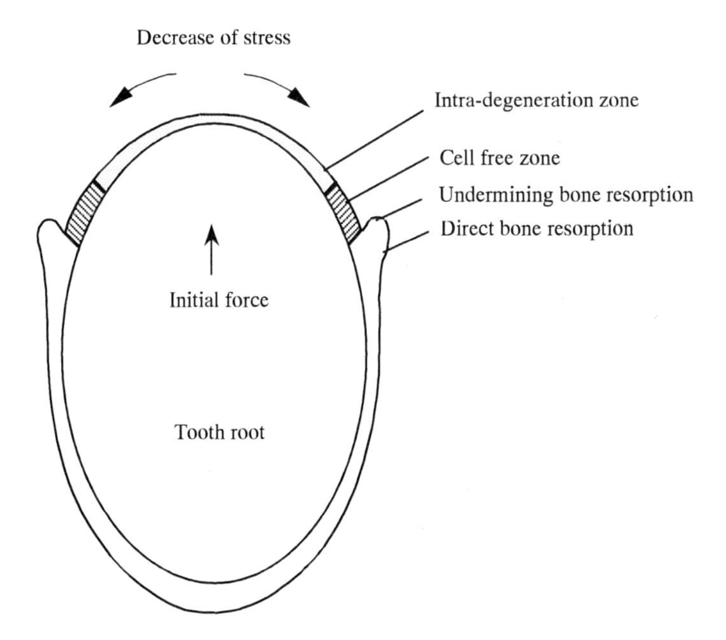

Fig. 18. Positional relation of degenerated tissue and bone resorption associated with stress distribution

At 4 days, both an IZ and a CFZ could be distinguished. In the electron microscopic observations, the interfiber space was filled with electron-dense substances and there were no empty gaps in the IZ, while some gaps could be seen in the CFZ. This indicates that pressure distribution was different in these two areas. Further, large gaps had appeared at the peripheral area of the compressed periodontium at 1 day. Considering the flow of tissue fluid, these results suggest the route of formation of the CFZ, which is without pyknotic cells or debris. Degenerated periodontal cells and debris are carried away by tissue fluid flow through the interfiber gaps, which exist under a less compressed condition than in the IZ. This is supported by observations in the first experiment. The reduction of the degenerated area was observed mainly in the IZ, and even at 28 days, the IZ and CFZ could be discriminated in the degenerated tissue. This implies that the CFZ gradually takes the place of the IZ as the pressure decreases in the process of accommodation.

As described earlier, orthodontic tooth movement is the biomechanical phenomenon in which tissue releases internal stress and accommodates to external forces. To clarify the mechanism of tooth movement, both a mechanical approach on a biological basis and a biological approach related to a mechanical stimulus should be conducted. In studies of mechanical simulation of orthodontic tooth movement, interest has been mainly in stress distribution in the periodontal tissue [32–36]. The finite-element method has been employed here, which is more effective on the basis of recent computer technology, but problems still exist in similarities to living substances. Efforts are now concentrated on a determination of material parameters and model construction on a biological basis.

Yamamoto et al. [37] developed a specimen-specific finite-element model by tracing histological images of an experimentally moved tooth and feeding these into a computer. The relationship between osteoclast appearance and stress distribution around a tooth was investigated with this model. Because stress distribution is greatly influenced by the morphological details of the periodontal tissue, this method is highly useful in terms of the similarity to anatomical structures. However, the effect of changing material parameters, such as elastic modulus and Poisson's ratio, should be considered. We have observed that periodontal fibers were gradually compressed from 6h to 1 day after the initial force was applied to the tooth. Electron microscopy has shown that there were no empty gaps in the IZ at 4 days. These findings suggest that the material parameters of the periodontium change with time from the initial force application, and that these may be influenced by tissue changes in the accommodation process.

To investigate the relationship between stress distribution and histological events that occur during a cycle of this accommodation process, different material parameters of the periodontium should be fed into a computer using different conditions. In addition, it should be also noted that rear bone resorption occurs in the alveolar bone marrow space and its openings, which are reduced with age. Because the extent of this type of bone resorption is strongly related to the anatomical structures of periodontal tissue, this seems to be the major factor in the differences in the speed of tooth movement among individual subjects. The histological study reported here provides details of the basic requirements for the biomechanical approach to orthodontics. Further investigation will be conducted to solve problems related to entering the biological information into a computer with regard to the variety of anatomical structures and the nonlinear elasticity of the periodontium.

Acknowledgment. This research was supported by a Grant-in-Aid for Scientific Research on Priority Area from the Ministry of Education, Science and Culture (Biomechanics, no. 04237102).

References

1. Reitan K (1951) The initial tissue reaction incident to orthodontic tooth movement as related to the influence of function. Acta Odontol Scand 6:1–240
2. Sandstedt C (1904) Some contributions to the theory of regulating tooth movement (in German). Nord Tandliäkere Tidskr Ht 1, 2, 4
3. Schwarz AM (1932) Tissue changes incident to tooth movement. Int Orthod Cong 18:331
4. Oppenheim A (1944) A possibility for physiologic orthodontic movement. Am J Orthod Oral Surg 30:277–345
5. Waldo CM (1953) Method for the study of tissue response to tooth movement. J Dent Res 32:690–691
6. Macapanpan LC, Weinmann JP, Brodie AG (1954) Early tissue changes following tooth movement in rats. Angle Orthod 24:79–95
7. Reitan K (1957) Some factors determining the evaluation of forces in orthodontics. Am J Orthod Dentofacial Orthop 43:32–45
8. Reitan K (1960) Tissue behavior during orthodontic tooth movement. Am J Orthod Dentofacial Orthop 46:881–900
9. Reitan K (1964) Effects of force magnitude and direction of tooth movement on different alveolar bone types. Angle Orthod 34:244–255
10. Reitan K (1967) Clinical and histologic observations on tooth movement during and after orthodontic treatment. Am J Orthod Dentofacial Orthop 53:721–745
11. Nakamura S (1967) An experimental study on the periodontal vasculature from orthodontic viewpoint. Part I: Responses of the periodontal vasculature to orthodontic pressure (in Japanese). J Stomatol Soc Jpn 62:342–348
12. Anthony A (1969) Force-induced changes in the vascularity of the periodontal ligament. Am J Orthod Dentofacial Orthop 55:5–11
13. Azuma M (1970) Study on histologic changes of periodontal membrane incident to experimental tooth movement. Bull Tokyo Med Dent Univ 17:149–178
14. Reitan K, Kvam E (1971) Comparative behaviour of human and animal tissue during experimental tooth movement. Angle Orthod 41:1–14
15. Rygh P (1972) Ultrastructural cellular reactions in pressure zone of rat molar periodontium incident to orthodontic tooth movement. Acta Odontol Scand 30:575–593
16. Rygh P (1974) Elimination of hyalinized periodontal tissues associated with orthodontic tooth movement. Scand J Dent Res 82:57–73

17. Koga M (1974) Histologic study on the periodontal structures incident to experimental tooth movement in rats. Light-microscopic and electron-microscopic investigations (in Japanese). J Tokyo Dent Coll Soc 74:498–557
18. Ichinokawa M (1975) Histologic study on the periodontal structures incident to experimental tooth movement in rats. Light and electron microscopic investigations according to the labeling with lead acetate (in Japanese). J Tokyo Dent Coll Soc 75:1435–1472
19. Okumura E (1982) Light and electron microscopic study of multinucleated giant cells related with the resorption of hyalinized tissues (in Japanese). J Jpn Orthod Soc 41:531–555
20. Kazama T (1989) Electron microscopic study of the periodontal tissues in the pressure side by intermittent forces (in Japanese). Tsurumi Univ Dent J 15:87–108
21. Satoh Y, Ishikawa H, Nakamura S, Wakita M (1995) Time dependent changes of periodontal tissue at pressure side incident to orthodontic tooth movement (in Japanese). J Jpn Orthod Soc 54:177–192
22. Takahashi N (1994) Ultrastructural study of the periodontal ligament at the pressure side incident to orthodontic tooth movement—mechanism of cell-free zone formation (in Japanese). Hokkaido J Dent Sci 15:108–128
23. Kaneko T (1994) Three-dimensional situation of periodontal tissue at pressure side incident to orthodontic tooth movement (in Japanese). Jpn J Oral Biol 36:170–186
24. Walker DG (1975) Control of bone resorption by hematopoietic tissue. J Exp Med 142:651–663
25. Fischman DA, Hay ED (1962) Origin of osteoclasts from mononuclear leucocytes in regenerating newt limbs. Anat Rec 143:329–337
26. Horton MA, Rimmer EF, Moore A, Chambers TJ (1985) On the origin of osteoclast: the cell surface phenotype of rodent osteoclasts. Calcif Tissue Int 37:46–50
27. Baron R, Tran V, Nefussi JR, Vignery A (1986) Kinetic and cytochemical identification of osteoclast precursors and their differentiation into mutinucleated osteoclast. Am J Pathol 122:368–378
28. Udagawa N, Takahashi N, Akatsu T, Tanaka H, Suda T (1990) Origin of osteoclasts: mature monocytes and macrophages are capable of differentiating into osteoclasts under a suitable microenvironment prepared by bone marrow-derived stromal cells. Proc Natl Acad Sci USA 85:7260–7264
29. Schwarz AM (1928) Ueber die Bewegung belastere Zahne. Z Stomatol 26:40–83
30. Kondo K (1969) A study of blood circulation in the periodontal membrane by electrical impedance plethysmography (in Japanese). J Stomatol Soc Jpn 36:20–42
31. Kurashima K (1963) The viscoelastic properties of the periodontal membrane and alveolar bone (in Japanese). J Stomatol Soc Jpn 30:361–385
32. Tanne K (1983) Stress induced in the periodontal tissue at the initial phase of the application of various types of orthodontic force (in Japanese). J Osaka Univ Dent Soc 28:209–261
33. Tanne K (1987) Three-dimensional finite element analysis for stress in the periodontal tissue by orthodontic forces. Am J Orthod Dentofacial Orthop 92:499–505
34. Middleton J, Jones ML, Wilson AN (1990) Three-dimensional analysis of orthodontic tooth movement. J Biomed Eng 12:319–327
35. Andersen KL, Mortensen HT, Pedersen EH, Melsen B (1991) Determination of stress levels and profiles in the periodontal ligament by means of an improved three-dimensional finite element model for various types of orthodontic and natural force system. J Biomed Eng 13:293–303

36. Juan C, Alberto S, Juan A, David S, Manuel V (1993) Initial stress induced in periodontal tissue with diverse degrees of bone loss by an orthodontic force: tridimensional analysis by means of the finite element method. Am J Orthod Dentofacial Orthop 104:448–454
37. Yamamoto K, Satoh Y, Nishihira M, Morikawa H, Ishikawa H, Nakamura S, Wakita M (1994) Finite element analysis of stress around a moved tooth and correlation with osteoclast distribution. In: Annual international conference of the IEEE engineering in medicine and biology society, vol 16, November 1994, Baltimore

Subject Index